Fungi: Diversity, Ecology and Biotechnological Applications

Fungi: Diversity, Ecology and Biotechnological Applications

Edited by Donald Vincent

SYRAWOOD
PUBLISHING HOUSE
New York

Published by Syrawood Publishing House,
750 Third Avenue, 9th Floor,
New York, NY 10017, USA
www.syrawoodpublishinghouse.com

Fungi: Diversity, Ecology and Biotechnological Applications
Edited by Donald Vincent

International Standard Book Number: 978-1-64740-391-1 (Hardback)

Cataloging-in-publication Data

Fungi : diversity, ecology and biotechnological applications / edited by Donald Vincent.
p. cm.
Includes bibliographical references and index.
ISBN 978-1-64740-391-1
1. Fungi. 2. Fungi--Biotechnology. 3. Mushrooms. 4. Bioengineering. 5. Biotechnology.
6. Biodiversity. I. Vincent, Donald.

TP248.27.F86 F86 2023
660.62--dc23

TABLE OF CONTENTS

Preface VII

Chapter 1 **Perstraction of Intracellular Pigments through Submerged Fermentation of *Talaromyces* spp. in a Surfactant Rich Media: A Novel Approach for Enhanced Pigment Recovery** 1
Lourdes Morales-Oyervides, Jorge Oliveira, Maria Sousa-Gallagher, Alejandro Méndez-Zavala and Julio Cesar Montañez

Chapter 2 **An Italian Research Culture Collection of Wood Decay Fungi** 13
Carolina Elena Girometta, Annarosa Bernicchia, Rebecca Michela Baiguera, Francesco Bracco, Simone Buratti, Marco Cartabia, Anna Maria Picco and Elena Savino

Chapter 3 **Microscopic Analysis of Pigments Extracted from Spalting Fungi** 32
Sarath M. Vega Gutierrez and Sara C. Robinson

Chapter 4 **Basidiomycetes Associated with *Alnus glutinosa* Habitats in Andros Island (Cyclades, Greece)** 47
Elias Polemis, Vassiliki Fryssouli, Vassileios Daskalopoulos and Georgios I. Zervakis

Chapter 5 **Utilization of High Performance Liquid Chromatography Coupled to Tandem Mass Spectrometry for Characterization of 8-*O*-methylbostrycoidin Production by Species of the Fungus *Fusarium*** 69
Mark Busman

Chapter 6 **Ecology, Phylogeny, and Potential Nutritional and Medicinal Value of a Rare White "Maitake" Collected in a Mediterranean Forest** 77
Maria Letizia Gargano, Georgios I. Zervakis, Omoanghe S. Isikhuemhen, Giuseppe Venturella, Roberta Calvo, Anna Giammanco, Teresa Fasciana and Valeria Ferraro

Chapter 7 **Biodiversity of Pigmented Fungi Isolated from Marine Environment in La Réunion Island, Indian Ocean: New Resources for Colored Metabolites** 89
Mireille Fouillaud, Mekala Venkatachalam, Melissa Llorente, Helene Magalon, Pascale Cuet and Laurent Dufossé

Chapter 8 **Does Arbuscular Mycorrhiza Determine Soil Microbial Functionality in Nutrient-Limited Mediterranean Arid Ecosystems?** 111
Neji Mahmoudi, Teresa Dias, Mosbah Mahdhi, Cristina Cruz, Mohamed Mars and Maria F. Caeiro

Chapter 9 **Carotenoid Biosynthesis in *Fusarium*** 126
Javier Avalos, Javier Pardo Medina, Obdulia Parra-Rivero, Macarena Ruger-Herreros, Roberto Rodríguez-Ortiz, Dámaso Hornero-Méndez and María Carmen Limón

Chapter 10 **Fungal Diversity in the Phyllosphere of *Pinus heldreichii* H. Christ—An Endemic and High-Altitude Pine of the Mediterranean Region** **142**
Jelena Lazarević and Audrius Menkis

Chapter 11 **Production and New Extraction Method of Polyketide Red Pigments Produced by Ascomycetous Fungi from Terrestrial and Marine Habitats** **158**
Juliana Lebeau, Mekala Venkatachalam, Mireille Fouillaud, Thomas Petit, Francesco Vinale, Laurent Dufossé and Yanis Caro

Chapter 12 **News from the Sea: A New Genus and Seven New Species in the Pleosporalean Families Roussoellaceae and Thyridariaceae** **179**
Anna Poli, Elena Bovio, Lucrezia Ranieri, Giovanna Cristina Varese and Valeria Prigione

Chapter 13 **Combinatorial Biosynthesis of Novel Multi-Hydroxy Carotenoids in the Red Yeast *Xanthophyllomyces dendrorhous*** **197**
Hendrik Pollmann, Jürgen Breitenbach, Hendrik Wolff, Helge B. Bode and Gerhard Sandmann

Chapter 14 **Genetic Structure and Phylogeography of *Tuber magnatum* Populations** **207**
Beatrice Belfiori, Valentina D'Angelo, Claudia Riccioni, Marco Leonardi, Francesco Paolocci, Giovanni Pacioni and Andrea Rubini

Chapter 15 **Assessment of the Dyeing Properties of the Pigments Produced by *Talaromyces* spp.** **221**
Lourdes Morales-Oyervides, Jorge Oliveira, Maria Sousa- Gallagher, Alejandro Méndez Zavala and Julio Cesar Montañez

Chapter 16 **Phylogenetic Characterization of *Botryosphaeria* Strains Associated with *Asphondylia* Galls on Species of Lamiaceae** **230**
Beata Zimowska, Sylwia Okoń, Andrea Becchimanzi, Ewa Dorota Krol and Rosario Nicoletti

Permissions

List of Contributors

Index

PREFACE

Fungi are eukaryotic microorganisms and can occur as yeasts, molds or as a combination of both forms. Mycology refers to the branch of biology involved in the study of fungi. It is concerned with the study of genetic and biochemical properties of fungi, as well as their taxonomy, benefits and harmful effects. There are three ways of classifying fungi based on their ability to reproduce asexually, sexually or by a combination of both. Fungi have several applications in a variety of industries such as food and pesticide. Yeast is a unicellular fungus used for making bread and other food items. The antibiotic known as penicillin is also obtained from a fungus named Penicillium notatum. This book provides significant information on fungi to help develop a good understanding of its diversity, ecology and biotechnological applications. It will also provide interesting topics for research which interested readers can take up.

Significant researches are present in this book. Intensive efforts have been employed by authors to make this book an outstanding discourse. This book contains the enlightening chapters which have been written on the basis of significant researches done by the experts.

Finally, I would also like to thank all the members involved in this book for being a team and meeting all the deadlines for the submission of their respective works. I would also like to thank my friends and family for being supportive in my efforts.

Editor

Perstraction of Intracellular Pigments through Submerged Fermentation of *Talaromyces* spp. in a Surfactant Rich Media: A Novel Approach for Enhanced Pigment Recovery

Lourdes Morales-Oyervides [1,2], Jorge Oliveira [1], Maria Sousa-Gallagher [1], Alejandro Méndez-Zavala [2] and Julio Cesar Montañez [2,*]

[1] School of Engineering, University College Cork, Cork, Ireland; lourdesmorales@uadec.edu.mx (L.M.-O.); j.oliveira@ucc.ie (J.O.); M.deSousaGallagher@ucc.ie (M.S.-G.)
[2] Department of Chemical Engineering, Universidad Autónoma de Coahuila, Saltillo 25280, Mexico; alejandro.mendez@uadec.edu.mx
* Correspondence: julio.montanez@uadec.edu.mx.

Abstract: A high percentage of the pigments produced by *Talaromyces* spp. remains inside the cell, which could lead to a high product concentration inhibition. To overcome this issue an extractive fermentation process, perstraction, was suggested, which involves the extraction of the intracellular products out of the cell by using a two-phase system during the fermentation. The present work studied the effect of various surfactants on secretion of intracellular pigments produced by *Talaromyces* spp. in submerged fermentation. Surfactants used were: non-ionic surfactants (Tween 80, Span 20 and Triton X-100) and a polyethylene glycerol polymer 8000, at different concentrations (5, 20, 35 g/L). The highest extracellular pigment yield (16 OD_{500nm}) was reached using Triton X-100 (35 g/L), which was 44% higher than the control (no surfactant added). The effect of addition time of the selected surfactant was further studied. The highest extracellular pigment concentration (22 OD_{500nm}) was achieved when the surfactant was added at 120 h of fermentation. Kinetics of extracellular and intracellular pigments were examined. Total pigment at the end of the fermentation using Triton X-100 was 27.7% higher than the control, confirming that the use of surfactants partially alleviated the product inhibition during the pigment production culture.

Keywords: fungal pigments; perstraction; surfactant; *Talaromyces*

---

1. Introduction

In recent years, consumer trends have been oriented to natural products, especially in the food industry. Consumer expectations have led the food processing industry to improve processes in order to deliver high quality-products [1]. To meet the requirements for safer food and also satisfy consumer preferences, scientists are searching for new food additives and new food processing methods [2]. The application of food additives will depend on their function in foods and can be classified as preservatives, nutritional additives, flavouring agents, texturising agents and colouring agents [3]. Natural colourants are obtained from sources like plants [4,5], insects [6] and microorganisms [7]. However, application of microbial pigments still represents a major challenge to biotechnology due to lower extraction or production yields. Dufosse et al. (2014) [8] emphasised in their fungal pigments review, the crucial role that filamentous fungi are currently playing as microbial cell factories, mainly due to the attractive range of stable bio-colourants that they are able to synthesise under controlled conditions. The most-well documented pigment producer fungi is *Monascus* [9]. However, in the last years, pigment producing fungi from other species have gained more attention due to the mycotoxin (citrinin) produced by *Monascus* [10].

Fungi belonging to the genus *Talaromyces* (formerly *Penicillium*) have attracted the attention of scientists due to their high pigment production yields [11,12] and moreover, due to the high thermal stability [13], antioxidant properties [14], antibacterial properties [15] and the absence of toxicity of these pigments [15,16].

However, their successful industrial application will not only depend on their safety or added value properties. The economics of the process need to be assessed early in the design stage before going through more expensive research, such as the scale-up of the process or the characterisation of the molecules. There are various strategies to achieve a cost-effective process such as using cheaper substrates to reduce raw material costs, optimising process conditions to achieve high yields or implementing an efficient product recovery. From these, developing an efficient downstream processing seems to be a rapid strategy to maximise product recovery and to reduce costs.

Pigments produced by *Talaromyces* are excreted out of the cell. However, most remain inside the cell, adding several unit operations to the downstream processing such as cell disruption, extraction of intracellular products, followed by the cell debris removal. Furthermore, a high intracellular pigment concentration could lead to product concentration inhibition. To overcome this issue researchers have been applying an extractive fermentation process, "perstraction", where the intracellular products are transported out of the cell by using a two-phase system during the fermentation, where the fermentation media represents one of the phases and the other should be an extractive solution. This process has been used in microalgae cultures [17,18] and is also known as "milking".

Ziolkowska and Simon, (2014) [17] reported that "milking algae" is a new successful technology that allows for the reduction of process costs by the extraction of intracellular products, increasing yields and reduction of unit operations. Two aspects are of significant importance for the selection of a suitable extractant for the perstraction of an intracellular product [19]: the biocompatibility of the solvent with the microorganism allowing cell growth, and that the extractant should induce permeability of the cell wall. Recent studies have reported the application of non-ionic surfactant as an effective extractant on the perstraction of intracellular pigments produced by the fungus *Monascus* [20,21]. Wang et al.(2013) [22] reported that the use of non-ionic surfactant Triton X-100 highly increased the pigment production and the extracellular/intracellular pigment ratio by *Monascus* in submerged fermentation. A specific characteristic of the non-ionic aqueous solution is the formation of the cloud point system.

The addition of non-ionic surfactant to an aqueous solution forms micelles and when this solution is above the surfactant cloud point, the micelle aqueous solution separates into two phases, a surfactant rich phase (coacervate phase) and a dilute phase [23]. The application of the cloud point system has been reviewed as an advantageous pre-concentration step prior to purification in many bioprocesses, such as extraction of fatty acids from microalgae cultures [18]; extractive fermentation of proteins produced by bacteria [24]; production of L-phenylacetylcarbinol by *Sacharomyces cerevisiae* [25] and in exporting intracellular pigments produced by *Monascus* into its fermentation broth [20,22,26].

The application of this concept to submerged fermentation of *Talaromyces* spp. has not been studied, and therefore, the objective of this work was to analyse the effect of various surfactants, at different concentrations, on secreting the intracellular pigments produced by *Talaromyces* spp. into the fermentation broth. The relevance of the precise time of addition of the surfactant was further studied in order to define conditions for maximum extracellular pigments recovery. In addition, the kinetics of pigment production, biomass and substrate consumption during *Talaromyces* spp. culture using non-ionic surfactant micelle aqueous solution were analysed.

2. Materials and Methods

2.1. Microorganism

Talaromyces spp. was used for the production of red pigments (DIA-UAdeC). The purified strain had been previously isolated and characterised as *Penicillium purpurogenum* GH2 [27,28]. *Penicillium purpurogenum* has, however, been transferred to *Talaromyces* spp. [29]. The strain was maintained on PDA (Potato dextrose agar) slants at 4 °C and sub-cultured periodically.

2.2. Culture Media

The PDA medium was prepared with a concentration of 39.0 g/L (Bioxon, Mexico). The medium Potato Dextrose Broth (PDB medium, ATCC medium: 336) was prepared by boiling 0.3 kg of finely diced potatoes in 500 mL of water until thoroughly cooked; then the potatoes were filtered through cheesecloth and water was added to the filtrate to complete a volume of 1.0 L. Finally, 20.0 g of glucose is added before sterilisation. The Czapek-dox modified medium [30] consisted in (g/L): D-xylose 15.0, $NaNO_3$ 3.0, $MgSO_4 \cdot 7H_2O$ 0.5, $FeSO_4 \cdot 7H_2O$ 0.1, K_2HPO_4 1.0, KCl 1.0 and ethanol 20.0.

2.3. Inoculum Preparation

PDB medium was used for the inoculum preparation. Flasks (125 mL) containing 25 mL of PDB medium were sterilised and then inoculated with a spore suspension (1×10^5 spores/mL) of *Talaromyces* spp. previously incubated for 5 days. The flasks were then incubated at 30 °C for 84 h in an orbital shaker (Innova 94, New Brunswick Scientific, Edison, NJ, USA) at 200 rpm [11,31].

2.4. Cultivation Conditions

Czapek-dox medium was used for pigment production. The initial pH of the Czapek-dox modified medium was adjusted to 5 before sterilising by using 0.22 m sterile membranes (Millipore, Billerica, MA, USA).

Four different surfactants were tested, assessing their effect on the pigment production and growth in submerged fermentation of *Talaromyces* spp. The surfactants analysed were Tween 80, Span 20, Triton X-100 and polyethylene glycerol polymer PEG 8000, at 3 different concentrations (5, 20, 35 g/L). Surfactants were added at the beginning of fermentation. Extractive fermentation was carried out in 125 mL Erlenmeyer flasks containing 25 mL of medium. Flasks were inoculated with a mycelial suspension 10% (v/v). The inoculated flasks were incubated at 30 °C in an orbital shaker (Innova 94, New Brunswick Scientific, Edison, NJ, USA) at 200 rpm for 8 days. Control experiments were also performed with no surfactant added.

2.5. Analytical Methods

Extracellular pigment recovery was performed according to the methodology reported by Méndez-Zavala et al. (2011) [32]. Each sample was centrifuged at 8000 rpm for 20 min at 4 °C (Sorvall, Primo R Biofuge Centrifugation Thermo, Waltham, MA, USA). The supernatant was then filtered through a 0.45 μm cellulose filter (Millipore, USA). Recovered mycelia were rinsed with distilled water until the supernatant was cleared and used for further extraction of intracellular pigments [20]. Mycelia were soaked in 25 mL of a 70% (v/v) ethanol aqueous solution which represented the initial fermentation broth volume. Extraction was carried out on an orbital shaker (Inova 94, New Brunswick Scientific, USA) at 200 rpm at 30 °C for 1 h. The concentration of red pigments (extracellular and intracellular) was quantified indirectly by simply measuring the optical density at 500 nm using a spectrophotometer (Cary 50, UV-Visible Varian, Palo Alto, CA, USA). The biomass concentration was determined using the gravimetric method. The analysis of substrate consumption was determined by quantifying the total sugar content using the method reported by [33]. The experiments were replicated three times.

2.6. Cloud Point Extraction

Extracellular pigment was separated into a two-phase cloud point system: a surfactant-rich phase and a dilute phase [18]. Pigments-surfactant solution was heated until the formation of these two phases (70 °C) and until reaching the equilibrium (30 min).

An estimation of the pigments that remained in each phase (diluted and coacervate) was done by performing a spectral analysis within the visible wave range (Cary 50, UV-Visible Varian). Various authors have expressed fungal pigments concentration as optical density units at the absorption maxima for a specific wavelength: 400–420 (yellow), 450–470 (orange) and red (490–510) [12,20,26,34,35].

2.7. Kinetic Parameters

Pigment extraction productivity ($P_{Y/t}$, OD/h), biomass productivity ($P_{B/t}$, g/L/h) and yield of product per unit of biomass ($Y_{Y/B}$, OD·L/h) were determined by the following equations, [36]:

$$P_{\frac{Y}{t}} = \frac{Y_{\infty} - Y_o}{t_{\infty} - t_o} \tag{1}$$

$$P_{\frac{B}{t}} = \frac{B_{\infty} - B_o}{t_{\infty} - t_o} \tag{2}$$

$$Y_{\frac{Y}{B}} = \frac{Y_{\infty} - Y_o}{B_{\infty} - B_o} \tag{3}$$

where Y_o and B_o correspond to the yield of intracellular pigment (OD_{500nm}) and biomass (g/L) at the time (t_o, min) when the surfactant (Triton X-100) was added; Y_{∞} and B_{∞} are the yield of pigments (OD_{500nm}) and biomass (g/L) at the end of the fermentation (t_{∞}, min).

2.8. Data Analysis

Kinetic parameters were set from experimental data. Statistical analyses were made with Statistica 7.0 software (StatSoft, Tulsa, OK, USA).

3. Results and Discussion

3.1. Effect of Surfactants on Secretion of Intracellular Pigments into Fermentation Broth (Screening)

As a first step towards developing a non-ionic surfactant micelle aqueous solution as a fermentation media, a screening of the non-ionic surfactants most commonly used in bioprocess was performed [22,23]. Selection of a suitable surfactant was based on the highest pigment production and extracellular/intracellular pigment ratio. Figure 1 shows the extracellular and intracellular pigments obtained after eight days of fermentation for each surfactant, evaluated at three different concentrations (5, 20 and 35 g/L).

Results for the control media indicated that a high extracellular pigment yield was reached and nearly 40% of the total pigment was still remaining inside of the cell. *Talaromyces* spp. was capable of growing in all surfactants studied; however, pigment production was strongly affected by the addition of surfactant at the beginning of the fermentation. The addition of PEG 8000 caused a reduction in pigment production by more than 50%, in comparison with the control media. This finding limits the use of PEG as a water soluble polymer in aqueous two-phase systems (ATPS) on in-situ product removal processes [37]. When Span 20 was added, the total pigment level was reduced by 80%, in comparison with control media. It was also observed that mycelia morphology changed (dispersed mycelia) and led to a higher biomass growth. Similar behaviour (higher cell growth) was reported [38] on β-Carotene production by *Blakeslea trispora* (Mating type, + and −) with the addition of Span 20. Biomass increment may be due to the change in morphology, i.e., fragmented mycelium led to a major aeration, conditions that allowed higher growth, but were not optimal for pigment production. The addition of Tween 80 at low concentration (5 g/L) caused a reduction of only less than 10% of the total pigment, however, the extracellular/intracellular pigment ratio was increased from 1.72 (control) to 1.79. This may be an indication of the permeability induced on the cell wall by Tween 80; however, increasing the levels of this surfactant resulted in a reduction in both biomass and total pigment, showing the low biocompatibility of this surfactant with *Talaromyces* spp. Similarly, Zhang et al. (2013) [39] reported that the addition of a high concentration of Tween 80 during the production of Antrodin C in submerged fermentation by *A. camphorata* led to massive damage on the cell membrane. When Triton X-100 surfactant was added (35 g/L), an enhancement of extracellular pigment production level (44%) was observed. Total pigment production (extracellular and intracellular) was not statistically different from the control experiments (without addition of surfactant); however, extracellular/intracellular pigment ratio was increased from 1.73 (control) to 11.75. Biomass reduction with elevated concentration of Triton X-100 may indicate that this non-ionic surfactant has a low biocompatibility with the strain, although it may also be related to the secretion of the intracellular pigments. A cell with an elevated concentration of pigment presents a higher weight than an "empty cell". Triton X-100 has been successfully applied on the secretion of intracellular pigments of *Monascus* [20,26,40], similarly, Triton X-100 has presented high biocompatibility with microbial cell allowing modification of membrane to induce permeability of certain desired metabolites [21,41]. Based on the above results, Triton X-100 (3.5%) was selected for subsequent studies.

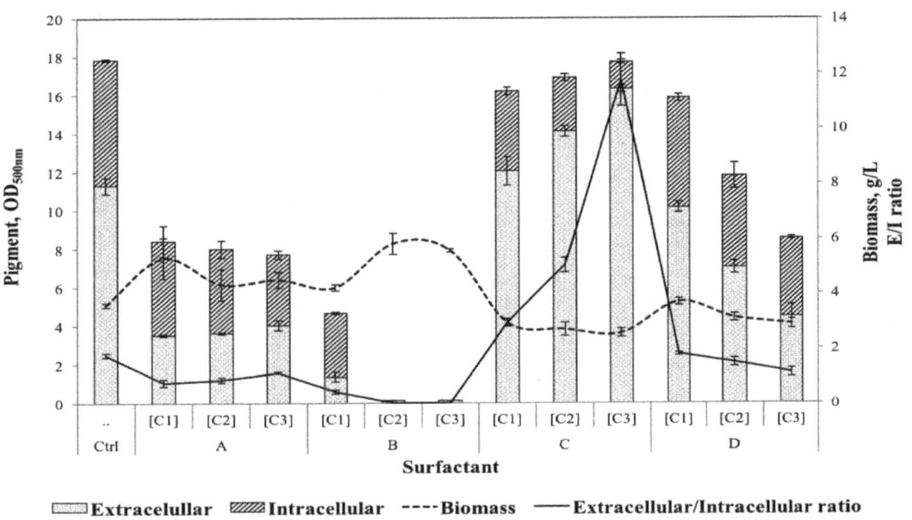

Figure 1. Screening of surfactants' effect on extracellular and intracellular pigment yield, extracellular/intracellular pigment ratio and growth. Ctrl) Without surfactant, (**A**) polyethylene glycerol polymer PEG 8000; (**B**) Span 20; (**C**) Triton X-100 and (**D**) Tween 80. [Ci, g/L]: C1, 5; C2, 20; C3, 35.

3.2. Effect of Addition Time of Triton X-100 during the Fermentation

In the previous experiments, surfactants were added at the beginning of the fermentation. Many authors have added surfactants at different stages of the fermentation to enhance microbial production [39,41]; however, the best time for surfactant addition will vary depending on the microorganism. In order to fully comprehend the effect of Triton X-100 on biomass growth and to enhance pigments production by *Talaromyces* spp., 35 g/L was added at different fermentation times (0, 24, 48, 72, 96, 120 and 144 h). Results for final biomass, extracellular and intracellular pigments are presented in Figure 2.

Biomass was reduced by 20% when Triton X-100 was added in the interval of 0–3 days and decayed to 50% when it was added at later stages of the fermentation, in comparison with the control. Wang et al. (2013) [22] reported that the addition of Triton X-100 to pigments production by *Monascus* at later stages of the fermentation process resulted in higher biomass growth. In contrast, in this study a consequent addition time of Triton X-100 did not represent better conditions to biomass growth; thus, the results obtained were not sufficient to demonstrate a noticeable Triton X-100 inhibitory effect on the cell. A time course study for each addition time could give more understanding of the Triton X-10 harmful effect on the cell growth, i.e., extended lag phase, reduced maximum biomass and reduced growth rate. However, the main objective of this study was to enhance pigment production and to increase extracellular/intracellular pigment ratio. Figure 2 shows that the excretion of intracellular pigments was considerably enhanced with the addition of Triton X-100; extracellular pigment increased gradually with the time of addition reaching the maximum extracellular pigment (22.1 ± 1.2 OD_{500nm}) when Triton X-100 was added at 120 h of fermentation, which corresponds to an increment of 90%, as compared to the control. Furthermore, extracellular/intracellular pigment ratio increased from 2.5 ± 0.1 to 37.9 ± 5.2 for control and Triton X-100 addition at 120 h of fermentation process, respectively. These results are in agreement with previous studies reported on *Monascus pigments* [20,26,40], where it was determined that adding Triton X-100 at a late stage of the fermentation process stimulated intracellular pigment excretion. Based on the above results, the addition of Triton X-100 at 120 h of fermentation at a concentration of 35 g/L was selected for further studies.

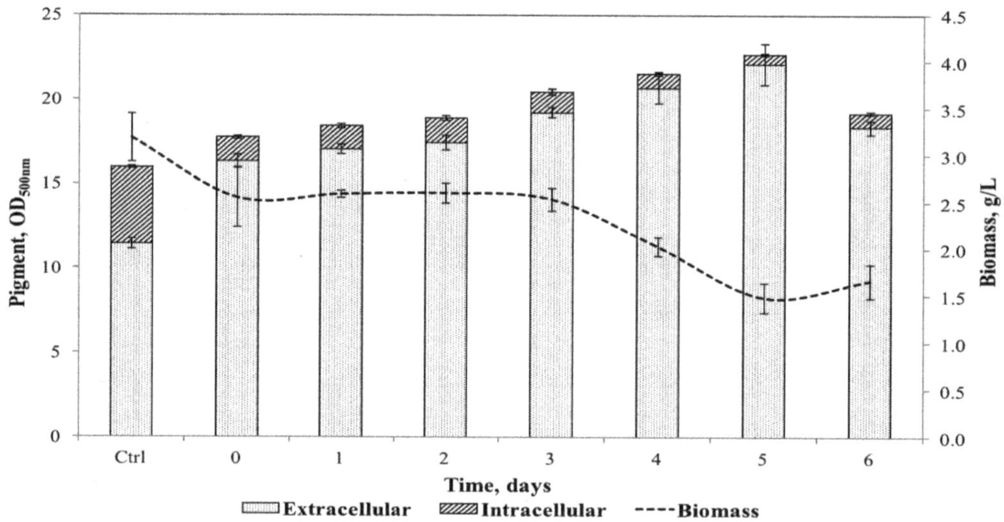

Figure 2. Effect of the addition time of Triton X-100 (35 g/L) on extracellular and intracellular pigment yield and biomass growth.

3.3. Pigments Production Kinetics of Talaromyces spp. by Using Triton X-100

Kinetics of biomass growth, pigment production and substrate consumption with and without the addition of Triton X-100 at 120 h of fermentation are depicted in Figure 3 (a and b respectively). Figure 3a shows that there is no lag phase for biomass growth; the microorganism started to grow (exponential phase) from 0 to 120 h, then reached a short stationary phase (120 to 150 h), after which biomass weight decayed. There was a slight relation between growth and pigment production.

Production of pigments started at 24 h and increased gradually until the end of fermentation. It is interesting to note that there is a correlation between extracellular, intracellular pigments and biomass decay. Extracellular and intracellular pigment profiles incremented in a similar pattern until 120 h. After 120 h of submerged fermentation, intracellular pigments decreased correspondingly with the biomass decay and further increment of the extracellular pigments. This behaviour led us to hypothesise that when the fungus reaches the stationary phase (120 h), it stops producing pigments and the further increase of extracellular pigments is partially due to the excretion of intracellular pigments out the cell. This excretion of intracellular pigments may contribute to the biomass decay after 120 of fermentation.

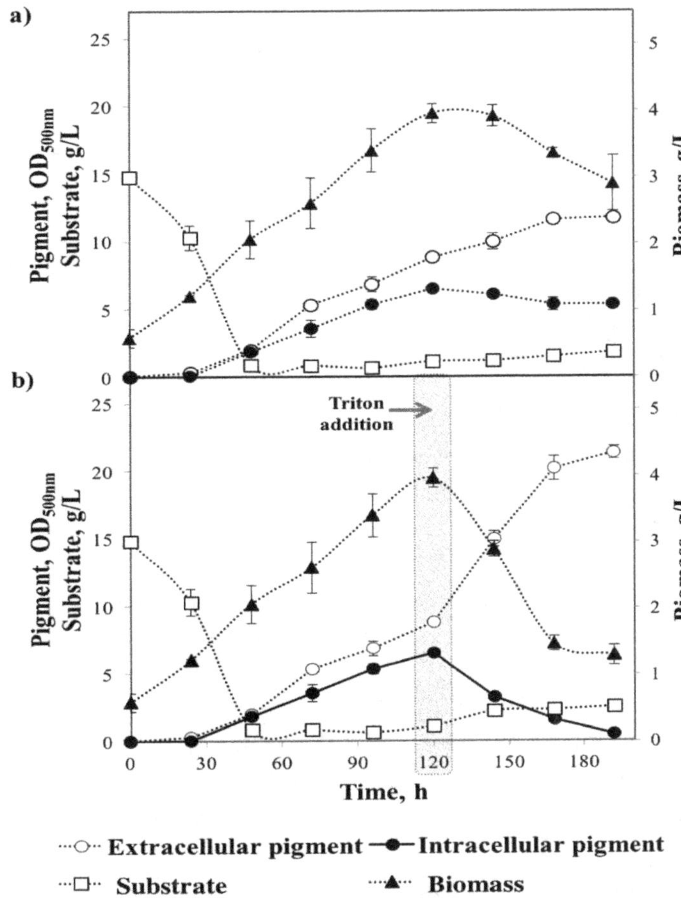

Figure 3. Kinetics of extracellular and intracellular pigment production, substrate consumption and biomass growth. (**a**) Without surfactant; (**b**) Triton X-100 added at 120 h of fermentation.

At 120 h of fermentation process, the fungus had reached the stationary phase and thus the addition of Triton X-100 did not cause biomass growth inhibition; this was the reason for the highest extracellular yields being obtained when Triton X-100 was added at this stage. The above results suggested that the surfactant only induced cell permeability and allowed intracellular pigments excretion. Previous work on excretion of *Monascus pigments* have suggested that the addition of Triton X-100 increases the permeability and fluidity of the cell due to the surfactant effect on cell membrane lipids [22]. Furthermore, Deive et al. (2009) [42] reported that the addition of surfactants increased the lipids solubility on the cell membrane changing the cell mass transfer.

The major effect on the cell of the addition of Triton X-100 to the fermentation process could be reflected in the elevated increment of extracellular pigments. Extracellular pigments rose from 8.8 ± 0.2 to 14.9 ± 0.56 OD_{500nm} after 24 h of Triton X-100 addition, which showed this increment as being 5.3 times higher than the control. Furthermore, the extracellular/intracellular pigment ratio was highly increased after 24 h Triton X-100 addition, in comparison with the control, from 1.6 to 4.6. A remarkable observation is that after Triton X-100 addition (120 h), the correlation between extracellular pigments accumulation, intracellular pigments decrease and biomass decay was similar in comparison to the control. This suggests that when the fungus has reached the stationary phase, extracellular pigment increment is only due to the release of intracellular pigments contributing to the higher reduction in biomass weight observed when using Triton X-100 than with the control.

In order to correlate the extraction of intracellular pigments to biomass decay; the pigment extraction productivity ($P_{Y/t}$, OD/h) and biomass productivity ($P_{B/t}$, g/L/h) was obtained after the addition of Triton X-100 for both kinetics (control and perstraction processes). Table 1 shows the parameters obtained. $P_{Y/t}$ and $P_{B/t}$ were 400% higher when using Triton X-100 in comparison with control (negative signs indicate intracellular pigments reduction and biomass decay).

Table 1. Intracellular realising (K_I), biomass decay (K_B) rates from 120 h to 168 h of fermentation kinetics and K_I/ K_B ratio. (**A**) Control, without surfactant; (**B**) Addition of Triton X-100.

Kinetics	Rate, k		Ratio
	Intracellular, OD_{500}/h	Biomass, g/L/h	K_I/K_B, OD_{500}, L/g
A	-0.023 ± 0.001	-0.0125 ± 0.002	1.90 ± 0.19
B	-0.103 ± 0.004	-0.052 ± 0.004	1.97 ± 0.10

The yield of intracellular product released per unit of biomass was calculated at 120 h of fermentation (time of Triton addition) for both kinetics. Statistical analysis showed that there were no significant differences (95% confidence level) between the $Y_{Y/B}$ obtained in the control and after the addition of Triton X-100. In both cases the cell was releasing approximately 1.90 OD_{500nm} per gram of biomass per litre.

Biomass decay (cell death) has represented a critical concept in microbiology and its estimation is continuously approached by different kinetic models [43]. Van Bodegom (2007) [44] mentioned that there are some inconsistencies related to cell death, because its quantification does not distinguish between biomass losses related to cell lysis or to the actual transport phenomena of intracellular products outside of the cell. In order to understand the mechanism of pigments release and biomass decay relationship, more studies need to be done, due to the variety of components present in the cell.

Regarding substrate consumption, there was no effect when Triton X-100 was added, as the addition occurred after substrate depletion. At the end of the fermentation process (192 h), extracellular pigments were enhanced above 80% with the addition of Triton X-100 at the end of the exponential growth phase (120 h). Visual kinetics for extracellular pigments enhancement are illustrated in Figure 4. Similar results were obtained by Wang et al. (2013) [22] who reported an increment of 88.4% of pigments produced by *Monascus* by adding Triton X-100 at a late stage of the submerged fermentation. Furthermore, total pigment (extracellular pigment plus intracellular pigment) at the end of the fermentation was 27.7% higher with Triton X-100 compared to control.

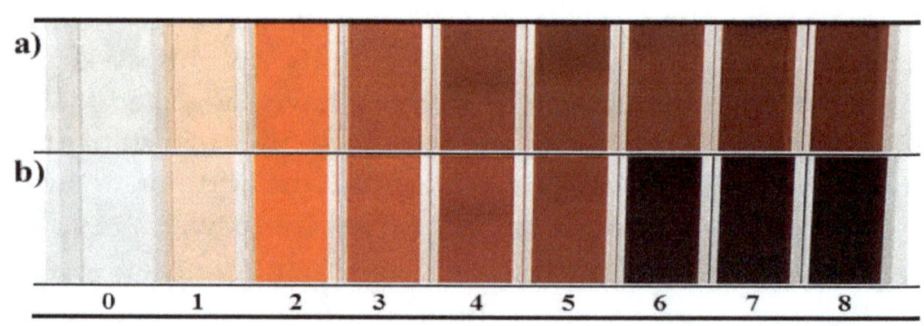

Figure 4. Visual kinetics of colour enhancement. X-axis in days. (**a**) Without surfactant; (**b**) Triton X-100 added at 120 h of fermentation.

3.4. Pigments Partitioning in Cloud Point System

The inner figure in Figure 5 shows the phase separation of the micelle aqueous solution when it reached a temperature above its cloud point (70 °C). It can be seen that the extracellular pigments obtained were separated between the dilute and coacervate phase.

Spectral analysis (Figure 5) shows a maximum absorbance at 414 and 500 nm for all the analysed samples. According to previous studies for *Talaromyces* pigments [12,34,45] these peaks correspond to yellow and red, respectively. Optical density values for each peak are also shown in Figure 5.

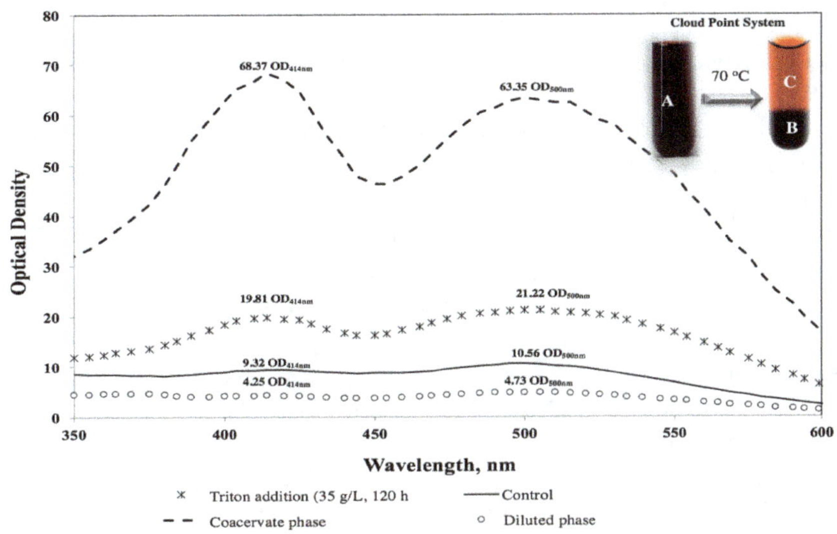

Figure 5. Spectral analysis within the visible wave range performed on the extracellular pigments obtained. Inner Figure: (**A**) the addition of Triton X-100 (35 g/L) at 120 h of fermentation, (**B**) coacervate phase, (**C**) diluted phase.

Red pigments concentration was much higher in the coacervate phase than in the diluted phase (13 times higher). Yellow pigments showed more affinity with the coacervate phase than red pigments, as its concentration was 16 times higher in the coacervate than in the diluted phase. The concentration factor, defined as the ratio of the pigment concentration in the coacervate phase to the concentration in the micelle aqueous solution [24], was 2.98 and 3.45 for red and yellow pigments, respectively. Hu et al. (2012) [20] confirmed the same affinity to the coacervate phase of pigments produced by *Monascus* in a cloud point system by Thin-layer chromatography (TLC) analysis.

4. Conclusions

In this study, applying a biphasic fermentation system was shown as a novel and effective bioprocess strategy for an in-situ extraction of intracellular pigments during fermentation, increasing the overall production. Among the studied surfactants, nonionic surfactant Triton X-100 was the most suitable for an in-situ extraction of *Talaromyces* pigments. Under the optimal perstraction conditions, the maximum extracellular pigment yield was 80% higher than the control. Total pigment enhancement (27.7%) confirmed that the addition of Triton X-100 had at least partially alleviated the product inhibition.

Maximising product recovery was evident, and it was demonstrated that the application of the cloud point system could be an advantageous pre-concentration step prior to purification of the pigments produced by *Talaromyces* spp. However, more studies are needed at large scale aired-bioreactors to assess if foaming issues must be addressed.

Also, a techno-economic analysis of the process using perstraction is required to verify if the extraction of pigments during the fermentation reduces, in fact, the costs related to the downstream processing. In any case, using a surfactant rich media can substitute the use of organic solvents for the cell disruption and extraction processes during the downstream processing.

Acknowledgments: Author L. Morales-Oyervides acknowledges CONACyT-México for the financial support provided for conducting her PhD studies (215490-2011).

Author Contributions: Julio Cesar Montañez and Jorge Oliveira conceived and designed the experiments; Lourdes Morales-Oyervides performed the experiments; Jorge Oliveira and Maria Sousa-Gallagher analysed the data; Alejandro Méndez-Zavala and Julio Cesar Montañez contributed reagents/materials/analysis tools; Lourdes Morales-Oyervides wrote the paper.

References

1. Grunert, K.G. Food quality and safety: Consumer perception and demand. *Eur. Rev. Agric. Econ.* **2005**, *32*, 369–391. [CrossRef]
2. El-Wahab, H.M.; Moram, G.S. Toxic effects of some synthetic food colorants and/or flavor additives on male rats. *Toxicol. Ind. Health* **2013**, *29*, 224–232. [CrossRef] [PubMed]
3. Carocho, M.; Barreiro, M.F.; Morales, P.; Ferreira, I.C. Adding molecules to food, pros and cons: A review on synthetic and natural food additives. *Compr. Rev. Food Sci. Food Saf.* **2014**, *13*, 377–399. [CrossRef]
4. Onslow, M.W. *The Anthocyanin Pigments of Plants*; Cambridge University Press: Cambridge, UK, 2014.
5. Boo, H.O.; Hwang, S.J.; Bae, C.S.; Park, S.H.; Heo, B.G.; Gorinstein, S. Extraction and characterization of some natural plant pigments. *Ind. Crop. Prod.* **2012**, *40*, 129–135. [CrossRef]
6. Borges, M.E.; Tejera, R.L.; Díaz, L.; Esparza, P.; Ibáñez, E. Natural dyes extraction from cochineal (*Dactylopius coccus*). New extraction methods. *Food Chem.* **2012**, *132*, 1855–1860. [CrossRef]
7. Tuli, H.S.; Chaudhary, P.; Beniwal, V.; Sharma, A.K. Microbial pigments as natural color sources: Current trends and future perspectives. *J. Food Sci. Technol.* **2015**, *52*, 4669–4678. [CrossRef] [PubMed]
8. Dufosse, L.; Fouillaud, M.; Caro, Y.; Mapari, S.A.; Sutthiwong, N. Filamentous fungi are large-scale producers of pigments and colorants for the food industry. *Curr. Opin. Biotechnol.* **2014**, *26*, 56–61. [CrossRef] [PubMed]
9. Feng, Y.; Shao, Y.; Chen, F. Monascus pigments. *Appl. Microbiol. Biotechnol.* **2012**, *96*, 1421–1440. [CrossRef] [PubMed]
10. Hajjaj, H.; Blanc, P.; Groussac, E. Kinetic analysis of red pigment and citrinin production by *Monascus ruber* as a function of organic acid accumulation. *Enzym. Microb. Technol.* **2000**, *27*, 619–625. [CrossRef]
11. Morales-Oyervides, L.; Oliveira, J.C.; Sousa-Gallagher, M.J.; Méndez-Zavala, A.; Montañez, J.C. Selection of best conditions of inoculum preparation for optimum performance of the pigment production process by *Talaromyces* spp. using the Taguchi method. *Biotechnol. Prog.* **2017**. [CrossRef] [PubMed]
12. Santos-Ebinuma, V.C.; Roberto, I.C.; Simas Teixeira, M.F.; Pessoa, A. Improving of red colorants production by a new *Penicillium purpurogenum* strain in submerged culture and the effect of different parameters in their stability. *Biotechnol. Prog.* **2013**, *29*, 778–785. [CrossRef] [PubMed]
13. Morales-Oyervides, L.; Oliveira, J.C.; Sousa-Gallagher, M.J.; Méndez-Zavala, A.; Montañez, J.C. Effect of heat exposure on the colour intensity of red pigments produced by *Penicillium purpurogenum* GH2. *J. Food Eng.* **2015**, *164*, 21–29. [CrossRef]
14. Dhale, M.A.; Vijay-Raj, A.S. Pigment and amylase production in *Penicillium* sp NIOM-02 and its radical scavenging activity. *Int. J. Food Sci. Technol.* **2009**, *44*, 2424–2430. [CrossRef]
15. Teixeira, M.; Teixeira, M.F.; Martins, M.; Caldas, J.; Kirsh, L.; Fernandes, O.; Fernandes, O.C.; Carneiro, A.; De Conti, R.; Durán, N. Amazonian biodiversity: Pigments from *Aspergillus* and *Penicillium*-characterizations, antibacterial activities and their toxicities. *Curr. Trends Biotechnol. Pharm.* **2012**, *6*, 300–311.
16. Sopandi, T.; Wardah, W. Sub-Acute toxicity of pigment derived from *Penicillium resticulosum* in mice. *Microbiol. Indones.* **2012**, *6*, 35–41. [CrossRef]
17. Ziolkowska, J.R.; Simon, L. Recent developments and prospects for algae-based fuels in the US. *Renew. Sustain. Energy Rev.* **2014**, *29*, 847–853. [CrossRef]
18. Glembin, P.; Kerner, M.; Smirnova, I. Cloud point extraction of microalgae cultures. *Sep. Purif. Technol.* **2013**, *103*, 21–27. [CrossRef]
19. Qureshi, N.; Maddox, I.S. Reduction in butanol inhibition by perstraction, utilization of concentrated Lactose/Whey permeate by *Clostridium acetobutylicum* to enhance butanol fermentation Economics. *Food Bioprod. Process.* **2005**, *83*, 43–52. [CrossRef]
20. Hu, Z.; Zhang, X.; Wu, Z.; Qi, H.; Wang, Z. Perstraction of intracellular pigments by submerged cultivation of *Monascus* in nonionic surfactant micelle aqueous solution. *Appl. Microbiol. Biotechnol.* **2012**, *94*, 81–89. [CrossRef] [PubMed]
21. Zhang, J.; Wang, Y.-L.; Lu, L.-P.; Zhang, B.-B.; Xu, G.-R. Enhanced production of Monacolin K by addition of precursors and surfactants in submerged fermentation of *Monascus purpureus* 9901. *Biotechnol. Appl. Biochem.* **2014**, *61*, 202–207. [CrossRef] [PubMed]
22. Wang, Y.; Zhang, B.; Lu, L.; Huang, Y.; Xu, G. Enhanced production of pigments by addition of surfactants in submerged fermentation of *Monascus purpureus* H1102. *J. Sci. Food Agric.* **2013**, *93*, 3339–3344. [CrossRef] [PubMed]

23. Wang, Z. The potential of cloud point system as a novel two-phase partitioning system for biotransformation. *Appl. Microbiol. Biotechnol.* **2007**, *75*, 1–10. [CrossRef] [PubMed]
24. Pan, T.; Wang, Z.; Xu, J.H.; Wu, Z.; Qi, H. Extractive fermentation in cloud point system for lipase production by *Serratia marcescens* ECU1010. *Appl. Microbiol. Biotechnol.* **2010**, *85*, 1789–1796. [CrossRef] [PubMed]
25. Zhang, W.; Wang, Z.; Li, W.; Zhuang, B.; Qi, H. Production of L-phenylacetylcarbinol by microbial transformation in polyethylene glycol-induced cloud point system. *Appl. Microbiol. Biotechnol.* **2008**, *78*, 233–239. [CrossRef] [PubMed]
26. Kang, B.; Zhang, X.; Wu, Z.; Qi, H.; Wang, Z. Effect of pH and nonionic surfactant on profile of intracellular and extracellular *Monascus* pigments. *Process Biochem.* **2013**, *48*, 759–767. [CrossRef]
27. Cruz-Hernández, M; Rodráguez, R. Aislamiento y caracterización morfológica de cepas microbianas degradadotas de taninos. In Proceedings of the XXII Annual Meeting of the Mexican Academy of Chemical Engineering (AMIDIQ), Mazatlan, Mexico, 9 September 2001; pp. 71–72.
28. Espinoza-Hernández, T. Caracterización Morfológica, Fisiológica y Molecular de Tres Cepas Fúngicas Productoras de Pigmentos. Bachelor's Thesis, Universidad Autónoma de Coahuila, Saltillo, Mexico, 2004.
29. Frisvad, J.C.; Yilmaz, N.; Thrane, U.; Rasmussen, K.B.; Houbraken, J.; Samson, R.A. Talaromyces atroroseus, a new species efficiently producing industrially relevant red pigments. *PLoS ONE* **2013**, *8*, e84102. [CrossRef] [PubMed]
30. Méndez-Zavala, A. Identificación de Factores que Afectan la Producción de Pigmentos por *Penicillium purpurogenum* GH2 y Obtención de Perfiles Cromatográficos. Master's Thesis, Universidad Autónoma de Coahuila, Saltillo, Mexico, 2011.
31. Morales-Oyervides, L.; Oliveira, J.C.; Sousa-Gallagher, M.J.; Méndez-Zavala, A.; Montañez, J.C. Quantitative assessment of the impact of the type of inoculum on the kinetics of cell growth, substrate consumption and pigment productivity by *Penicillium purpurogenum* GH2 in liquid culture with an integrated stochastic approach. *Food Bioprod. Process.* **2015**, *96*, 221–231. [CrossRef]
32. Méndez-Zavala, A.; Pérez, C.; Montañez, J.C.; Martínez, G.; Aguilar, C.N. Red pigment production by *Penicillium purpurogenum* GH2 is influenced by pH and temperature. *J. Zhejiang Univ. Sci. B* **2011**, *12*, 961–968. [CrossRef] [PubMed]
33. Dubois, M.; Gilles, K.; Hamilton, J. Colorimetric method for determination of sugars and related substances. *Anal. Chem.* **1956**, *28*, 300–306. [CrossRef]
34. Kang, B.; Zhang, X.; Wu, Z.; Qi, H.; Wang, Z. Isolation of natural red colorants from fermented broth using ionic liquid-based aqueous two-phase systems. *J. Ind. Microbiol. Biotechnol.* **2013**, *40*, 507–516.
35. Babitha, S.; Soccol, C.R.; Pandey, A. Effect of stress on growth, pigment production and morphology of *Monascus* sp. in solid cultures. *J. Basic Microbiol.* **2007**, *47*, 118–126. [CrossRef]
36. Velmurugan, P. Studies on the Production and Dyeing Properties of Water Soluble Pigments from Filamentous Fungi. Ph.D. Thesis, Bharathiar University, Coimbatore, India, 2008.
37. Dafoe-Samper, T. Properties of Absorbent Polymer Extractants for the Selective Removal of Target Molecules from Fermentation Systems. Ph.D. Thesis, Queen's University, Kingston, ON, Canada, 2014.
38. Kim, S.W.; Seo, W.T.; Park, Y.H. Enhanced production of β-carotene from *Blakeslea trispora* with Span 20. *Biotechnol. Lett.* **1997**, *19*, 557–560. [CrossRef]
39. Zhang, H.; Xia, Y.J.; Wang, Y.L.; Zhang, B.B.; Xu, G.R. Coupling use of surfactant and in situ extractant for enhanced production of Antrodin C by submerged fermentation of *Antrodia camphorata*. *Biochem. Eng. J.* **2013**, *79*, 194–199. [CrossRef]
40. Hu, Z.; Zhang, X.; Wu, Z.; Qi, H.; Wang, Z. Export of intracellular *Monascus* pigments by two-stage microbial fermentation in nonionic surfactant micelle aqueous solution. *J. Biotechnol.* **2012**, *162*, 202–209. [CrossRef] [PubMed]
41. Dhamole, P.B.; Wang, Z.; Liu, Y.; Wang, B.; Feng, H. Extractive fermentation with non-ionic surfactants to enhance butanol production. *Biomass Bioenergy* **2012**, *40*, 112–119. [CrossRef]
42. Deive, F.J.; Carvalho, E.; Pastrana, L.; Rúa, M.L.; Longo, M.A.; Sanroman, M.A. Strategies for improving extracellular lipolytic enzyme production by *Thermus thermophilus* HB27. *Bioresour. Technol.* **2009**, *100*, 3630–3637. [CrossRef] [PubMed]
43. Mavituna, F.; Sinclair, C. *Practical Fermentation Technology*; John Wiley & Sons: West Sussex, UK, 2008; p. 402.

44. Van Bodegom, P. Microbial maintenance: A critical review on its quantification. *Microb. Ecol.* **2007**, *53*, 513–523. [CrossRef] [PubMed]
45. Patil, S.; Sivanandhan, G.; Thakare, D. Effect of physical and chemical parameters on the production of Red exopigment from *Penicillium purpurogenum* isolated from spoilt onion and study of its antimicrobial activity. *Int. J. Curr. Microbiol. Appl. Sci.* **2015**, *4*, 599–609.

2

An Italian Research Culture Collection of Wood Decay Fungi

Carolina Elena Girometta [1,*], Annarosa Bernicchia [2], Rebecca Michela Baiguera [1], Francesco Bracco [1], Simone Buratti [1], Marco Cartabia [1,3], Anna Maria Picco [1] and Elena Savino [1]

[1] Department of Earth and Environmental Sciences (DSTA), University of Pavia, 27100 Pavia, Italy; rebeccamichela.baiguera01@universitadipavia.it (R.M.B.); francesco.bracco@unipv.it (F.B.); simone.buratti01@universitadipavia.it (S.B.); marco.cartabia01@universitadipavia.it (M.C.); annamaria.picco@unipv.it (A.M.P.); elena.savino@unipv.it (E.S.)

[2] School of Agriculture and Veterinary Medicine, University of Bolognavia, Guidotti 39, 40134 Bologna, Italy; corticia.polypores@gmail.com

[3] MOGU S.r.l. Via San Francesco d'Assisi 62, 21020 Inarzo (VA), Italy

* Correspondence: carolinaelena.girometta@unipv.it

Abstract: One of the main aims of the University of Pavia mycology laboratory was to collect wood decay fungal (WDF) strains in order to deepen taxonomic studies, species distribution, officinal properties or to investigate potential applications such as biocomposite material production based on fungi. The Italian Alps, Apennines and wood plains were investigated to collect Basidiomycota basidiomata from living or dead trees. The purpose of this study was to investigate the wood decay strains of the Mediterranean area, selecting sampling sites in North and Central Italy, including forests near the Ligurian and Adriatic seas, or near the Lombardy lakes. The isolation of mycelia in pure culture was performed according to the current methodology and the identity of the strains was confirmed by molecular analyses. The strains are maintained in the Research Culture Collection MicUNIPV of Pavia University (Italy). Among the 500 WDF strains in the collection, the most interesting isolates from the Mediterranean area are: *Dichomitus squalens* (basidioma collected from *Pinus pinea*), *Hericium erinaceus* (medicinal mushroom), *Inocutis tamaricis* (white-rot agent on *Tamarix* trees), *Perenniporia meridionalis* (wood degrader through Mn peroxidase) and *P. ochroleuca*. In addition, strains of species related to the Mediterranean climate (e.g., *Fomitiporia mediterranea* and *Cellulariella warnieri*) were obtained from sites with a continental-temperate climate.

Keywords: wood decay fungi (WDF); culture collection; fungal strain; host; Italy; morphological and molecular identification

---

1. Introduction

Wood decay fungi provide an extraordinary model both for pure and applied research, as well as a food or medicinal mushroom resource.

From an ecological point of view, wood decay fungi have a fundamental role, since they are important degraders of lignocelluloses. Heterogeneity in degradation strategies consists in different enzymatic pools, conditions for secretion and catalysis, alternative non-enzymatic pathways and strategy-switch depending on environmental conditions [1,2]. Consistently, wood decay fungi shift from necrotrophism to pure saprotrophism, sometimes at an intraspecific level.

Systematic revisions based on multi-locus or genomic approach have revealed an extremely complex scenario concerning both biochemical features and morphology. Similar degradation strategies and similar morphologies are widespread, even among phylogenetically distant taxa, whereas the same taxonomic group may include species displaying different strategies and different morphologies.

As a whole, wood decayers appear to be a pivotal model in the study of the evolutionary relationships of both Dikarya and extra-Dikarya taxa [3–5]. Consistently, the incipient molecular-based biogeography of wood decayers seems to display distribution patterns strongly affected by preferred host species (trees or shrubs), which are apparently followed throughout [6].

Wood decay fungi include several edible species as well as species reported to be the source of bioactive compounds, related to either primary (e.g., β-glucans) or secondary metabolism (terpenoids, phenolics, acids, superior alcohols, etc.). Structural diversity, occurrence and distribution among taxa, synthesis stimulation factors and correlation to growth stage, bioactivity pathways, standardization of products and crude extracts are the main current topics under investigation [7–9].

Due to their relatively easy reproduction in culture, several other applications are being developed on wood decay fungi: degradation of organic pollutants and bioremediation [10]; pretreatment of biomasses for production of sugars and bioethanol [11,12]; production of enzymes for industrial purposes, namely Mn peroxidases, laccases, cellulases and hemicellulases [13]; and bioadsorption and bioaccumulation of metal ions either in living or dead biomass [14–16].

Last but not least, necrotrophic wood decay means a loss of harvest in forestry and woody cultures, whereas in public and private green areas it means destabilization and consequently risk for people and objects [17].

Culture collections are an important reference, since availability and exchange of authenticated, quality-guaranteed pure cultures are increasingly needed by researchers at an international level [18,19]. Above all, tests on different species and strains (at intraspecific level) are required since biochemical differences are often not negligible [20–22].

Actually, only a few research centers can afford to structure their culture collection in conformity to international guidelines provided by the World Federation for Culture Collections (WFCC) [23,24]. The strains maintained in many universities or research centers can be considered an important source of experimental material, even without a WFCC certification. Since small uncertified collectors are geographically widespread, their contribution may be significantly representative of local ecosystems and biodiversity [25].

The Mycology Laboratory at the Botanical Garden of Pavia University (Italy) has a long tradition of the isolation, identification and preservation of fungal strains in various areas of mycology. This is supported by the numerous publications from the middle of the last century [26] up to now. Currently, the fungal strains collection is named MicUNIPV and each working group preserves and enriches the collection.

Although the definition of ecotypes is usually hard, a remarkable intraspecific variability is well documented in several fungal species and it may be particularly true for rare species, whose populations are supposed to be more isolated [27]. This highlights the value of Italian territory for fungal biodiversity and the great potentiality for research [28].

Italy has a wide variety of climates and morphologies, both due to its remarkable latitudinal range (about 13°) and structural-topographic complexity, including the presence of four different seas and two main mountain chains. According to the official maps of MATTM (Ministero dell'Ambiente e della Tutela del Territorio e del Mare), 28 different phytoclimatic classes are recognized, five of which are specifically referred to as Mediterranean [29], also taking into account the biogeographic reference map suggested by Rivas-Martínez [30]. Nevertheless, the pluri-millennial stratification of human impact has made it difficult to distinguish between actual and potential ecosystem features. As a consequence, the classification and mapping of either Italian ecoregions or phytoclimates provide a tool for the comprehension of biodiversity instead of a strict map of biodiversity itself [31].

The present article reports the results obtained by the researchers of the Laboratory of Mycology in DSTA-University of Pavia (Italy) who continuously collect new cultures of wood decay fungi, focusing on fungal biodiversity of species related to the Mediterranean area and climates.

2. Materials and Methods

2.1. Sampling Sites and Field Work

Basidiomata were mostly (but not exclusively) collected in North and Central Italy. Sampling stratification was selectively applied, i.e., specific areas have been more frequently and strictly examined than others and sampling effort was not equal among different species [32]. The different environments examined are resumed hence:

(a) highly-fragmented marginal woodlands and shrublands placed into an agricultural landscape, particularly referring to vegetation surrounding the hydrographic network, including major lakes;
(b) mountain continuous woodlands and shrublands, both managed and unmanaged;
(c) woody cultures (e.g., poplar plantations and vineyards), tree rows and hedges in agricultural landscape;
(d) urban and suburban environments (tree rows, parks, private and public gardens).

Environments a, c, and d are mostly related to basal altitudinal belt and upper hill altitudinal belt in Po Plain, Apennines and Prealps (lower mountain thermal belt), as well as Adriatic, Tirrenian and Ligurian coasts; Environment b is mostly related to the lower and upper montane belt in the North and Central Apennines.

The basidiomata were completely or partially harvested by knife, gently brushed to eliminate debris and stored in paper bags until laboratory operations. The collecting sites were geolocalized, and the host species and general features were detected.

2.2. Experimental Procedures

Basidiomata identification was carried out by macro and micro-morphological analysis [6,33,34]; stereo and light microscopy were performed by Zeiss Axioplan and Zeiss Stemi 2000-C.

According to Stalpers [35] and Gams et al. [36], as well as Stamets [37], isolation of mycelia in pure culture was obtained in sterile conditions by inoculating small portions of the basidioma context into Petri dishes containing MEA medium and antibiotic (malt extract 2% + agar 1.5% + cloramphenicol 50 ppm). The incubation was carried out at 24 °C in the dark and each strain growth was checked constantly for a month. Based on the above, all the mentioned strains are to be regarded as dikaryotic.

Besides the morphological checks, molecular identifications of isolates were carried out on mycelia cultured in liquid medium (malt extract 2%). DNA was extracted from lyophilized mycelia by Nucleospin Plant II kit (Macherey-Nagel). Amplification by Polymerase Chain Reaction (PCR) used the primer pair ITS1 (19bp) and ITS4 (20bp)—that is, Internal Transcribed Spacer of ribosomal DNA; this region has been widely used for different fungal taxa [38,39]. PCR protocol exploited Dream Taq Mastermix (Promega) and was performed in a thermocycler, as reported in Table 1.

Table 1. Thermocycling protocol for PCR.

Step	Aim	T (°C)	Duration	Cycle Repetitions
I	Denaturation	95	5 min	
	Denaturation	95	30 s	
II	Annealing	50	45 s	35
	Elongation	72	1 min	
III	Final elongation	72	10 min	

The qualitative checking of DNA (5 μL/sample) was performed both after extraction and amplification by DNA run (30 min, 100 V) on electrophoretic gel (1% agarose). SYBR Safe-DNA Gel Stain (Invitrogen) was used as an intercalant; GeneRuler 1kb (Thermo Scientific, Waltham - USA) was used as a ladder; BlueJuice (Invitrogen) was used as a gel loading buffer. The imaging was performed by Gel Doc (Biorad, Berkeley, CA, USA).

ExoSAP-IT (Applied Biosystems, Foster City, CA, USA) was used for the purification of amplification products. According to the suggested protocol, the sample/ExoSAP ratio was 5:2 μL; the reaction was carried out in a thermocycler in two steps—15 min at 37 °C and 15 min at 80 °C.

The sequencing was ordered to Macrogen (The Netherlands). Sequence analysis was performed by Sequencher 5.0 Demo. The sequences were finally matched with the ones available in the molecular identification facility of Mycobank [40].

Strains in pure culture were stored by different methods:

(a) on malt extract agar (MEA) in a aPetri plate at 3 °C;

(b) in a glass tube corked with cotton at room temperature;

(c) colonized paper discs in demineralized water at 4 °C;

(d) at −80 °C in glycerol (selected strains only).

Periodic checking and refreshment of cultures was performed to avoid contamination and devitalization.

The strains are maintained in the Fungal Research Culture Collection (MicUNIPV) of Department of Earth and Environmental Sciences of University of Pavia (Italy); each strain is included in a private database with all the information regarding sampling sites, data of collection and ecological notes.

3. Results and Discussion

MicUNIPV includes species related to plant pathology, soil, extreme environments, fresh and marine water, monuments and cultural heritage. As previously mentioned, different working groups within the Laboratory of Mycology (DSTA-University of Pavia) are engaged in the management, preservation and improvement of each MicUNIPV section. The section regarding wood decay species has up to now achieved 500 strains belonging to 110 different species [41–43]. The broad focus on wood decay led us to include in this section species related to different applications such as nutraceuticals, forest pathology, wood degradation and biocomposite materials.

The distribution of most species exceeds the Mediterranean area; nevertheless, several of them also display wide spatial gaps among stations and clear heterogeneity in host preference depending on the geographic location of the population.

Here, we present the species that have a distribution strongly related to the Mediterranean region and/or Southern Europe and/or warm climates, according to Ryvarden and Melo [6] and Bernicchia [33,34,44]. The species related to the Mediterranean diversity are reported in Table 2 and the most peculiar are discussed below.

Table 2. Selected Italian strains from MicUNIPV related to the Mediterranean area. Phytoclimate class as in [29].

Mic UNIPV ID	Species	Authors	Locality	Municipality	Host	Phytoclimate Class
D.con.1	*Daedaleopsis confragosa*	(Bolton) J. Schröt.	Dormelletto	Dormelletto (NO)	*Unidentified broadleaf*	mesotemperate/humid supratemperate
D.con.2	*Daedaleopsis confragosa*	(Bolton) J. Schröt.	Pian Porcino	Bagno di Romagna (FC)	*Unidentified broadleaf*	hyperhumid supratemperate/ultrahyperhumid
D.q.1	*Daedalea quercina*	(L.) Pers.	R.N. Bosco Giuseppe Negri	Pavia (PV)	*Quercus robur*	humid supratemperate/subhumid
D.q.2	*Daedalea quercina*	(L.) Pers.	Cono di Volo Malpensa	Gallarate (VA)	*Quercus rubra*	mesotemperate/humid supratemperate
D.q.3	*Daedalea quercina*	(L.) Pers.	Fosso dell'Oca	Rovescala (PV)	*Quercus petraea*	humid supratemperate/subhumid
D.sq.1	*Dichomitus squalens*	(P. Karst.) D.A. Reid	Pineta di San Vitale	Ravenna (RA)	*Pinus pinea*	supratemperate/humid-subhumid mesotemperate
D.sq.2	*Dichomitus squalens*	(P. Karst.) D.A. Reid	Ispra, Lungolago	Ispra (VA)	*Cedrus sp.*	mesotemperate/humid supratemperate
D.tric.1	*Daedaleopsis tricolor*	(Bull.) Bondartsev and Singer	Rio Bardonezza	Santa Maria della Versa (PV)	*Prunus avium*	humid supratemperate/subhumid
Des.t.1	*Desarmillaria tabescens*	(Scop.) R.A. Koch and Aime	RNIS Bosco Siro Negri	Zerbolò (PV)	*Quercus robur*	humid supratemperate/subhumid
Fm.i.1	*Fomitopsis iberica*	Melo and Ryvarden	Via Montello	Varese (VA)	*Corylus avellana*	humid supratemperate/hyperhumid
Fm.i.2	*Fomitopsis iberica*	Melo and Ryvarden	Villa Baragiola	Varese (VA)	*Abies alba*	humid supratemperate/hyperhumid
Fm.i.3	*Fomitopsis iberica*	Melo and Ryvarden	Via Tasso	Varese (VA)	*Cedrus deodara*	humid supratemperate/hyperhumid
Fm.i.4	*Fomitopsis iberica*	Melo and Ryvarden	Via S. Francesco	Inarzo (VA)	*Betula pendula*	humid supratemperate/hyperhumid
Fm.i.5	*Fomitopsis iberica*	Melo and Ryvarden	Villa Toeplitz	Varese (VA)	*Fagus sylvatica*	humid supratemperate/hyperhumid
Fm.i.6	*Fomitopsis iberica*	Melo and Ryvarden	Villa Mylius	Varese (VA)	*Fagus sylvatica*	humid supratemperate/hyperhumid
Fm.m.1	*Fomitiporia mediterranea*	M. Fisch.	RNIS Bosco Siro Negri	Zerbolò (PV)	*Quercus robur*	humid supratemperate/subhumid
Fm.m.2	*Fomitiporia mediterranea*	M. Fisch.	RNIS Bosco Siro Negri	Zerbolò (PV)	*Quercus robur*	humid supratemperate/subhumid
Fm.m.3	*Fomitiporia mediterranea*	M. Fisch.	RNIS Bosco Siro Negri	Zerbolò (PV)	*Hedera helix*	humid supratemperate/subhumid
Fm.m.4	*Fomitiporia mediterranea*	M. Fisch.	Rio Bardonezza	Santa Maria della Versa (PV)	*Robinia pseudoacacia*	humid supratemperate/subhumid
Fm.m.5	*Fomitiporia mediterranea*	M. Fisch.	Comiso	Ragusa (RG)	*Cistus sp.*	mesomediterranean/subhumid-dry thermomediterranean
Fm.m.6	*Fomitiporia mediterranea*	M. Fisch.	Cono di Volo Malpensa	Gallarate (VA)	*Quercus rubra*	mesotemperate-humid supratemperate
Fm.m.7	*Fomitiporia mediterranea*	M. Fisch.	R.N. Torbiere del Sebino	Provaglio d'Iseo (BS)	*Corylus avellana*	mesotemperate-humid supratemperate
Fm.m.8	*Fomitiporia mediterranea*	M. Fisch.	Villa Augusta	Varese (VA)	*Fagus sylvatica*	humid supratemperate - hyperhumid
Fm.m.9	*Fomitiporia mediterranea*	M. Fisch.	Olgiate Comasco	Olgiate Comasco (CO)	*Actinidia chinensis*	mesotemperate-humid supratemperate

Table 2. *Cont.*

Mic UNIPV ID	Species	Authors	Locality	Municipality	Host	Phytoclimate Class
Fm.m.10	*Fomitiporia mediterranea*	M. Fisch.	Pradone nord	Rovescala (PV)	*Vitis vinifera*	humid supratemperate - subhumid
Fm.m.11	*Fomitiporia mediterranea*	M. Fisch.	Pradone nord	Rovescala (PV)	*Vitis vinifera*	humid supratemperate - subhumid
Fm.m.12	*Fomitiporia mediterranea*	M. Fisch.	Pradone nord	Rovescala (PV)	*Vitis vinifera*	humid supratemperate - subhumid
Fm.m.13	*Fomitiporia mediterranea*	M. Fisch.	Pradone nord	Rovescala (PV)	*Vitis vinifera*	humid supratemperate - subhumid
Fm.m.14	*Fomitiporia mediterranea*	M. Fisch.	Pradone nord	Rovescala (PV)	*Vitis vinifera*	humid supratemperate - subhumid
Fm.m.15	*Fomitiporia mediterranea*	M. Fisch.	Pradone nord	Rovescala (PV)	*Vitis vinifera*	humid supratemperate - subhumid
Fm.m.16	*Fomitiporia mediterranea*	M. Fisch.	Pradone nord	Rovescala (PV)	*Vitis vinifera*	humid supratemperate - subhumid
Fm.m.17	*Fomitiporia mediterranea*	M. Fisch.	Unipv_polo scientifico via Ferrata	Pavia (PV)	*Salix alba*	humid supratemperate - subhumid
G.adsp.1	*Ganoderma adspersum*	(Schulzer) Donk	Orto Botanico	Pavia (PV)	*Quercus sp.*	humid supratemperate - subhumid
G.adsp.2	*Ganoderma adspersum*	(Schulzer) Donk	R.N. Bosco Giuseppe Negri	Pavia (PV)	*Populus nigra*	humid supratemperate - subhumid
G.adsp.3	*Ganoderma adspersum*	(Schulzer) Donk	R.N. Bosco Giuseppe Negri	Pavia (PV)	*Unidentified broadleaf*	humid supratemperate - subhumid
G.adsp.4	*Ganoderma adspersum*	(Schulzer) Donk	Parco della Vernavola	Pavia (PV)	*Alnus glutinosa*	humid supratemperate - subhumid
G.adsp.5	*Ganoderma adspersum*	(Schulzer) Donk	RNIS Bosco Siro Negri	Zerbolò (PV)	*Unidentified broadleaf*	humid supratemperate - subhumid
G.adsp.6	*Ganoderma adspersum*	(Schulzer) Donk	Rio Bardonezza	Rovescala (PV)	*Quercus robur*	humid supratemperate - subhumid
G.adsp.7	*Ganoderma adspersum*	(Schulzer) Donk	Madonna del Bocco	Santa Margherita Staffora (PV)	*Quercus cerris*	supratemperate - humid mesotemperate
G.adsp.8	*Ganoderma adspersum*	(Schulzer) Donk	Morina	Rovescala	*Quercus petraea*	humid supratemperate - subhumid
G.adsp.9	*Ganoderma adspersum*	(Schulzer) Donk	RNIS Bosco Siro Negri	Zerbolò (PV)	*Quercus robur*	humid supratemperate - subhumid
G.adsp.10	*Ganoderma adspersum*	(Schulzer) Donk	RNIS Bosco Siro Negri	Zerbolò (PV)	*Populus nigra*	humid supratemperate - subhumid
G.adsp.11	*Ganoderma adspersum*	(Schulzer) Donk	RNIS Bosco Siro Negri	Zerbolò (PV)	*Quercus robur*	humid supratemperate - subhumid
G.adsp.12	*Ganoderma adspersum*	(Schulzer) Donk	Ca′ del Bosco	Nibbiano Val Tidone (PC)	*Quercus cerris*	humid supratemperate - hyperhumid
G.adsp.13	*Ganoderma adspersum*	(Schulzer) Donk	Montebolone	Pavia (PV)	*Cedrus atlantica*	humid supratemperate - subhumid
G.adsp.14	*Ganoderma adspersum*	(Schulzer) Donk	Cascina Scova	Pavia (PV)	*Unidentified broadleaf*	humid supratemperate - subhumid
G.adsp.15	*Ganoderma adspersum*	(Schulzer) Donk	Ticino	Torre d'Isola (PV)	*Quercus sp.*	humid supratemperate - subhumid
G.adsp.17	*Ganoderma adspersum*	(Schulzer) Donk	unknown	Bologna (BO)	*Unidentified broadleaf*	supratemperate/humid mesotemperate-subhumid
G.car.1	*Ganoderma carnosum*	Pat.	Foreste Casentinesi	Poppi (AR)	*Abies alba*	hyperhumid supratemperate/ultrahyperhumid

Table 2. *Cont.*

Mic UNIPV ID	Species	Authors	Locality	Municipality	Host	Phytoclimate Class
G.pf.1	*Ganoderma pfeifferi*	Bres.	Prati di Tivo	Pietracamela (TE)	*Fagus sylvatica*	supratemperate/hyperhumid mesotemperate/humid
H.e.1	*Hericium erinaceus*	(Bull.) Pers.	Colle Ciupi	Siena (SI)	*Quercus ilex*	subhumid mesotemperate/humid
H.e.2	*Hericium erinaceus*	(Bull.) Pers.	Castello di Belcaro	Siena (SI)	*Quercus ilex*	subhumid mesotemperate/humid
H.e.3	*Hericium erinaceus*	(Bull.) Pers.	Strada per Castello di Belcaro	Siena (SI)	*Quercus ilex*	subhumid mesotemperate/humid
H.e.4	*Hericium erinaceus*	(Bull.) Pers.	Strada per Castello di Belcaro	Siena (SI)	*Quercus ilex*	subhumid mesotemperate/humid
H.e.5	*Hericium erinaceus*	(Bull.) Pers.	Colle Val d'Elsa	Colle Val d'Elsa (SI)	*Quercus ilex*	subhumid mesotemperate/humid
I.t.1	*Inocutis tamaricis*	(Pat.) Fiasson and Niemelä	Apani	Brindisi (BR)	*Tamarix gallica*	thermomediterranean/mesomediterranean/dry inframediterranean/subhumid
I.t.2	*Inocutis tamaricis*	(Pat.) Fiasson and Niemelä	Ostia Lido	Roma (RM)	*Tamarix gallica*	subhumid mesomediterranean
L.s.1	*Laetiporus sulphureus*	(Bull.) Murrill	RNIS Bosco Siro Negri	Zerbolò (PV)	*Quercus robur*	humid supratemperate - subhumid
L.s.2	*Laetiporus sulphureus*	(Bull.) Murrill	RNIS Bosco Siro Negri	Zerbolò (PV)	*Quercus robur*	humid supratemperate - subhumid
L.s.3	*Laetiporus sulphureus*	(Bull.) Murrill	RNIS Bosco Siro Negri	Zerbolò (PV)	*Quercus robur*	humid supratemperate - subhumid
L.s.4	*Laetiporus sulphureus*	(Bull.) Murrill	RNIS Bosco Siro Negri	Zerbolò (PV)	*Quercus robur*	humid supratemperate - subhumid
L.s.5	*Laetiporus sulphureus*	(Bull.) Murrill	Pietragavina	Varzi (PV)	*Castanea sativa*	supratemperate - hyperhumid mesotemperate - humid
L.s.6	*Laetiporus sulphureus*	(Bull.) Murrill	Cono di Volo Malpensa	Gallarate (VA)	*Quercus sp.*	mesotemperate-humid supratemperate
L.w.1	*Cellulariella warnieri*	(Durieu and Mont.) Zmitr. and V. Malysheva	R.N. Bosco Giuseppe Negri	Pavia (PV)	*Quercus robur*	humid supratemperate - subhumid
L.w.2	*Cellulariella warnieri*	(Durieu and Mont.) Zmitr. and V. Malysheva	Bosco del Cecco	Santa Maria della Versa (PV)	*Ulmus minor*	humid supratemperate - subhumid
L.w.3	*Cellulariella warnieri*	(Durieu and Mont.) Zmitr. and V. Malysheva	Rio Marsinola-Fracion	Rovescala (PV)	*Populus nigra*	humid supratemperate - subhumid
L.w.4	*Cellulariella warnieri*	(Durieu and Mont.) Zmitr. and V. Malysheva	RNIS Bosco Siro Negri	Zerbolò (PV)	*Quercus robur*	humid supratemperate - subhumid
L.w.5	*Cellulariella warnieri*	(Durieu and Mont.) Zmitr. and V. Malysheva	RNIS Bosco Siro Negri	Zerbolò (PV)	*Robinia pseudoacacia*	humid supratemperate - subhumid
L.w.6	*Cellulariella warnieri*	(Durieu and Mont.) Zmitr. and V. Malysheva	Bosco Sforza nord-Rio Bardonezza	Ziano Piacentino (PC)	*Ulmus minor*	humid supratemperate - subhumid
P.f.1	*Perenniporia fraxinea*	(Bull.) Ryv.	R.N. Bosco Giuseppe Negri	Pavia (PV)	*Populus nigra*	humid supratemperate - subhumid
P.f.2	*Perenniporia fraxinea*	(Bull.) Ryv.	Via Scala	Pavia (PV)	*Celtis australis*	humid supratemperate - subhumid
P.f.3	*Perenniporia fraxinea*	(Bull.) Ryv.	Rio Bardonezza	Rovescala (PV)	*Salix alba*	humid supratemperate - subhumid

Table 2. *Cont.*

Mic UNIPV ID	Species	Authors	Locality	Municipality	Host	Phytoclimate Class
P.f.4	*Perenniporia fraxinea*	(Bull.) Ryv.	Via Ubaldo degli Ubaldi	Pavia (PV)	*Unidentified broadleaf*	humid supratemperate - subhumid
P.f.5	*Perenniporia fraxinea*	(Bull.) Ryv.	Via Borgo Calvenzano	Pavia (PV)	*Platanus x hispanica*	humid supratemperate - subhumid
P.f.6	*Perenniporia fraxinea*	(Bull.) Ryv.	Via Borgo Calvenzano	Pavia (PV)	*Platanus x hispanica*	humid supratemperate - subhumid
P.f.7	*Perenniporia fraxinea*	(Bull.) Ryv.	Bosco Giuseppe Negri	Pavia (PV)	*Populus nigra*	humid supratemperate - subhumid
P.f.8	*Perenniporia fraxinea*	(Bull.) Ryv.	Bosco Giuseppe Negri	Pavia (PV)	*Populus nigra*	humid supratemperate - subhumid
P.f.9	*Perenniporia fraxinea*	(Bull.) Ryv.	Cascina Venara	Zerbolò (PV)	*Populus alba*	humid supratemperate - subhumid
P.f.10	*Perenniporia fraxinea*	(Bull.) Ryv.	Pizzofreddo	Santa Maria della Versa (PV)	*Unidentified broadleaf*	mesomediterranean - humid thermotemperate - subhumid
P.f.11	*Perenniporia fraxinea*	(Bull.) Ryv.	Parco della Vernavola	Pavia (PV)	*Robinia pseudoacacia*	humid supratemperate - subhumid
P.f.12	*Perenniporia fraxinea*	(Bull.) Ryv.	Parco della Vernavola	Pavia (PV)	*Robinia pseudoacacia*	humid supratemperate - subhumid
P.f.13	*Perenniporia fraxinea*	(Bull.) Ryv.	Parco della Vernavola	Pavia (PV)	*Robinia pseudoacacia*	humid supratemperate - subhumid
P.f.14	*Perenniporia fraxinea*	(Bull.) Ryv.	Parco della Vernavola	Pavia (PV)	*Robinia pseudoacacia*	humid supratemperate - subhumid
P.f.15	*Perenniporia fraxinea*	(Bull.) Ryv.	Parco della Vernavola	Pavia (PV)	*Robinia pseudoacacia*	humid supratemperate - subhumid
P.f.16	*Perenniporia fraxinea*	(Bull.) Ryv.	Parco della Vernavola	Pavia (PV)	*Robinia pseudoacacia*	humid supratemperate - subhumid
P.f.17	*Perenniporia fraxinea*	(Bull.) Ryv.	Parco della Vernavola	Pavia (PV)	*Robinia pseudoacacia*	humid supratemperate - subhumid
P.f.18	*Perenniporia fraxinea*	(Bull.) Ryv.	Parco della Vernavola	Pavia (PV)	*Robinia pseudoacacia*	humid supratemperate - subhumid
P.f.19	*Perenniporia fraxinea*	(Bull.) Ryv.	Via D. Alighieri 25	Illasi (VR)	*Olea europaea*	humid supratemperate - subhumid
P.f.20	*Perenniporia fraxinea*	(Bull.) Ryv.	via Mascherpa	Castelvetro Piacentino (PC)	*Populus alba*	humid supratemperate - subhumid
P.f.21	*Perenniporia fraxinea*	(Bull.) Ryv.	Lungolago	Mergozzo (VCO)	*Robinia pseudoacacia*	humid supratemperate/hyperhumid
P.f.22	*Perenniporia fraxinea*	(Bull.) Ryv.	Cascina Scova	Pavia (PV)	*Robinia pseudoacacia*	humid supratemperate - subhumid
P.f.23	*Perenniporia fraxinea*	(Bull.) Ryv.	RNIS Bosco Siro Negri	Zerbolò (PV)	*Quercus robur*	humid supratemperate - subhumid
P.f.24	*Perenniporia fraxinea*	(Bull.) Ryv.	RNIS Bosco Siro Negri	Zerbolò (PV)	*Populus nigra*	humid supratemperate - subhumid
P.f.25	*Perenniporia fraxinea*	(Bull.) Ryv.	Santa Sofia	Torre d'Isola (PV)	*Quercus robur*	humid supratemperate - subhumid
P.f.26	*Perenniporia fraxinea*	(Bull.) Ryv.	Cono di Volo Malpensa	Gallarate (VA)	*Unidentified broadleaf*	mesotemperate - humid supratemperate
P.f.27	*Perenniporia fraxinea*	(Bull.) Ryv.	Viale Gorizia - Mura Spagnole	Pavia (PV)	*Celtis australis*	humid supratemperate - subhumid
P.f.28	*Perenniporia fraxinea*	(Bull.) Ryv.	Fossone - Bosco della Fame	Rovescala (PV)	*Populus nigra*	humid supratemperate - subhumid
P.m.1	*Perenniporia meridionalis*	Decock and Stalpers	R.N. Regionale Piramidi di Zone	Zone (BS)	*Quercus robur*	hyperhumid supratemperate/humid
P.m.2	*Perenniporia meridionalis*	Decock and Stalpers	Sormano	Castellina in Chianti (SI)	*Olea europaea*	supratemperate/humid mesotemperate/hyperhumid

Table 2. *Cont.*

Mic UNIPV ID	Species	Authors	Locality	Municipality	Host	Phytoclimate Class
P.och.1	*Perenniporia ochroleuca*	(Berk.) Ryvarden	Belcaro	Siena (SI)	*Quercus ilex*	subhumid mesotemperate/humid
P.och.2	*Perenniporia ochroleuca*	(Berk.) Ryvarden	Le Manie	Savona (SV)	*Quercus ilex*	mesomediterranean/dry-subhumid thermomediterranean
Ph.c.1	*Phellinus contiguus*	(Pers.) Pat.	Rio Bardonezza	Ziano Piacentino (PC)	*Robinia pseudoacacia*	humid supratemperate - subhumid
Ph.tor.1	*Phellinus torulosus*	(Pers.) Bourdot and Galzin	Ticino	Torre d'Isola (PV)	*Prunus avium*	humid supratemperate - subhumid
Pl.e.1	*Pleurotus eryngii*	(DC.) Quél.	Aidomaggiore	Aidomaggiore (OR)	*unidentified*	subhumid mesomediterranean
Pl.e.2	*Pleurotus eryngii*	(DC.) Quél.	Spadafora	Spadafora (ME)	*unidentified*	thermomediterranean/subhumid mesomediterranean
Punc.s.1	*Punctularia strigosozonata*	(Schwein.) P.H.B. Talbot	Bosco di Bauli'	Palazzolo Acreide (SR)	*Quercus sp.*	mesomediterranean/dry thermomediterranean-subhumid
Sp.p.1	*Spongipellis pachyodon*	(Pers.) Kotl. and Pouzar	Rio Marsinola-Fracion	Rovescala (PV)	*Prunus avium*	humid supratemperate - subhumid
Sp.p.2	*Spongipellis pachyodon*	(Pers.) Kotl. and Pouzar	Civezza	Imperia (IM)	*Quercus pubescens*	mesomediterranean/dry thermomediterranean-subhumid

The species reported in Table 2 represent about one in five of the comprehensive collection of wood decay species in MicUNIPV. The temperate region and Mediterranean region in Italy are reciprocally intersected and several phytoclimates are represented based on both thermal–pluviometrical parameters and floristic–vegetational ones. According to our field observations, this has the consequence that several species can be found in different phytoclimates and on different hosts, whereas a minor fraction is strictly related to one or few hosts.

Daedaleopsis confragosa and *D. tricolor* are easily distinguished by morphology; nevertheless, ITS sequences are important to discriminate the species strain. *D. tricolor* seems more common in Central and Southern Europe; in Italy, it has been reported in seven out of 20 regions [45]. Our strain (MicUNIPV D.tric.1) comes from the lower Apennines in Pavia Province; other field observations suggest that *Prunus avium* is the favourite host of *D. tricolor* in North and Central Italy. The strain has not been characterized yet, although pharmacological effects have been reported [46].

Daedalea quercina has been reported in 11/20 Italian regions [45]; as expected, all MicUNIPV strains were isolated from *Quercus* spp. Nevertheless, strain MicUNIPV D.q.1 efficiently colonized poplar wood chips and confirmed that this species is a typical brown rot agent [47].

Despite being apparently cosmopolitan, *Desarmillaria tabescens* is strictly related to *Quercus* in warm climates, where it behaves as a secondary pathogen [48]. *D. tabescens* has been reported in 16/20 Italian Regions [45]. Accordingly, our strain was isolated from roots of *Q. robur* in RNIS Bosco Siro Negri (Pavia, Italy), which is a significant, unmanaged residue of typical forest of the western Po Plain.

Fomitopsis iberica is a rare species, reported in three Italian regions [45]. All the strains in MicUNIPV were isolated close to Varese lakes, either on broadleaves or conifers.

Ganoderma is represented in MicUNIPV by seven species: *G. adspersum*, *G. applanatum*, *G. carnosum*, *G. pfeifferi*, *G. lucidum*, *G. resinaceum* and *G. valesiacum*. This genus has been intensely studied due to its wide range of secondary metabolites, including several bioactive compounds [49]. According to Ryvarden and Melo [6], *Ganoderma* is one of the most difficult genera to identify at species level. As reported in Table 2, we obtained strains of *G. adspersum*, confirming that is a southern species in Europe [50]. *G. carnosum* is usually located in the *G. lucidum* complex due to its morphological similarity, despite it showing clear differences in host relationship. Our strain was isolated from its type-locality in Italy, i.e., a forest of *Abies alba*, that is likely to be its preferred host in South Europe [33]. Molecular identification by ITS region met difficulties in discriminating *G. carnosum* from *G. tsugae* and *G. oregonensis*; this topic presents questions about the real interspecific diversity within this conifer-related species in *Ganoderma*. Strains in pure culture will thus help us to investigate both the molecular and morphological nested diversity in this complex. An analogous problem concerns *G. pfeifferi*, as it partially shares its trophic niche with *G. lucidum* and *G. resinaceum*.

L. sulphureus is particularly related to *Quercus* according to our observations in North Italy, although *Castanea sativa* is also represented. Interestingly, *C. sativa* is also one of the favourite hosts of *Phellinus torulosus*, according to field observations, despite strain MicUNIPV Ph.tor.1 being isolated from *P. avium*.

Genus *Perenniporia* is represented in MicUNIPV by *P. fraxinea*, *P. meridionalis* and *P. ochroleuca*. According to our field observation, *P. fraxinea* is more common and widespread than expected, particularly in urban areas. We have focused our attention in indentifying strains, which, to date, number 27. Some of the isolated strains have been used for population studies and tests on heavy metal bioaccumulation [16,51].

Pleurotus eryngii is a typically Mediterranean species as well as its herbaceous hosts in *Apiaceae*. Consistently, the MicUNIPV Pl.e.1 and Pl.e.2 strains were isolated in properly Mediterranean areas (Sardinia and Sicily).

Punctularia strigosozonata is a rare, poorly studied species, typically related to the Mediterranean area; its resupinate morphology increases the difficulty in achieving pure isolates.

Spongipellis pachyodon has a mainly central–southern distribution in Europe; according to Onofri et al. [45], in Italy it is known in five out of 20 Regions, not including either Lombardy or Liguria.

This species is reported as uncommon but locally abundant; regarding this, our field observations suggest that the population in the Pavia-Piacenza Apennines is particularly related to *P. avium*.

Further species listed in Table 2 are reported below in more detail owing to their taxonomic controversy or potential applications.

3.1. *Cellulariella Warnieri (Durieu and Mont.) Zmitr. and V. Malysheva*

3.1.1. Background

As detailed in Table 2, the basidiomata of some species were collected in the Mediterranean area and others that are known to prefer warm environment, even if they were collected in continental or temperate zones. An example is *Cellulariella warnieri*, a poorly investigated species related to warm climates, according to Bernicchia and Gorjón [44]; despite not strictly being related to the Mediterranean region, Ryvarden and Melo [6] reported it as a southern and rare species.

The notable scarcity of data about this species has probably contributed to its uncertain systematic and taxonomic status. Currently it is reported as: *Lenzites warnieri* Durieu and Mont. by Mycobank [40], *C. warnieri* by Index Fungorum [52] and *Trametes warnieri* (Durieu and Mont.) Zmitr., Wasser and Ezhov by Ryvarden and Melo [6]. The latter indication is suggested also by Justo and Hibbett [53] based on a five marker-based phylogenetic classification of *Trametes*. Significantly, only 108 records for this species have been reported by the GBIF (Global Biodiversity Information Facility) [54] and only 10 sequences are available in GenBank, almost half of them being critical as they are reported from South East Asia [55]. Further analyses on a more representative number of strains are thus needed to clarify the position of *C. warnieri*.

Strain MicUNIPV L.w.1 was tested for the evaluation of lignocellulolytic activity and resulted in a very low production of Mn peroxydase and lack of lignin peroxydase, whereas cellulase and hemicellulase had the highest presentation among the species under examination [47]. This was also confirmed when testing the effect of its colonization on *Medicago sativa* for pre-treatment, as cellulose and hemicellulose were preferentially removed [56].

3.1.2. MicUNIPV WDF Strains Results

According to our field observations, localities are distributed as small local clusters which are very scattered in turn. Thus, the strains MicUNIPV L.w.1, L.w.4 and L.w.5 were collected from Po plain areas (Pavia and RNIS Bosco Siro Negri), whereas strains the MicUNIPV L.w.2, L.w.3 and L.w.6 were collected from hill area (Oltrepo Pavese).

The six strains of MicUNIPV were collected from *Quercus*, *Ulmus*, *Populus* and *Robinia*; interestingly, our field observations pointed out some preference for *Ulmus*, which was not previously reported as a host in Italy. As expected, all the strains were isolated from individuals behaving as saprotrophs. It should be noted that, since *C. warnieri* develops basidiomata in late autumn but it releases spores in spring, the basidioma itself remains vital even at low temperatures and under the snow.

3.2. *Dichomitus squalens (P. Karst.) D.A. Reid*

3.2.1. Background

Dichomitus squalens is a model species for studies about the selectivity of white rot and its enzymatic basis [57–60]. Despite being reported all throughout the boreal emisphere, it appears scattered and is commonly found in the northern parts of Europe, North America and Asia [61]; the GBIF [54] places the wide majority of records in the Fennoscandian region. The host relationship is apparently controversial and surprising: Ryvarden and Melo [6] assumed *Pinus* as the only European host species, whereas Bernicchia and Gorjón [44] recorded *Picea abies* for the Italian sample and Niemelä [62] assigned most samples to *Pinus* and a smaller fraction to *Picea abies* in Białowieża Forest (Poland / Belarus). Nevertheless, it should be noted that American samples have been reported on six different genera in *Pinaceae*. Furthermore, young basidiomata of *D. squalens* are easily misidentified due to the close morphological resemblance with *Neoantrodia serialis* (Fr.) Audet and related species. Consistently, a remarkable intraspecific variability in growth and enzyme profiles was revealed by testing different monokarya strains [61]. This is also consistent with the numerous mating types deriving from tetrapolarity [61].

3.2.2. MicUNIPV WDF Strains Results

Strains MicUNIPV D.sq.1 and MicUNIPV D.sq.2 were recovered from *Pinus pinea* and *Cedrus* sp. respectively near the Adriatic Sea and Varese Lake. The Italian strains have not yet been investigated for their enzymatic properties, so they may provide an additional tool to explore the diversity in degradation potential of this selective decayer.

3.3. Hericium Erinaceus (Bull.) Pers.

3.3.1. Background

Hericium erinaceus (Bull.) Pers. is one of the most famous cultivated medicinal species in the world; a wide range of peculiar compounds, both related to primary (e.g., β-glucans) and secondary metabolism (e.g., erinacines and hericenones) have been up to now characterized and screened for bioactivity [7,63–66]. GBIF [54] places the wide majority of *H. erinaceus* sites in Europe, North America and North Eastern Asia. According to the phylogenetic study by Cesaroni et al. [67], a subclade containing European and American ITS sequences is well distinguished from the Asian clade. Despite relying on ITS region only, these data suggest the possibility to differentiate *H. erinaceus* strains also by the phylogeographic structure. Notwithstanding the scarcity of available data for Asian samples, *H. erinaceus* apparently has a quite broad trophic niche including several host species in *Fagaceae,* and *Aceraceae* to lesser extent, and particularly showing a preference for *Quercus* all throughout its distribution area [37]. Consistently, European samples have mainly been recovered from *Quercus* and *Fagus*, the former likely being the exclusive host in Italy and the only known host in North Africa [34,68]. Strain MicUNIPV H.e.2 was analyzed for the production of erinacine A and hericenes (presumably A, B, C, D). Thus, a complete quali-quantitative comparison of these selected metabolites was provided throughout different growth stages but within the same strain, which is a powerful tool for the standardization of bioactive products [69].

Strain MicUNIPV H.e.1 was selected to test the effect of oral supplementation on mice memory. The results indicate an improvement in recognition memory and induction of hyppocampal and cerebellar neurogenesis during aging. This strain has therefore contributed to pointing out which areas are directly involved in the neuroactivity of *H. erinaceus* compounds, highlighting which type of memory is increased [70].

3.3.2. MicUNIPV WDF Strains Results

Accordingly, all four strains in the MicUNIPV collection were recovered in the municipality of Siena from *Q. ilex*, that is, a featuring species in the flora of Mediterranean area often forming homogeneous woodlands. It is noteworthy that the Mediterranean basin hosts a great variety of *Quercus* species, whose phylogenetic and systematic relationships are still controversial, with particular concern to the *Q. ilex* group [71–73].

3.4. Inocutis Tamaricis (Pat.) Fiasson and Niemelä

3.4.1. Background

The relationship of *Inocutis tamaricis* with *Tamarix* is apparently so strict to be regarded as a discriminant character in identification [6,33,44]. Although the genus *Tamarix* consists of 72 accepted species in Europe, Asia and Africa [74], *I. tamaricis* is restricted to the Mediterranean basin and

Macaronesia. Here, it grows on different *Tamarix* species according to their availability but shows a preference for *T. gallica* [75,76]. Consistently, strains MicUNIPV I.t.1 and I.t.2 were both isolated from *T. gallica*. As a whole, the intra-familiar phylogeny of *Hymenochaetaceae* is still to be clarified; multiple revisions have tried to point out nested diversity within polyphyletic taxa, such as *Inonotus* [77,78]. Thus, the genus *Inocutis* is nowadays accepted to be distinct from *Inonotus* itself, as formerly suggested by Fiasson and Niemelä [79]. Interestingly, the type-species for *Inocutis* is *I. rheades* (Pers.) Fiasson and Niemelä, which is morphologically very similar to *I. tamaricis* and is mostly distinguished by host and distribution [6]. Thus, *I. tamaricis* may be regarded as the Mediterranean counterpart of *I. rheades*. As a whole, only 12 sequences have been up to now deposited in GenBank [55] as belonging to this species, some of which are lacking data to assess their effective reliability. Further sequences from the Mediterranean area, equipped with information about host and geographic origin, are needed to support studies about intrageneric diversity in *Inocutis* in the light of biogeographic patterns.

3.4.2. MicUNIPV WDF Strains Results

Strains MicUNIPV I.t.1 and I.t.2 were both isolated from *T. gallica*, forming in both cases ornamental rows along the sea coast.

3.5. *Fomitiporia Mediterranea M. Fisch.*

3.5.1. Background

As mentioned for *Inonotus*, the genus *Phellinus* is increasingly revealing its hidden diversity; recognized as being polyphyletic, several species have been distributed into other genera, such as *Fomitiporia* Murrill. *Fomitiporia mediterranea* is a peculiar example due to its morphology, being actually indistinguishable from *P. punctatus*. According to Fischer [80], these two species also show differences in growth rate at selected temperatures and mating behaviour. The same study provides strong evidence for dichotomy in host selection by *F. mediterranea* depending on biogeography, i.e., this species grows on several tree species in Italy [81,82], whereas north of the Alps it apparently grows on *Vitis vinifera* exclusively [83,84].

It should be considered that misidentification with *P. punctatus* has probably led to the underestimation of *F. mediterranea* in the Mediterranean area [85]. Analogously, Polemis et al. [86] suggested that the relationship with *P. pseudopunctatus* A. David, Dequatre and Fiasson should be reconsidered as well, enclosing the latter in *F. mediterranea* clade. It may be observed that the *P. pseudopunctatus* is apparently more related to the South Mediterranean region and climates [87,88]. Further analyses on strains from different geographic origins and hosts are thus needed to clarify both phylogenetic relationships and biogeographic patterns. As a whole, this species complex is characterized by intense necrotrophic white rot; *F. mediterranea* in particular is regarded as one of the main agents responsible for wood rot in *V. vinifera*, *Corylus avellana* and *Olea europaea* [82,83,85].

3.5.2. MicUNIPV WDF Strains Results

All of the 17 Italian strains up to now attained by the Laboratory of Mycology DSTA–University of Pavia were assigned to *F. mediterranea* instead of *P. punctatus* and recovered from different substrates in North Italy. Even within one province (Pavia), *F. mediterranea* was located on five hosts, namely *Q. robur*, *Hedera helix* (State Natural Strict Reserve Bosco Siro Negri), *Salix alba* (University of Pavia courtyard), *R. pseudocacacia* and *V. vinifera* (Oltrepo Pavese hills).

The identified strains thus provide a tool to deepen pathology dynamics and different susceptibility depending on host species and cultivar.

3.6. Perenniporia meridionalis Decock and Stalpers

3.6.1. Background

Genus *Perenniporia* Murrill sensu lato is large, cosmopolitan and supposed to be polyphyletic, and thus is in need of further phylogenetic analysis based on sequences from different species [6]. In turn, an example of intrageneric complexity is provided by *P. meridionalis*, within *P. medulla-panis* (Jacq.) Donk group. Actually, the complete revision by Decock and Stalpers [89] arose the doubt that several records, as well as specimens in herbaria, are to be referred to *P. meridionalis* instead of *P. medulla-panis* (or closely related species), particularly when coming from the Mediterranean area. Up to now, only a partial and fragmentary investigation into the intra-generic diversity in *Perenniporia* has been carried out from a molecular and phylogenetic point of view [90]. It is noteworthy that no sequences at all are available in GenBank by the name of *P. meridionalis*, whereas 40 sequences are referred to *P. medulla-panis* [55]. Strains from culture collections are thus needed as basic material for this purpose. According to the indications by both Bernicchia and Gorjón [44], as well as Ryvarden and Melo [6], *P. meridionalis* is particularly related to *Quercus* (more than *P. medulla-panis*), the holotype having been isolated from *Q. ilex* in Sardinia [62,89].

3.6.2. MicUNIPV WDF Strains Results

Strains MicUNIPV P.m.1 and P.m.2 strains were respectively isolated from *Q. robur* (North Italy, near a lake) and *Q. ilex* (Central Italy). Besides considerations on biodiversity, *P. meridionalis* has a great applicative potential. MicUNIPV P.m.1 showed remarkable selectivity as a white rot agent and versatility when inoculated onto unusual substrates such as grass. The selective removal of lignin by this species contemporarily relies on high activity for Mn peroxidases and very low for one cellulase; the final delignification in the substrate is clear both in thermogravimetric analysis and FTIR spectroscopy [12,47].

3.7. Perenniporia ochroleuca (Berk.) Ryvarden

3.7.1. Background

Perenniporia ochroleuca is another example of the unsolved intra-generic diversity within *Perenniporia*. This species is suspected to hide a complex, and transfer to *Truncospora* Pilát ex Pilát has thus been suggested [91]. According to the same authors, the Iberian/Macaronesian clade gives *T. atlantica* Spirin and Vlasák, whereas the status of Australian samples is more uncertain, which would mainly belong to *T. ochroleuca*. Nevertheless, the new taxonomy has not yet been fully accepted, neither by Mycobank [40] nor by Index Fungorum [52]. These hypotheses therefore need to be supported by entering further sequences into the phylogenetic analyses from an exhaustive geographic range. *P. ochroleuca* was reported by Bernicchia and Gorjón [44] and Ryvarden and Melo [6] as tropical and growing on several hosts, whereas in Europe it is particularly related to the Mediterranean area. Nevertheless, Bernicchia and Gorjón [44] report a range of typically Mediterranean hosts, whereas Ryvarden and Melo [6] also include host plants whose distribution exceeds the Mediterranean area to include samples from the coasts of South England and Wales. Further phylogenetic analyses focused on the Mediterranean region versus the adjacent Atlantic ones are needed to test the monophyly of the proposed *T. atlantica*.

3.7.2. MicUNIPV WDF Strains Results

Both strains MicUNIPV P.och.1 and P.och.2 were isolated from *Q. ilex* in Central Italy and the Ligurian west coast, respectively.

4. Conclusions

At the moment, MicUNIPV, the fungal research culture collection of University of Pavia (Italy), maintains 500 strains from wood decay species. Examples particularly correlated to the Mediterranean area were discussed and their roles in accomplished research were mentioned in this study.

Culture collections of wood decay fungi are an important tool both for systematic and applied studies. Strains in pure culture are more easily and reliably identified and analyzed for metabolic activities and competitivity. The environmental features of the strain origin place have often been underestimated; nevertheless, the diversity of wood decay fungi strongly depends on biogeography and is related to host distribution. This also highlights the need for an investigation including a wider concept of the Mediterranean region than one strictly limited by climate or phytoclimate classification, i.e., even continental regions surrounding the Mediterranean area contribute to the explanation of Mediterranean diversity.

The Laboratory of Mycology DSTA–University of Pavia (Italy) has up to now successfully collaborated with both researchers from other universities and amateurs in order to increase the diversity richness and geographic origin range of strains, as well as to enter these strains in original pure and applied research such as MATER and CE4WE (grants from Cariplo Foundation and Regione Lombardia).

Author Contributions: Conceptualization, E.S. and C.E.G.; Methodology, E.S., A.B., A.M.P., C.E.G., R.M.B. and M.C.; Investigation, E.S., A.B., C.E.G., R.M.B., S.B. and M.C. Resources, E.S., A.M.P. and F.B.; Data Curation, E.S., A.B. and C.E.G.; Writing – Original Draft Preparation, C.E.G., S.B.; Writing – Review and Editing, C.E.G., E.S., A.B., A.M.P.; Supervision, E.S., A.M.P.; Project Administration, E.S.; Funding Acquisition, E.S., F.B. All authors have read and agreed to the published version of the manuscript.

Acknowledgments: The State Natural Strict Reserve (RNIS) Bosco Siro Negri – Zerbolò for support and authorization for sampling; the Botanical Garden of Pavia (Orto Botanico di Pavia) for authorization for sampling. The authors are also grateful to C. Perini, University of Siena (Italy), Valentina Cesaroni and Fabio Savino for providing some original specimens.

References

1. Liers, C.; Arnstadt, T.; Ullrich, R.; Hofrichter, M. Patterns of lignin degradation and oxidative enzyme secretion by different wood-and litter-colonizing basidiomycetes and ascomycetes grown on beech-wood. *FEMS Microbiol. Ecol.* **2011**, *78*, 91–102. [CrossRef] [PubMed]
2. Van den Brink, J.; de Vries, R.P. Fungal enzyme sets for plant polysaccharide degradation. *Appl. Microbiol. Biotechnol.* **2011**, *91*, 1477. [CrossRef] [PubMed]
3. Riley, R.; Salamov A., A.; Brown, D.W.; Nagy, L.G.; Floudas, D.; Held, B.W.; Levasseur, A.; Lombard, V.; Morin, E.; Otillar, R.; et al. Extensive sampling of basidiomycete genomes demonstrates inadequacy of the white-rot/brown-rot paradigm for wood decay fungi. *Proc. Natl. Acad. Sci. USA* **2014**, *111*, 9923–9928. [CrossRef]
4. Hibbett, D.S.; Bauer, R.; Binder, M.; Giachini, A.J.; Hosaka, K.; Justo, A.; Larsson, E.; Larsson, K.H.; Lawrey, J.D.; Miettinen, O.; et al. Agaricomycetes. In *the Mycota*, 2nd ed.; McLaughlin, D.J., Spatafora, J.W., Eds.; Springer: Berlin, Germany, 2014; Volume 7, pp. 373–429. [CrossRef]
5. Nagy, L.G.; Riley, R.; Tritt, A.; Adam, C.; Daum, C.; Floudas, D.; Sun, H.; Yadav, J.S.; Pangilinan, J.; Larsson, K.H.; et al. Comparative genomics of early-diverging mushroom-forming fungi provides insights into the origins of lignocellulose decay capabilities. *Mol. Biol. Evol.* **2015**, *33*, 959–970. [CrossRef] [PubMed]
6. Ryvarden, L.; Melo, I. *Poroid Fungi of Europe*, 2nd ed.; Fungiflora: Oslo, Norway, 2017.
7. Ma, B.J.; Shen, J.W.; Yu, H.Y.; Ruan, Y.; Wu, T.T.; Zhao, X. Hericenones and erinacines: Stimulators of nerve growth factor (NGF) biosynthesis in Hericium erinaceus. *Mycology* **2010**, *1*, 92–98. [CrossRef]

8. Dresch, P.; Rosam, K.; Grienke, U.; Rollinger, J.M.; Peintner, U. Fungal strain matters: Colony growth and bioactivity of the European medicinal polypores Fomes fomentarius, Fomitopsis pinicola and Piptoporus betulinus. *AMB Express* **2015**, *5*, 4. [CrossRef]
9. Angelini, P.; Girometta, C.; Tirillini, B.; Moretti, S.; Covino, S.; Cipriani, M.; D'Ellena, E.; Angeles, G.; Federici, E.; Savino, E.; et al. A comparative study of the antimicrobial and antioxidant activities of Inonotus hispidus fruit and their mycelia extracts. *Int. J. Food Prop.* **2019**, *22*, 768–783. [CrossRef]
10. Pozdnyakova, N.N. Involvement of the ligninolytic system of white-rot and litter-decomposing fungi in the degradation of polycyclic aromatic hydrocarbons. *Biotechnol. Res. Int.* **2012**, *2012*. [CrossRef]
11. Giles, R.L.; Galloway, E.R.; Zackeru, J.C.; Naithani, V.; Parrow, M.W. Two stage fungal biopulping solubilizes lignocellulosic carbohydrates without supplemental enzymatic hydrolysis. *Int. Biodeter. Biodegradation* **2014**, *86*, 265–271. [CrossRef]
12. Girometta, C.; Zeffiro, A.; Malagodi, M.; Savino, E.; Doria, E.; Nielsen, E.; Buttafava, A.; Dondi, D. Pretreatment of alfalfa stems by wood decay fungus Perenniporia meridionalis improves cellulose degradation and minimizes the use of chemicals. *Cellulose* **2017**, *24*, 3803–3813. [CrossRef]
13. Mendonça Maciel, M.J.; Ribeiro, H.C.T. Industrial and biotechnological applications of ligninolytic enzymes of the basidiomycota: A review. *Electron. J. Biotechnol.* **2010**, *13*, 14–15. [CrossRef]
14. Gadd, G.M. Biosorption: Critical review of scientific rationale, environmental importance and significance for pollution treatment. *J. Chem. Technol. Biot.* **2009**, *84*, 13–28. [CrossRef]
15. Javaid, A.M.N.A.; Bajwa, R.; Manzoor, T. Biosorption of heavy metals by pretreated biomass of Aspergillus niger. *Pak. J. Bot.* **2011**, *43*, 419–425.
16. Sturini, M.; Girometta, C.; Maraschi, F.; Savino, E.; Profumo, A. A Preliminary Investigation on Metal Bioaccumulation by Perenniporia fraxinea. *B Environ. Contam. Tox* **2017**, *98*, 508–512. [CrossRef]
17. Arciniegas, A.; Prieto, F.; Brancheriau, L.; Lasaygues, P. Literature review of acoustic and ultrasonic tomography in standing trees. *Trees* **2014**, *28*, 1559–1567. [CrossRef]
18. CBS-KNAW Collections. Available online: www.cbs.knaw.nl (accessed on 29 December 2019).
19. All Russian Collection of Microorganisms—VKM. Available online: www.vkm.ru (accessed on 29 December 2019).
20. Baldrian, P.; Gabriel, J. Intraspecific variability in growth response to cadmium of the wood-rotting fungus Piptoporus betulinus. *Mycologia* **2002**, *94*, 428–436. [CrossRef]
21. Pawlik, A.; Janusz, G.; Dębska, I.; Siwulski, M.; Frąc, M.; Rogalski, J. Genetic and metabolic intraspecific biodiversity of Ganoderma lucidum. *BioMed Res. Int.* **2015**, *2015*. [CrossRef]
22. Zeng, Z.; Sun, H.; Vainio, E.J.; Raffaello, T.; Kovalchuk, A.; Morin, E.; Duplessis, S.; Asiegbu, F.O. Intraspecific comparative genomics of isolates of the Norway spruce pathogen (Heterobasidion parviporum) and identification of its potential virulence factors. *BMC Genom.* **2018**, *19*, 220. [CrossRef]
23. WFCC—World Federation for Culture Collections. Available online: www.wfcc.info (accessed on 29 December 2019).
24. MIRRI—Microbial Resorurce Research Infrastructure. Available online: www.mirri.org (accessed on 29 December 2019).
25. Gargano, M.L. Mycotheca of edible and medicinal mushrooms at herbarium SAF as a potential source of nutraceuticals and cultivated mushrooms. *Int. J. Med. Mushrooms* **2018**, *20*, 405–409. [CrossRef]
26. Caretta, G. Micologo a Pavia. In *Raffaele Ciferri Scienziato Versatile e Critico, Cisalpino*; Istituto Editoriale Universitario: Milano, Italy, 2000; pp. 51–54.
27. Lowry, D.B. Ecotypes and the controversy over stages in the formation of new species. *Biol. J. Linn. Soc.* **2012**, *106*, 241–257. [CrossRef]
28. Angelini, P.; Compagno, R.; Arcangeli, A.; Bistocchi, G.; Gargano, M.L.; Venanzoni, R.; Venturella, G. Macrofungal diversity and ecology in two Mediterranean forest ecosystems. *Plant Biosyst.* **2016**, *150*, 540–549. [CrossRef]
29. Ministero Dell'ambiente e Della Tutela del Territorio e del Mare. Geoportale Nazionale—Nuovo Visualizzatore. Available online: http://www.pcn.minambiente.it/viewer/ (accessed on 28 December 2019).
30. Rivas-Martínez, S.; Penas, Á.; Díaz, T.E. Bioclimatic and Biogeographic Maps of Europe. 1:16.000.000 University of León, E-24071, Spain. 2004. Available online: http://www.globalbioclimatics.org/form/bg_mEd.htm (accessed on 28 December 2019).

31. Blasi, C.; Capotorti, G.; Copiz, R.; Guida, D.; Mollo, B.; Smiraglia, D.; Zavattero, L. Classification and mapping of the ecoregions of Italy. *Plant Biosyst.* **2014**, *148*, 1255–1345. [CrossRef]
32. Sutherland, W.J. *Ecological Census Techniques: A Handbook*, 2nd ed.; Cambridge University Press: Cambridge, UK, 2006.
33. Bernicchia, A. *Polyporaceae sl.*; Candusso: Alassio (SV), Italy, 2005.
34. Bernicchia, A.; Gorjón, S.P. *Corticiaceae sl.*; Candusso: Alassio (SV), Italy, 2010.
35. Stalpers, J.A. Identification of wood-inhabiting fungi in pure culture. *Stud. Mycol.* **1978**, *16*, 1–248.
36. Gams, W.; Hoekstra, E.S.; Aptroot, A. *CBS Course of Mycology*; Centraalbureau voor Schimmelcultures Baarn: Delft, NL, USA, 1998.
37. Tamets, P. *Growing Gourmet and Medicinal Mushrooms*; Ten Speed Press: Berkeley, CA, USA, 2011.
38. Toju, H.; Tanabe, A.S.; Yamamoto, S.; Sato, H. High-coverage ITS primers for the DNA-based identification of ascomycetes and basidiomycetes in environmental samples. *PLoS ONE* **2012**, *7*, e40863. [CrossRef]
39. Nilsson, R.H.; Hyde, K.D.; Pawlowska, J.; Ryberg, M.; Tedersoo, L.; Aas, A.B.; Alias, S.A.; Alves, A.; Anderson, C.L.; Antonelli, A.; et al. Improving ITS sequence data for identification of plant pathogenic fungi. *Fungal Divers* **2014**, *67*, 11–19. [CrossRef]
40. Mycobank. Available online: www.mycobank.org (accessed on 19 December 2019).
41. Altobelli, E.; Bernicchia, A.; Pecoraro, L.; Savino, E. Raccolta, isolamento e coltivazione di funghi poliporoidi con proprietà medicinali. *Micol. Ital.* **2012**, *41*, 3–10.
42. Savino, E.; Girometta, C.; Chinaglia, S.; Guglielminetti, M.; Rodolfi, M.; Bernicchia, A.; Perini, C.; Salerni, E.; Picco, A.M. Medicinal mushrooms in Italy and their ex situ conservation through culture collection. In Proceedings of the 8th International Conference on Mushroom Biology and Mushroom Products, New Delhi, India, 19–22 Novemver 2014; pp. 50–54.
43. Savino, E.; Girometta, C.; Miteva-Staleva, J.; Kostadinova, A.; Krumova, E. Wood decay macrofungi: Strain collection and studies about antioxidant properties. *Comptes Rendus l'Académie Bulg. Sci.* **2016**, *69*, 747–755.
44. Bernicchia, A.; Gorjón, S.P. *Polypores of Mediterranean Region*; S.P. Publishers: Bologna, Italy, 2019; in press.
45. Onofri, S.; Bernicchia, A.; Filipello Marchisio, V.; Padovan, F.; Perini, C.; Ripa, C.; Salerni, E.; Savino, E.; Venturella, G.; Vizzini, A.; et al. *Checklist dei Funghi Italiani Basidiomycetes, Basidiomycota*, 1st ed.; Carlo Delfino Editore: Sassari, Italy, 2005; pp. 1–380.
46. Bernicchia, A.; Fugazzola, M.A.; Gemelli, V.; Mantovani, B.; Lucchetti, A.; Cesari, M.; Speroni, E. DNA recovered and sequenced from an almost 7000 y-old Neolithic polypore, Daedaleopsis tricolor. *Mycol. Res.* **2006**, *110*, 14–17. [CrossRef]
47. Doria, E.; Altobelli, E.; Girometta, C.; Nielsen, E.; Zhang, T.; Savino, E. Evaluation of lignocellulolytic activities of ten fungal species able to degrade poplar wood. *Int. Biodeter Biodegrad.* **2014**, *94*, 160–166. [CrossRef]
48. Koch, R.A.; Wilson, A.W.; Séné, O.; Henkel, T.W.; Aime, M.C. Resolved phylogeny and biogeography of the root pathogen Armillaria and its gasteroid relative, Guyanagaster. *BMC Evol. Biol.* **2017**, *17*, 33. [CrossRef]
49. Lindequist, U.; Jülich, W.D.; Witt, S. Ganoderma pfeifferi—A European relative of Ganoderma lucidum. *Phytochemistry* **2015**, *114*, 102–108. [CrossRef] [PubMed]
50. Karadelev, M.; Rusevska, K.; Kajevska, I. Distribution and ecology of Genus Ganoderma (Ganodermataceae) in the Republic of Macedonia. In Proceedings of the International Conference on Biological and Enveronmental Sciences, Tirana, Albania, 26–28 September 2008; pp. 320–326.
51. Sillo, F.; Savino, E.; Giordano, L.; Girometta, C.; Astegiano, D.; Picco, A.M.; Gonthier, P. Analysis of genotypic diversity provides a first glimpse on the patterns of spread of the wood decay fungus Perenniporla fraxinea in an urban park in northern Italy. *J. Plant Pathol.* **2016**, *98*, 617–624.
52. Index Fungorum. Available online: www.indexfungorum (accessed on 19 December 2019).
53. Justo, A.; Hibbett, D.S. Phylogenetic classification of Trametes (Basidiomycota, Polyporales) based on a five-marker dataset. *Taxon* **2011**, *60*, 1567–1583. [CrossRef]
54. GBIF—Global Biodiversity Information Facility. Available online: www.gbif.org (accessed on 28 December 2019).
55. GenBank- NCBI (National Center for Biotechnology Information). Available online: https://www.ncbi.nlm.nih.gov (accessed on 19 December 2019).
56. Zeffiro, A.; Dondi, D.; Marconi, R.P.; Malagodi, M.; Girometta, C.; Bentivoglio, A.; Lazzaroni, S.; Savino, E.; Nielsen, E.; Buttafava, A. Sugar Production for bioethanol from alfalfa stems. Results, and comparative study with application of lignocellulolytic activities of novel fungal species. In Atti del Convegno, Proceedings

of the VI Workshop nazionale AICIng "Molecules and materials: Chemistry for engineering", Roma, Italy, 22–23 June 2015.

57. Floudas, D.; Binder, M.; Riley, R.; Barry, K.; Blanchette, R.A.; Henrissat, B.; Martínez, A.T.; Otillar, R.; Spatafora, J.W. The Paleozoic origin of enzymatic lignin decomposition reconstructed from 31 fungal genomes. *Science* **2012**, *336*, 1715–1719. [CrossRef] [PubMed]
58. Rytioja, J.; Hildén, K.; Di Falco, M.; Zhou, M.; Aguilar-Pontes, M.V.; Sietiö, O.M.; Tsang, A.; de Vries, R.P.; Mäkelä, M.R. The molecular response of the white-rot fungus Dichomitus squalens to wood and non-woody biomass as examined by transcriptome and exoproteome analyses. *Environ. Microbiol.* **2017**, *19*, 1237–1250. [CrossRef] [PubMed]
59. Daly, P.; López, S.C.; Peng, M.; Lancefield, C.S.; Purvine, S.O.; Kim, Y.M.; Zink, E.M.; Dohnalkova, A.; Singan, V.R.; Lipzen, A.; et al. Dichomitus squalens partially tailors its molecular responses to the composition of solid wood. *Environ. Microbiol.* **2018**, *20*, 4141–4156. [CrossRef]
60. López, S.C.; Peng, M.; Issak, T.Y.; Daly, P.; de Vries, R.P.; Mäkelä, M.R. Induction of genes encoding plant cell wall-degrading carbohydrate-active enzymes by lignocellulose-derived monosaccharides and cellobiose in the white-rot fungus Dichomitus squalens. *Appl. Environ. Microbiol.* **2018**, *84*, e00403-18. [CrossRef]
61. L'opez, S.C.; Theelen, B.; Manserra, S.; Issak, T.Y.; Rytioja, J.; Mäkelä, M.R.; de Vries, R.P. Functional diversity in Dichomitus squalens monokaryons. *IMA Fungus* **2017**, *8*, 17. [CrossRef]
62. Niemelä, T. *Polypores of the Białowieża Forest*; Białowieski Park Narodowy: Białowieża, Poland, 2013.
63. Kawagishi, H.; Shimada, A.; Shirai, R.; Okamoto, K.; Ojima, F.; Sakamoto, H.; Ishiguro, Y.; Furukawa, S. Erinacines A, B and C, strong stimulators of nerve growth factor (NGF)-synthesis, from the mycelia of Hericium erinaceum. *Tetrahedron Lett.* **1994**, *35*, 1569–1572. [CrossRef]
64. Friedman, M. Chemistry, nutrition, and health-promoting properties of Hericium erinaceus (lion's mane) mushroom fruiting bodies and mycelia and their bioactive compounds. *J. Agric. Food Chem* **2015**, *63*, 7108–7123. [CrossRef]
65. Brandalise, F.; Cesaroni, V.; Gregori, A.; Repetti, M.; Romano, C.; Orrù, G.; Botta, L.; Girometta, C.; Guglielminetti, M.L.; Savino, E.; et al. Dietary supplementation of Hericium erinaceus increases mossy fiber-CA3 hippocampal neurotransmission and recognition memory in wild-type mice. *Evid-Based Compl. Alt.* **2017**. [CrossRef] [PubMed]
66. Rossi, P.; Cesaroni, V.; Brandalise, F.; Occhinegro, A.; Ratto, D.; Perrucci, F.; Lanaia, V.; Girometta, C.; Orrù, G.; Savino, E. Dietary supplementation of lion's mane medicinal mushroom, Hericium erinaceus (Agaricomycetes), and spatial memory in wild-type mice. *Int. J. Med. Mushrooms* **2018**, *20*, 485–494. [CrossRef] [PubMed]
67. Cesaroni, V.; Brusoni, M.; Cusaro, C.M.; Girometta, C.; Perini, C.; Picco, A.M.; Rossi, P.; Salerni, E.; Savino, E. Phylogenetic Comparison between Italian and Worldwide Hericium Species (Agaricomycetes). *Int. J. Med. Mushrooms* **2019**, *21*, 943–954. [CrossRef]
68. Ouali, Z.; Sbissi, I.; Boudagga, S.; Rhaiem, A.; Hamdi, C.; Venturella, G.; Saporita, P.; Jaouani, A.; Gargano, M.L. First report of the rare tooth fungus Hericium erinaceus in North African temperate forests. *Plant Biosyst.* **2018**, *154*, 24–28. [CrossRef]
69. Corana, F.; Cesaroni, V.; Mannucci, B.; Baiguera, R.M.; Picco, A.M.; Savino, E.; Ratto, D.; Perini, C.; Kawagishi, H.; Girometta, C.E.; et al. Array of Metabolites in Italian Hericium erinaceus Mycelium, Primordium, and Sporophore. *Molecules* **2019**, *24*, 3511. [CrossRef] [PubMed]
70. Ratto, D.; Corana, F.; Mannucci, B.; Priori, E.C.; Cobelli, F.; Roda, E.; Ferrari, B.; Occhinegro, A.; Di Iorio, C.; De Luca, F.; et al. Hericium erinaceus Improves recognition memory and induces hippocampal and cerebellar neurogenesis in frail mice during aging. *Nutrients* **2019**, *11*, 715. [CrossRef]
71. Manos, P.S.; Doyle, J.J.; Nixon, K.C. Phylogeny, biogeography, and processes of molecular differentiation in Quercus subgenus Quercus (Fagaceae). *Mol. Phylogenet. Evol.* **1999**, *12*, 333–349. [CrossRef]
72. Simeone, M.C.; Piredda, R.; Papini, A.; Vessella, F.; Schirone, B. Application of plastid and nuclear markers to DNA barcoding of Euro-Mediterranean oaks (Quercus, Fagaceae): Problems, prospects and phylogenetic implications. *Bot. J. Linn. Soc.* **2013**, *172*, 478–499. [CrossRef]
73. Hubert, F.; Grimm, G.W.; Jousselin, E.; Berry, V.; Franc, A.; Kremer, A. Multiple nuclear genes stabilize the phylogenetic backbone of the genus Quercus. *Syst. Biodivers.* **2014**, *12*, 405–423. [CrossRef]
74. Kew Science—Plants of the World Online. Available online: http://www.plantsoftheworldonline.org/ (accessed on 19 December 2019).

75. Loizides, M. Diversity of wood-inhabiting aphyllophoraceous basidiomycetes on the island of Cyprus. *Mycotaxon* **2017**, *132*, 985–986.
76. Gargano, M.L. Nuovi ospiti di Inonotus tamaricis (Hymenochaetaceae) in Sicilia. *Inf. Bot. Ital.* **2010**, *42*, 319–321.
77. Wagner, T.; Fischer, M. Natural groups and a revised system for the European poroid Hymenochaetales (Basidiomycota) supported by nLSU rDNA sequence data. *Mycol. Res.* **2001**, *105*, 773–782. [CrossRef]
78. Wagner, T.; Fischer, M. Proceedings towards a natural classification of the worldwide taxa Phellinus s.l. and Inonotus s.l., and phylogenetic relationships of allied genera. *Mycologia* **2002**, *94*, 998–1016. [CrossRef] [PubMed]
79. Fiasson, J.L.; Niemelä, T. The Hymenochaetales: A revision of the European poroid taxa. *Karstenia* **1984**, *24*, 14–28. [CrossRef]
80. Fischer, M. A new wood-decaying basidiomycete species associated with esca of grapevine: Fomitiporia mediterranea (Hymenochaetales). *Mycol. Prog.* **2002**, *1*, 315–324. [CrossRef]
81. Ciccarone, C.; Graniti, A.; Schiaffino, A.; Marras, F. Molecular analysis of Fomitiporia mediterranea isolates from esca-affected grapevines in southern Italy. *Phytopathol. Mediterr.* **2004**, *43*, 268–272.
82. Pilotti, M.; Tizzani, L.; Brunetti, A.; Gervasi, F.; Di Lernia, G.; Lumia, V. Molecular identification of Fomitiporia mediterranea on declining and decayed hazelnut. *JPP* **2010**, *92*, 115–129.
83. Fischer, M. Biodiversity and geographic distribution of basidiomycetes causing esca-associated white rot in grapevine: A worldwide perspective. *Phytopathol. Mediterr.* **2006**, *45*, 30–42.
84. Kovács, C.; Sándor, E. The increasing importance of grapevine trunk diseases. *Int. J. Hortic. Sci.* **2016**, *22*, 21–30. [CrossRef]
85. Markakis, E.A.; Ligoxigakis, E.K.; Roussos, P.A.; Sergentani, C.K.; Kavroulakis, N.; Roditakis, E.N.; Koubouris, G.C. Differential susceptibility responses of Greek olive cultivars to Fomitiporia mediterranea. *Eur. J. Plant Pathol.* **2019**, *153*, 1055–1066. [CrossRef]
86. Polemis, E.; Dimou, D.M.; Fryssouli, V.; Zervakis, G.I. Diversity of saproxylic basidiomycetes in Quercus ilex woodlands of central and insular Greece. *Plant Biosyst.* **2019**, *153*, 385–397. [CrossRef]
87. Saitta, A.; Venturella, G. On the presence of Diplomitoporus lindbladii and Phellinus pseudopunctatus in Sicily (Southern Italy). *Bocconea* **2009**, *23*, 273–276.
88. Karadelev, M.; Rusevska, K.; Kost, G.; Kopanja, D.M. Checklist of macrofungal species from the phylum Basidiomycota of the Republic of Macedonia. *Acta Musei Maced. Sci. Nat.* **2018**, *21*, 23–112.
89. Decock, C.; Stalpers, J.A. Studies in Perenniporia: Polyporus unitus, Boletus medulla-panis, the nomenclature of Perenniporia, Poria and Physisporus, and a note on European Perenniporia with a resupinate basidiome. *Taxon* **2006**, *55*, 759–778. [CrossRef]
90. Zhao, C.L.; Cui, B.K.; Dai, Y.C. New species and phylogeny of Perenniporia based on morphological and molecular characters. *Fungal Divers.* **2013**, *58*, 47–60. [CrossRef]
91. Spirin, V.; Kout, J.; Vlasák, J. Studies in the Truncospora ohiensis–T. ochroleuca group (Polyporales, Basidiomycota). *Nova Hedwig.* **2015**, *100*, 159–175. [CrossRef]

3

Microscopic Analysis of Pigments Extracted from Spalting Fungi

Sarath M. Vega Gutierrez * and Sara C. Robinson

Department of Wood Science & Engineering, 119 Richardson Hall, Oregon State University, Corvallis, OR 97331, USA; sara.robinson@oregonstate.edu

* Correspondence: sarath.vega@oregonstate.edu.

Academic Editors: Laurent Dufossé, Yanis Caro and Mireille Fouillaud

Abstract: Pigments that are currently available in the market usually come from synthetic sources, or, if natural, often need mordants to bind to the target substrate. Recent research on the fungal pigment extracts from *Scytalidium cuboideum*, *Scytalidium ganodermophthorum*, *Chlorociboria aeruginosa*, and *Chlorociboria aeruginascens* have been shown to successfully dye materials, like wood, bamboo, and textiles, however, there is no information about their binding mechanisms. Due to this, a microscopic study was performed to provide information to future manufacturers interested in these pigments. The results of this study show that *S. ganodermophthorum* and *C. aeruginosa* form an amorphous layer on substrates, while *S. cuboideum* forms crystal-like structures. The attachment and morphology indicate that there might be different chemical and physical interactions between the extracted pigments and the materials. This possibility can explain the high resistance of the pigments to UV light and color fastness that makes them competitive against synthetic pigments. These properties make these pigments a viable option for an industry that demands natural pigments with the properties of the synthetic ones.

Keywords: extracted fungal pigments; spalting; microscopy; SEM; FIB

1. Introduction

The impact of synthetic dyes on the textile market is significant, in terms of the dyes' ecological impact [1], and most dyes in use today come from synthetic sources. For example, the commonly used red pigments of ferrite red oxide and Venetian red contain iron oxide, cadmium, and copper oxide. One of the most common yellow pigments, "lead chromate", and the popular green pigment, "chromium oxide", contain the aforementioned elements [2]. Other commonly used dyes, such as aniline dyes, are amino based, and are a derivate from petroleum [3].

Natural pigment and dye alternatives do exist, although they are not competitive in the market anymore for a number of reasons. Many of these natural dyes come from plants, lichens, and insects, and may exhibit issues with color fastness [4], adherence [4], UV stability [5], and toxicity [6], among others, making them less competitive compared with their renewable counterparts. Studies on bacterial-produced pigments have promising results for their use in textiles [7], although work with fungal pigments has been disappointing. Historic fungal pigment work focused on the use of fruiting bodies of some fungal species for dye extractions, but these dyes had the same issues as other natural dyes (low resistance to UV light, need for mordants). Lichens, likewise, have similar issues [8]. However, recent research by Hinsch showed the potential of spalting fungi pigments for their application in textiles in terms of their color fastness, crocking, and stability in UV light, all without mordants [9]. Recent research on a specific group of spalting fungi, fungi that produce extracellular pigments into wood, has shown that these pigments have the potential to equal, and in some instances even outperform, synthetic pigments [10].

Pigment-type spalting fungi are a select group of soft-rotting ascomycetes that have been shown to reliably dye a number of substrates, including wood [11], bamboo [12], and textiles [9,13]. These pigments have been found to be light fast, color fast [9], and UV light stable [9,10,14]. The specific pigments of interest in these spalting fungi are as follows: *Scytalidium cuboideum* (Sacc. And Ellis) Singler and Kang, which produces a red pigment called draconin red [9]. *Scytalidium ganodermophthorum* Kand, Singler, Y.W., Lee and S.H., Yun, produces an unidentified yellow pigment [9,15], and *Chlorociboria aeruginosa* (Oeder) Seaver and *Chlorociboria aeruginascens* (Nyl.) Kanouse ex C.S. Ramamurthi, Korf and L.R. Batra. produce xylindein, a naphthoquinone blue-green pigment [16,17].

The commercial viability of the extracted pigments obtained from spalting fungi, especially if they are to be used within the textile and wood markets, requires a better understanding of their microscopic characteristics. Previous microscopic studies have focused on naturally-produced fungal pigments. Tudor et al. focused mainly in melanin producing spalting fungi in natural form, finding that the pigments located mainly in rays and fibers. Blanchette et al. used transmission electron microscopy (TEM) to verify the presence of the fungus *Chlorociboria* sp. in art pieces of the 1400s, finding evidence of hyphae mostly in rays and vessels of the wood of the analyzed art pieces [18], and Michaelsen et al. used thin-layer chromatography and mass spectrometry to identify the pigments produced by *Chlorociboria* in art pieces of the sixteenth to eighteenth century [19]. These studies have focused mainly on the pigments produced naturally by the fungi in wood. No previous works have been done on extracted fungal pigments from these specific spalting fungi under laboratory growth conditions, likely due to the fact that the pigment extraction is relatively recent [20]. Most of the pigment extraction research has been done on the genus *Chlorociboria* [21], but it was not until Robinson et al. developed a standard method to extract the pigments from the genera *Chlorociboria* and *Scytalidium* with dichloromethane (DCM) [20] that pigment extraction from spalting fungi became a common method for working with fungal pigments.

These extracted pigments have a wide spectrum of possible applications in the wood finish and textile fields. Before this can be accomplished, however, it will be necessary to better understand the spalting pigments: how they look, how they interact, and where they are deposited on their substrates. In this study, the pigments of *Chlorociboria* sp., *S. ganodermophthorum*, and *S. cuboideum*, were characterized in terms of deposition on wood and textile substrates. The information gathered from this microscopic study will give a broader understanding of how the extracted fungal pigments work, and whether or not they effectively deposit directly onto substrates, such as wood and textiles. This is important because there are no prior studies on how extracted fungal pigments deposit onto materials on a microscopic level. How dyes deposit is critical to set a basis for future research and to gain an understanding of their potential ability to bind to, and stay on, potential substrates.

2. Materials and Methods

Different microscopy techniques were required for the different pigments studied. All of the techniques are detailed below.

2.1. Light Microscopy

Fourteen-millimeter cubes of cottonwood (*Populus trichocarpa* Torr. and A. Gray) were treated with 60 drops of extracted pigments of *S. cuboideum* UAMH 11517 (isolated from *Quercus* sp. in Memphis, TN), *S. ganodermophthorum* UAMH 10320 (isolated from oak wood logs in Gyeonggi Province, South Korea) and *C. aeruginosa* UAMH 11657 (isolated from a decaying hardwood log in Haliburton, ON, Canada) carried in dichloromethane (DCM). All of the pigments were standardized following the values set by Robinson et al. [11,22]. The pigment drops were applied in the standard amount on the cross-section of wood blocks following the protocol of Robinson et al. [11].

After application, the blocks were air dried for 48 h in a fume hood at 20 °C to evaporate the DCM, then cut into slices between 10–14 μm with a Spencer Buffalo microtome (Spencer Lens Co.,

Buffalo, NY, USA). The wood slides were mounted on VWR (VWR, Radnor, PA, USA) glass slides and covered with VWR #1 glass cover slides. The imaging was done with a Nikon Eclipse Ni-U equipped with a Nikon DS-Ri2 camera (Nikon Instruments Inc., Melville, NY, USA). Samples were analyzed with special focus on the rays and vessels of the samples. Coordinates were taken from the areas that showed a higher wood coloration by the pigments for further analysis.

2.2. *Confocal Microscopy*

The samples used for the light microscopy analysis were also used for their evaluation with confocal microscopy. A Zeiss LSM 780 NLO (Carl Zeiss Microscopy, LLC, Thornwood, NY, USA) was used for this analysis. The coordinates taken during the light microscopy analysis were used on this experiment, as they were the areas of interest. Each area of interest was analyzed with a set power of the argon laser (488 nm) at 0%–2%, and the filters were set from 405 to 633 nm. The wider use of filters was used to identify the optimum combination to try to distinguish the pigments from the wood. The images were obtained with the Zeiss ZEN software and analyzed with the Zeiss ZEN 2.3 lite freeware (Carl Zeiss Microscopy).

2.3. *Scanning Electron Microscopy (SEM)*

Silica squares (2 cm $\times$ 0.5 mm) were used as a base to apply extracted pigments as a control. Fifteen drops of each fungal pigment, solubilized in DCM at a standardized concentration were applied on them [11,22]. The DCM was allowed to evaporate between each drop. Each silica piece was fixed to a Ted Pella, Inc. (Ted Pella, Inc., Redding, CA, USA) aluminum stud of one centimeter in diameter for SEM. Ted Pella, Inc. double-coated carbon conductive tape was used to fix the samples to the aluminum studs.

Wood pieces of sugar maple and cottonwood (6 mm $\times$ 3 mm $\times$ 0.3 mm) were prepared, with the wider face oriented on the radial cut. Three pieces of each were treated on a similar way than the wood blocks, using 10 and 40 drops of each extracted fungal pigment, allowing the DCM to evaporate between drops.

Fabric squares of cotton and polyester (Testfabrics, Inc., West Pittston, PA, USA) of 5 mm $\times$ 5 mm were treated with 40 drops of extracted pigment of *S. cuboideum*, letting the DCM carrier to evaporate between drop applications. The selection of the pigment of *S. cuboideum* was done considering its characteristic morphology that would allow an easier identification of the pigments and the fibers.

The samples were mounted on Ted Pella, Inc. aluminum studs of one centimeter in diameter for SEM. A Ted Pella, Inc. double-coated carbon conductive tape was used to fix the samples to the studs. A sputter coating of gold-palladium was applied with a Cressington Sputter Coater 108 Auto (Cressington Scientific Instruments, Inc., Cranberry Twp, PA, USA) for giving the samples an enhanced optical contrast on all of the samples. The samples were exposed for 35 s to develop an even coating of 30–45 nm to avoid the electron charging of the samples and an enhanced optical contrast.

The samples were placed in the stage of an FEI QUANTA 600F environmental SEM (FEI Co., Hillsboro, OR, USA). The samples were viewed at an electron spot size of 4.5 to 5, and a high voltage (HV) between 10 to 20 kV.

2.4. *Energy-Dispersive X-Ray Spectroscopy (EDS)*

The extracted fungal pigment control samples were analyzed using the software EDAX Genesis (EDAX Inc., Mahwah, NJ, USA) for EDS analysis. For the analysis different sections of the samples were selected in the SEM and they were read using the X-ray detector. The software was calibrated for the detection of basic organic and inorganic elements. The limitation of the analysis is that it cannot detect hydrogen (H) due to its low molecular weight. For this reason, the analysis was focused on the detection of carbon (C), oxygen (O) and other elements that could have been present in the samples.

2.5. Focused Ion Beam (FIB)

The samples that showed higher saturation of the pigments on their surface during the SEM analysis were selected for performing an FIB cut. The selected samples were sputter coated with 10 nm of chromium oxide (Cr_2O_3) using a Varian Vacuum Evaporator VE10 (Agilent Technologies, Santa Clara, CA, USA). For the sputter coating, the bell of the sputter coater was opened and pieces of chromium (Cr) were placed in the holder connected to the back electrodes. On the stage below the holder, the aluminum studs with the wood samples were placed. Then the bell was closed and the vacuum system started. The first vacuum value required was 4×10^{-1} kPa, after reaching this value; the high-speed vacuum was activated to reach a pressure of 1×10^{-3} kPa. After obtaining the desired vacuum pressure, the stage where the samples were placed was set to a rotating speed of 10 rpm. After setting the vacuum and stage rotation, the electrodes where the holder with the Cr was connected was soaked at a power level of 2 for 60 s. Then, the power was increased to level 6 in 10 s. At this power level, the vaporization of the Cr and formation of the Cr_2O_3 started. The system remained at this level for 30 s to obtain the desired coating thickness. After 30 s, the system was turned off and the vacuum system was allowed to reach atmospheric pressure before being opened to retrieve the samples.

The coated samples were then placed in an FEI QUANTA 3D dual beam SEM/FIB (FEI Co., Hillsboro, OR, USA). Imaging of the samples was done to identify areas with higher saturation of pigments. On these areas, a rectangle of 20 μm × 3 μm was outlined. On this rectangle, two coating deposits were applied. The first coating was a layer of 20 μm × 3 μm × 100 nm of carbon (C). The second layer of coating was platinum (Pt) with the same size and depth as the carbon one. Both deposits were applied at a current of 5 Kv with 3.4 nA. After applying the deposits, the ion beam was turned on and the stage was tilted 52 degrees. The coated area was found with the ion beam and focused. There, a deposition of 1 μm of C and Pt was applied with a current of 5 Kv and 3.4 nA to protect the pre-coated area from the ion beam. Then the ion beam was activated to create a laddered cross-section of 15 μm × 1.5 μm × 5 μm at 5 Kv and 3 nA on the wood samples. The depth was enough to erode the protective coating layers, the pigment and the wood cell wall. After finishing the cross-section cutting, the borders were cleaned with the ion beam at a depth of 1 μm, to enhance the surface and clean debris generated during the cutting.

3. Results

3.1. Light Microscopy

The samples analyzed with light microscopy showed the pigments tended to accumulate in the wood vessels. Samples treated with *S. cuboideum* pigment extract showed filament-like structures accumulated on the vessels, especially in the helical thickenings of cottonwood, as shown in Figure 1.

The wood slides that contained *C. aeruginosa* pigments tended to accumulate in the intervessel pits (see Figure 2). Compared with the pigments from *S. cuboideum*, *C. aeruginosa* pigment formed what looks like an even layer on the vessel wall and it also covered the helical thickenings.

With light microscopy, the pigment of *S. ganodermophthorum* was visible on the vessel cell walls, and it also tended to accumulate on the helical thickenings (see Figure 3). Compared with the other two pigments, the yellow pigment tended to create a thicker layer on the vessel cell walls in contrast with the pigment from *C. aeruginosa*. It also accumulated between the helical thickenings, but compared to *S. cuboideum*, the pigment of *S. ganodermophthorum* did not show a characteristic shape.

Figure 1. Vessel cell from cottonwood with extracted pigments of *S. cuboideum* between the helical thickenings, pointed to with arrows. Picture taken with a Nikon Eclipse Ni-U at a magnification of 20×.

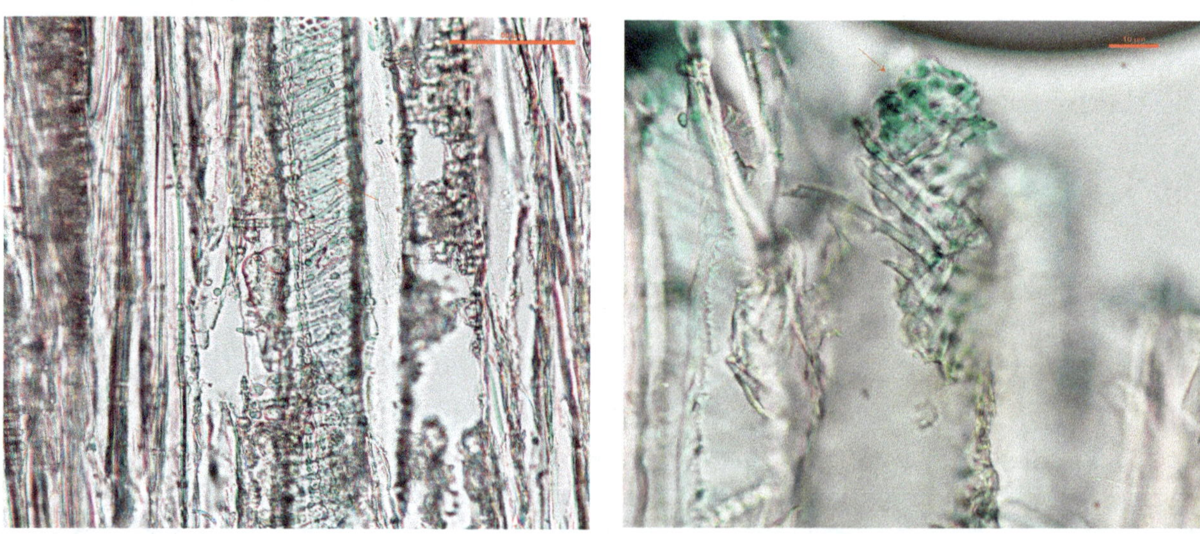

(**A**) (**B**)

Figure 2. (**A**) Vessel cell wall of cottonwood covered with extracted pigment, pointed to with arrows, from *C. aeruginosa* taken at a 20× magnification; (**B**) Detail of the vessel cell wall showing concentration of the pigment in the vessel pits at a 40× magnification. Picture taken with a Nikon Eclipse Ni-U.

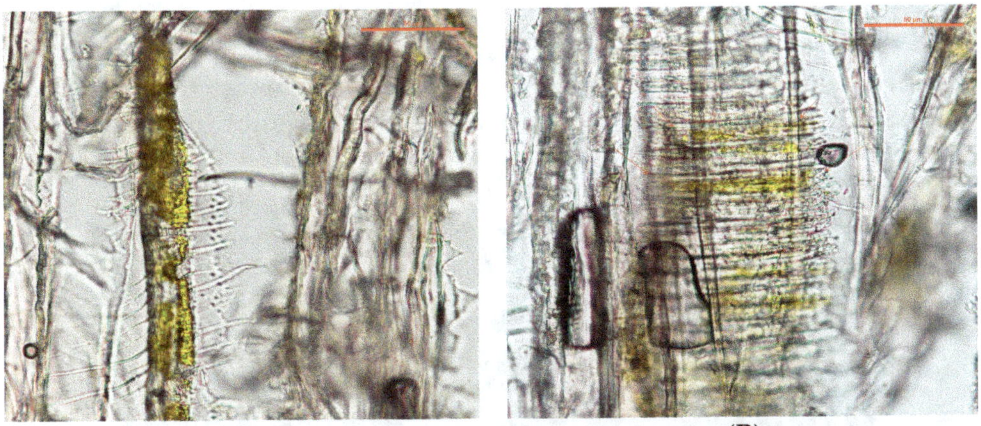

(A) (B)

Figure 3. (**A**) Vessels of cottonwood with extracted pigment of *S. ganodermophthorum* accumulating on the cell walls; (**B**) Pigments of *S. ganodermophthorum* between the helical thickenings, pointed to with arrows. Picture taken with a Nikon Eclipse Ni-U at 20× magnification.

3.2. Confocal Microscopy

With confocal microscopy, no difference between pigments and wood was visible due to the autofluorescence of wood and the broad-spectrum fluorescence of the extracted pigments at the wavelengths of 405, 488 and 561 nm. Images of the pure dry extracted pigments showed that they had textures. The red pigment had some peaks visible when analyzed at 561 nm. The yellow pigment showed an uneven texture when analyzed at 488 nm. The green pigment imaged at 488 nm showed an almost even surface. However, in general, this method showed to not be adequate for further analysis.

3.3. SEM on Dry Extracted Fungal Pigments

Results with SEM were the most promising. Silica squares with pure *S. cuboideum* pigment showed that they crystalize and form flower-like structures, as shown in Figure 4. This structure is composed by filament-like elements, which are consistent with the results obtained with light and confocal microscopy.

Figure 4. Crystal like structures from the extracted pigment of *S. cuboideum*. Picture taken with an FEI QUANTA 600 F.

The controls for *C. aeruginosa*, showed that the pigment has an amorphous structure, but it also showed two distinct areas, which can correspond to a molecular differences that affects the way that the electrons impact the compound (see Figure 5). Additionally, this pigment does not show a characteristic shape as the pigment from *S. cuboideum*, but it resembles a film.

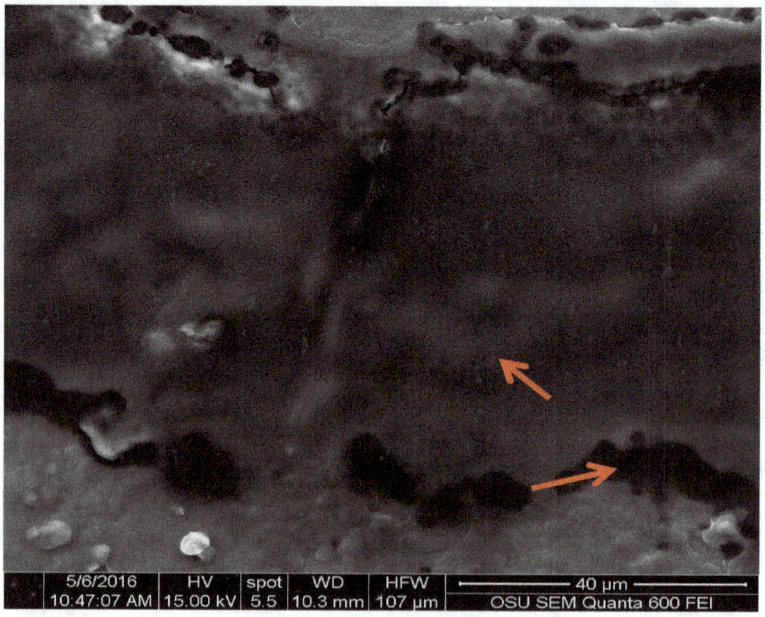

Figure 5. Extracted pigment of *C. aeruginosa* showing two reflective electron areas, pointed to with arrows. Picture taken with an FEI QUANTA 600 F.

The two electron reflective surface was also observed for *S. ganodermophthorum* (see Figure 6). The control also showed a more rugged surface, compared to the one of *C. aeruginosa*. This uneven surface was also observed with the light microscope and confocal microscopy, but with the use of SEM it is possible to have a more detailed view of the topography of this pigment.

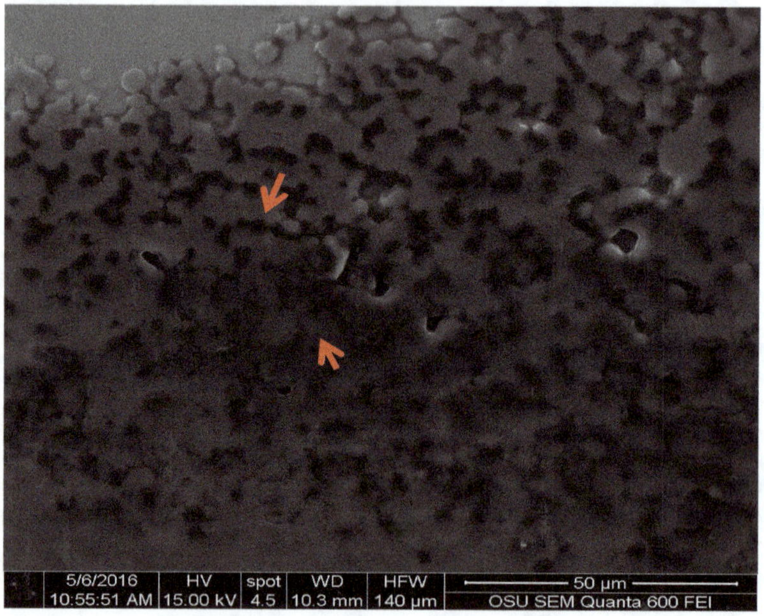

Figure 6. Extracted pigment of *S. ganodermophthorum* showing texture. Different electron reflectance is indicated with arrows. Picture taken with an FEI QUANTA 600 F.

3.4. SEM on Wood

For the wood samples, all three pigments were deposited on the surface of wood. For *S. cuboideum*, the crystal-like structures accumulated on top of the cell walls on the radial section, and on the extremes of the cell walls. They also tended to group as seen the dry pigment on silica where the pigment forms the flower-like structure (see Figure 7) that could reach up to an average of 81 μm of diameter.

Areas with less concentration of pigment presented smaller crystals on the surface of the wood cell walls shows small crystals forming on the surface of a vessel cell wall.

For *C. aeruginosa* and *S. ganodermophthorum*, the pigments formed an uneven film, covering the surface of all of the cell elements of the wood samples (see Figure 8).

Figure 7. Flower-like structures formed by the extracted pigment of *S. cuboideum* on the fiber cells of cottonwood. Picture taken with an FEI QUANTA 600 F.

Figure 8. Amorphous layer formed by the extracted pigment of *C. aeruginosa*. Picture taken with an FEI QUANTA 600 F.

3.5. SEM on Textiles

The fabrics showed that the pigment of *S. cuboideum* formed crystal-like structures that wrapped to the fiber surface of the polyester sample as shown on Figure 9.

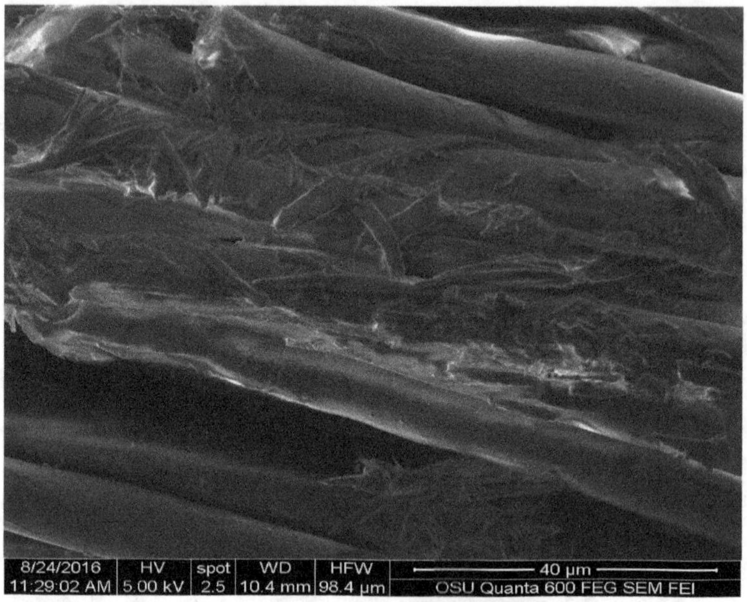

Figure 9. Pigment of *S. cuboideum* wrapping around polyester fibers. Picture taken with an FEI QUANTA 600 F.

The behavior of the pigment on cotton was similar to wood, tending to accumulate at the ends of the fibers and the pigment tended to form convoluted areas on top of the cotton fibers as shown in Figure 10.

Figure 10. Extracted pigment of *S. cuboideum* accumulating on top of a cotton fiber. Picture taken with an FEI QUANTA 600 F.

3.6. EDX Analysis

The EDX analysis showed that the elements contained by the three pigments were carbon (C) and oxygen (O). No data were recorded for hydrogen (H) due to the limitation of the machine to detect that element. This made the calculation of the elements show up to 100% between C and O. In general, the pigments showed similar amounts of C and O in their structure, and we can infer the presence of hydrogen (H) on the three of them (see Table 1).

Table 1. Percentage of carbon (C) and oxygen (O) present in the extracted fungal pigments.

Extracted Pigment	Carbon (C) %	Oxygen (O) %	Silica (Si) and Gold (Au) % from Coating and Substrate	Total
Draconin red from *Scytalidium cuboideum*	78%	20%	2%	100%
Yellow pigment from *Scytalidium ganodermophthorum*	79%	21%	0%	100%
Xylindein from *Chlorociboria aeruginosa*	77%	23%	0%	100%

3.7. FIB

FIB images were obtained for the three pigments. They showed that all of the three extracted pigments have a surface attachment to the wood and they do not penetrate the wood cell wall. The sample of *S. cuboideum* and *C. aeruginosa* showed a similar attachment to the surface of wood, consisting of irregular void spaces between attachment points. For *S. cuboideium,* the void spaces were smaller than the ones seen for *C. aeruginosa* (see Figures 11 and 12).

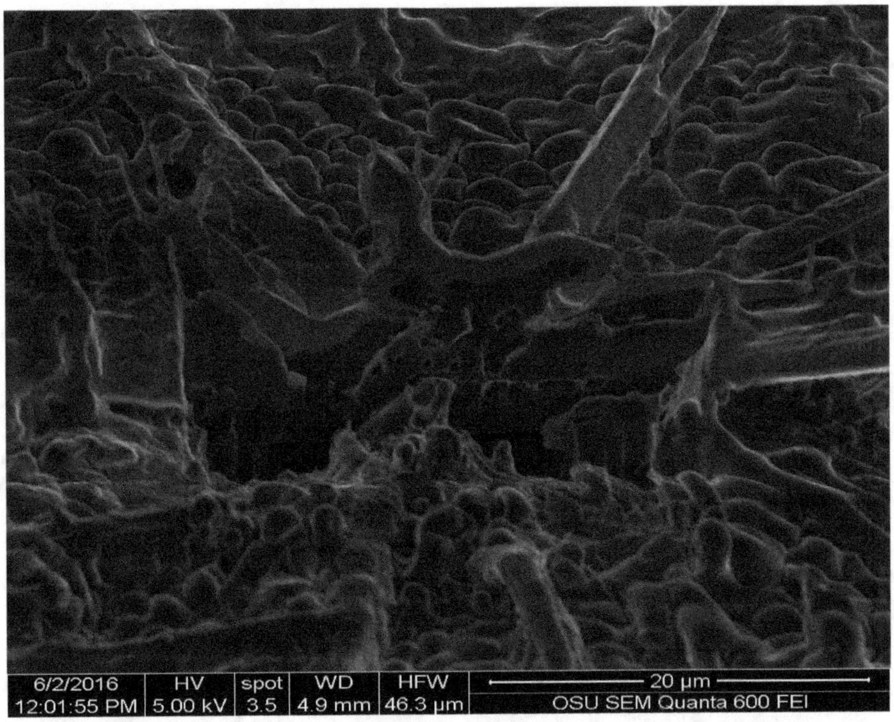

Figure 11. Cross-section of a crystal-like structure produced by *S. cuboideum* on cottonwood, showing void spaces between the attachment area of the pigment with the wood cell wall. Picture taken with an FEI QUANTA 3D dual beam SEM/FIB.

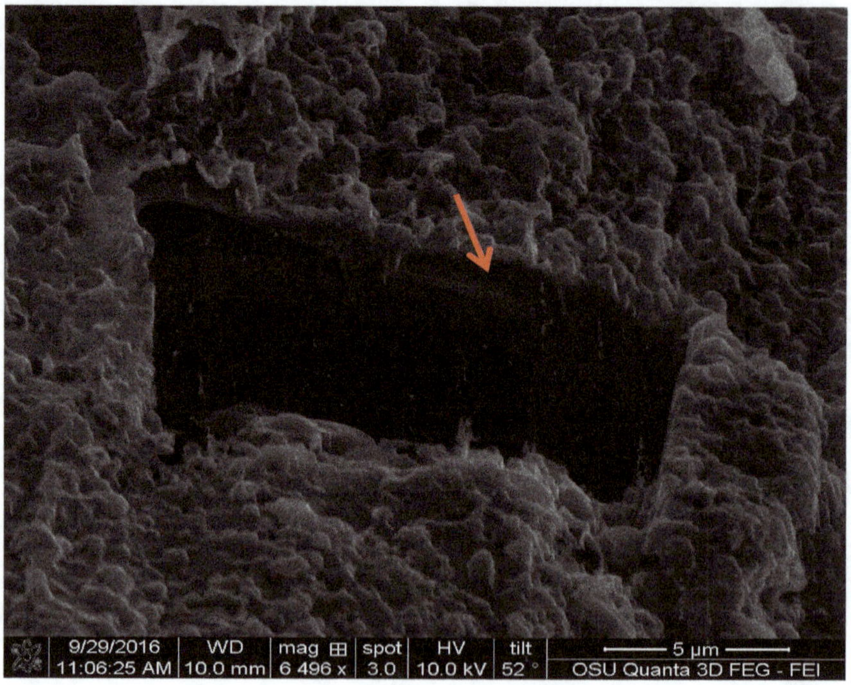

Figure 12. FIB cross-section of the extracted pigment of *C. aeruginosa* with cottonwood, showing large void spaces between the wood cell wall and the pigment, pointed to with an arrow. FEI QUANTA 3D dual beam SEM/FIB.

For *S. ganodermophthorum*, the attachment to wood was in two primary areas. In the first area it is possible to observe a continuous layer of pigment on top of the cell wall, and on the second area the surface showed a sponge-like texture (see Figure 13).

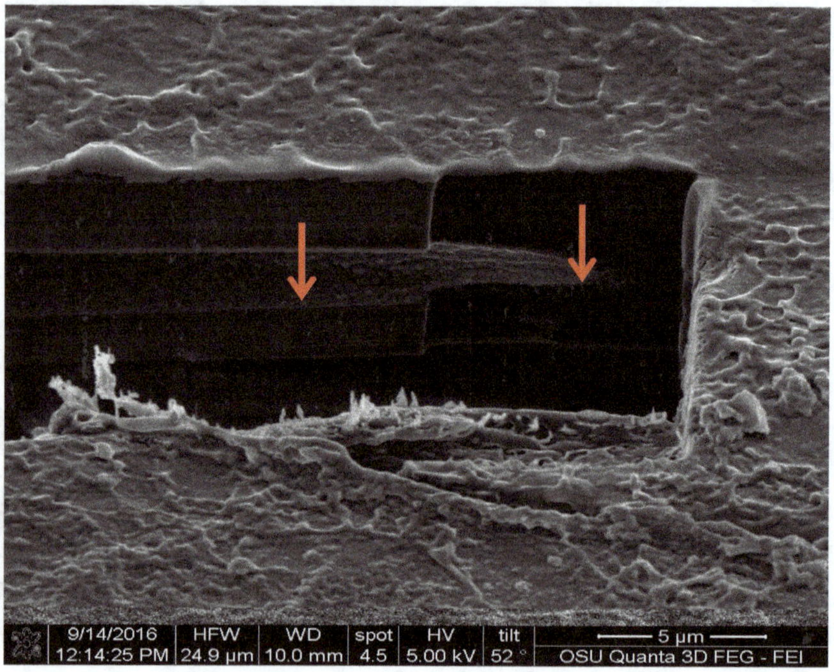

Figure 13. Cross-section of the area between the extracted pigment of *S. ganodermophthorum* and cottonwood. The left arrow shows an even attachment layer while the right arrow shows a sponge-like surface between the pigment and the wood. FEI QUANTA 3D dual beam SEM/FIB.

4. Discussion

The main distribution in wood of the extracted pigments was found in the vessels. The location of the extracted pigments contrast with naturally-secreted pigments. Naturally-secreted pigments will almost always be found in higher concentrations in wood ray cells, where the simple sugars are located [23]. Fungi develop and deposit their pigments in the ray cells of wood, as seen in the study by Tudor (melanin-producing fungi) [24] and Robinson [25] (pigmenting fungi, such as *C. aeruginosa*). Compared to other decay fungi, spalting fungi, specifically those that secrete extracellular pigments not classified as melanin, are Ascomycetes that prefer to first digest the available simple sugars in wood rays [23]. In contrast, extracted pigments, when applied by a dripping test, allocate in the vessels. This location is similar for wood coatings that are applied on the cross-section of wood [26], and likely has to do with the solutions moving through a path of least resistance.

The change of location of the pigments, from rays to vessels, could affect some of the inherent properties of the spalting fungal pigments, such as UV light resistance or color fastness. Were this the case, it could be due to differences in cellular structure. Wood ray cells can compartmentalize the pigments and hold higher concentrations of them due to their septa, in contrast to the wider and continuous vessel cells [27]. Additionally, the location of the rays in the wood can make the pigments less exposed to the effects of light, as they are organized in grouped structures. This is in contrast to the alignment of vessel elements, which, depending on the wood species, can be disperse in irregular patterns. However, recent studies on extracted fungal pigments focused on textiles [9,13] and UV light resistance [14], have shown that these pigments still retain the same properties as the ones naturally deposited by fungi.

Unfortunately the confocal microscopy method was ineffective with the samples due to the autofluorescence of wood (due to lignin and extractives) [28–31], and also because the pigments showed a similar fluorescence range. Usually, markers, like calcofluor white, are used in confocal microscopy to highlight the structures of fungal cells [32]. In the case of the extracted fungal pigments, there are no fungal cell walls where the marker can attach. This made the use of confocal microscopy not viable for further analysis of extracted fungal pigments.

Images obtained with SEM gave a detailed view of the extracted fungal pigments' topography. A study by Vega et al. on bamboo [12], showed that the pigments of *S. cuboideum* formed a rough surface on the hyphae cell wall, while microscopy on the genus *Chlorociboria* did not find a specific structure of the pigments within the hyphae and the colored areas [18]. The lack of a specific morphology was confirmed for the extracted pigments of *S. ganodermophthorum* and *C. aeruginosa*, but for *S. cuboideum*, crystal-like structures were observed. It was originally suspected that these crystals were calcium oxalates, which are commonly formed by fungi [33–35], but the EDS analysis showed no presence of calcium (Ca) in the samples. Other fungal pigments are also known to crystalize as shiraiachrome B [36], a red crystal with a perylenequinone nucleus, which differs from the described naphthoquinone composition of the red pigment draconin red from *S. cuboideum*, previously described by Golinsky et al., that only includes carbon, oxygen, and hydrogen in its structure [37]. Future studies could determine if the crystal-like structures identified on *S. cuboideum* pigments are organic single crystals or organic poly-crystals (aggrupation of single crystals) by determining if it has a diffraction pattern. This could be performed with a Kikuchi diffraction pattern analysis with a transmission electron microscope (TEM), which is broadly used for determining the crystal nature of materials [38] or with the accurate X-ray diffraction analysis [39]. If the aforementioned pigment is confirmed as a single crystal or poly-crystal, the synthesis of the pure pigment could be easier regardless of the crystal type. As a crystal compound, it would be possible to obtain a powder-like state pigment, which could allow a wider use of the pigment in diverse materials without the limitation of the solvent.

Specifically for xylindein (from *C. aeruginosa*), a compound known since the late 1800s [16,17,40–42], two clear areas were identified in the SEM images. These two areas correspond to a difference in the way that the material reflects electrons. A study by Saikawa et al. determined that xylindein produces two isomers, one that reflects the blue spectrum of light, while the second isomer reflects the yellow spectrum of light [40]. This characteristic in the xylindein molecule can explain the presence of the two areas where the electrons reflection differs. The images obtained from *S. ganodermophthorum* also presented a similar two-electron reflection area as *C. aeruginosa*, but the chemical identification of the yellow pigment is still not completed [9,15]. By the images obtained, it is possible to suspect that there might be two different molecules in the pigment, or possibly isomers.

For the textiles, *S. cuboideum* on polyester showed that the crystal-like structures wrap around the fibers of the material. This might confer better color stability on polyester than cotton by creating a more stable pigment deposition, which could make the pigments more resistant to physical and chemical abrasions. This was confirmed by Hinsch et al., who found that the red pigment performed better on polyester, which has a chemical structure similar to lignin, in terms of color fastness compared to cellulosic fibers, such as cotton [9]. Is possible to theorize that the pigments have a higher chemical affinity to lignin-like materials over the cellulose-like ones, mainly due to a higher amount of bonding sites on lignin-like compounds compared with the cellulose-like ones [27].

For the FIB analysis, the three pigments showed a similar attachment to wood. The pigments presented irregular attachment and void areas that looked similar to how adhesives attach to wood. It is possible that the pigments form similar hydrogen bonds to wood [27], and also mechanical interactions. Future research in this area should be able to confirm this using a nuclear magnetic resonance spectroscopy (NMR) or a proton nuclear magnetic resonance (HNMR) analysis.

5. Conclusions

The results obtained shed new light on how these extracted pigments deposit onto wood and textiles, and offer a view of how coating and dye manufacturers can visualize the possibilities of these fungal pigments. Extracted fungal pigments from *S. cuboideum*, *C. aeruginosa*, and *S. ganodermophthorum*, when applied on the cross-section of wood, are transported through the vessels of wood and have a surface attachment that has void and contact spaces. This may mean that there can be a chemical and mechanical interaction between the pigments and the materials. On textiles it was observed that the pigment of *S. cuboideum* wrapped around the polyester fiber, while the pigment behaved similarly on cotton as it did with wood. For base morphology, the red pigment from *S. cuboideum* forms a crystal-like structure. To confirm this, diffraction studies are needed. If a crystalline structure is confirmed, a broader use of fungal pigments can be conceived as the pigments could be turned into a powdered form that can then be applied to a wider variety of materials.

Additionally, this study determined the best methods to analyze the pigments microscopically. Scanning electron microscopy was the best method to visualize and identify the extracted fungal pigments on different materials. Light microscopy was useful to identify the allocation of the pigments in wood. Confocal microscopy was not effective to identify the attachment of pigments on wood, due to the autofluorescence of the material and the broad-range fluorescence of the pigments.

Acknowledgments: Oregon State University Electron Microscopy facilities Joan Hudson, Peter Eschbach, Teresa Sawyer, Oregon State University Center of Genome Research and Biocomputing Anne-Marie Girard Pohjanpelto. The authors gratefully acknowledge the financial support provided by the Walmart Manufacturing Innovation Fund for this work.

Author Contributions: Sarath M. Vega Gutierrez and Sara C. Robinson conceived and designed the experiments; Sarath M. Vega Gutierrez performed the experiments, analyzed the data, and wrote the paper.

References

1. Pointing, S. Feasibility of bioremediation by white-rot fungi. *Appl. Microbiol. Biotechnol.* **2001**, *57*, 20–33. [PubMed]
2. Preuss, H.P. *Pigments in Paint*; Noyes Data Corp.: Park Ridge, NJ, USA, 1974.
3. Carrubba, R.V.; Golden, R.L. Preparation of Anilines from Oxygenated Cyclohexanes. Google Patents US3347921A, 17 October 1967.

4. Oda, H. An attempt to improve the light fastness of natural dyes. Effect of UV absorbers on the photofading of red carthamin. *Jpn Home Econ. J.* **2002**, *53*, 271–277.
5. Erdoğrul, Ö.; Azirak, S. Review of the studies on the red yeast rice (*Monascus purpureus*). *Turk. Electron. J. Biotechnol.* **2004**, *2*, 37–49.
6. Ahmad, W.A.; Ahmad, W.Y.W.; Zakaria, Z.A.; Yusof, N.Z. Application of bacterial pigments as colorant. In *Application of Bacterial Pigments as Colorant*; Springer: Berlin, Germany, 2012; pp. 57–74.
7. Ferreira, E.S.; Hulme, A.N.; McNab, H.; Quye, A. The natural constituents of historical textile dyes. *Chem. Soc. Rev.* **2004**, *33*, 329–336. [CrossRef] [PubMed]
8. Hinsch, E.M.; Weber, G.; Chen, H.-L.; Robinson, S.C. Colorfastness of extracted wood-staining fungal pigments on fabrics: A new potential for textile dyes. *J. Text. Appar. Technol. Manag.* **2015**, *9*, 3.
9. Cristea, D.; Vilarem, G. Improving light fastness of natural dyes on cotton yarn. *Dyes Pigments* **2006**, *70*, 238–245. [CrossRef]
10. Hinsch, E.M. A Comparative Analysis of Extracted Fungal Pigments and Commercially Available Dyes for Colorizing Textiles. Mater's Thesis, Oregon State University, Corvallis, OR, USA, 2015.
11. Robinson, S.C.; Weber, G.; Hinsch, E.; Vega Gutierrez, S.M.; Pittis, L.; Freitas, S. Utilizing extracted fungal pigments for wood spalting: A comparison of induced fungal pigmentation to fungal dyeing. *J. Coat.* **2014**, *2014*, 1–8. [CrossRef]
12. Vega Gutierrez, S.; Vega Gutierrez, P.; Godinez, A.; Pittis, L.; Huber, M.; Stanton, S.; Robinson, S. Feasibility of coloring bamboo with the application of natural and extracted fungal pigments. *Coatings* **2016**, *6*, 37. [CrossRef]
13. Weber, G.; Chen, H.L.; Hinsch, E.; Freitas, S.; Robinson, S. Pigments extracted from the wood-staining fungi *Chlorociboria aeruginosa*, *Scytalidium cuboideum*, and *S. ganodermophthorum* show potential for use as textile dyes. *Color. Technol.* **2014**, *130*, 445–452. [CrossRef]
14. Beck, H.G.; Freitas, S.; Weber, G.; Robinson, S.C.; Morrell, J.J. Resistance of fungal derived pigments to ultraviolet light exposure. In *International Research Group in Wood Protection*; IRG/WP: St George, UT, USA, 2014.
15. Weber, G.; Boonloed, A.; Naas, K.M.; Koesdjojo, M.T.; Remcho, V.T.; Robinson, S.C. A method to stimulate production of extracellular pigments from wood-degrading fungi using a water carrier. *Curr. Res. Environ. Appl. Mycol.* **2016**, *6*, 218–230.
16. Edwards, R.L.; Kale, N. The structure of xylindein. *Tetrahedron* **1965**, *21*, 2095–2107. [CrossRef]
17. Blackburn, G.M.; Ekong, D.E.; Nielson, A.H.; Todd, L. Xylindein. *Chimia* **1965**, *19*, 208–212.
18. Blanchette, R.A.; Wilmering, A.M.; Baumeister, M. The use of green-stained wood caused by the fungus *Chlorociboria* in intarsia masterpieces from the 15th century. *Holzforschung* **1992**, *46*, 225–232. [CrossRef]
19. Michaelsen, H.; Unger, A.; Fischer, C.H. Blaugrüne färbung an intarsienhölzern des 16. Bis 18. Jahrhunderts. Available online: https://opus4.kobv.de/opus4-fhpotsdam/frontdoor/index/index/docId/1247 (accessed on 10 November 2016).
20. Robinson, S.C.; Hinsch, E.; Weber, G.; Freitas, S. Method of extraction and resolubilisation of pigments from *Chlorociboria aeruginosa* and *Scytalidium cuboideum*, two prolific spalting fungi. *Color. Technol.* **2014**, *130*, 221–225. [CrossRef]
21. Maeda, M.; Yamauchi, T.; Oshima, K.; Shimomura, M.; Miyauchi, S.; Mukae, K.; Sakaki, T.; Shibata, M.; Wakamatsu, K. Extraction of xylindein from *Chlorociboria aeruginosa* complex and its biological characteristics. *Bull. Nagaoka Univ. Technol.* **2003**, *25*, 105–111.
22. Robinson, S.C.; Hinsch, E.; Weber, G.; Leipus, K.; Cerney, D. Wood colorization through pressure treating: The potential of extracted colorants from spalting fungi as a replacement for woodworkers' aniline dyes. *Materials* **2014**, *7*, 5427–5437. [CrossRef]
23. Zabel, R.A.; Morrell, J.J. *Wood Microbiology. Decay and Its Prevention*; Harcourt Brace Jovanovich, Academic Press, Inc.: New York, NY, USA, 1992.
24. Tudor, D.; Robinson, S.C.; Sage, T.L.; Krigstin, S.; Cooper, P.A. Microscopic investigation on fungal pigment formation and its morphology in wood substrates. *Open Mycol. J.* **2014**, *8*, 174–186. [CrossRef]
25. Robinson, S.C.; Michaelsen, H.; Robinson, J.C. *Spalted Wood. The History, Science and Art of a Unique Material*, 1st ed.; Schiffer Publishing, Ltd: Atglen, PA, USA, 2016; p. 287.
26. De Meijer, M.; Thurich, K.; Militz, H. Comparative study on penetration characteristics of modern wood coatings. *Wood Sci. Technol.* **1998**, *32*, 347–365. [CrossRef]

27. Shmulsky, R.; Jones, D. *Forest Products and Wood Science*, 6th ed.; John Wiley & Sons: Hoboken, NJ, USA, 2011.
28. Billinton, N.; Knight, A.W. Seeing the wood through the trees: A review of techniques for distinguishing green fluorescent protein from endogenous autofluorescence. *Anal. Biochem.* **2001**, *291*, 175–197. [CrossRef] [PubMed]
29. Olmstead, J.A.; Gray, D.G. Fluorescence emission from mechanical pulp sheets. *J. Photochem. Photobiol. A Chem.* **1993**, *73*, 59–65. [CrossRef]
30. Olmstead, J.; Gray, D. Fluorescence spectroscopy of cellulose, lignin and mechanical pulps: A review. *J. Pulp Pap. Sci.* **1997**, *23*, J571–J581.
31. Albinsson, B.; Li, S.; Lundquist, K.; Stomberg, R. The origin of lignin fluorescence. *J. Mol. Struct.* **1999**, *508*, 19–27. [CrossRef]
32. Hickey, P.C.; Swift, S.R.; Roca, M.G.; Read, N.D. Live-cell imaging of filamentous fungi using vital fluorescent dyes and confocal microscopy. *Methods Microbiol.* **2004**, *34*, 63–87.
33. Whitney, K.D.; Arnott, H.J. Calcium oxalate crystal morphology and development in agaricus bisporus. *Mycologia* **1987**, 180–187. [CrossRef]
34. Dutton, M.V.; Evans, C.S.; Atkey, P.T.; Wood, D.A. Oxalate production by basidimycetes, including the white rot species *Coriolus versicolor* and *Phanerochaete chrysosporium*. *Appl. Microbiol. Biotechnol.* **1993**, *39*, 5–10. [CrossRef]
35. Dutton, M.V.; Kathiara, M.; Gallagher, I.M.; Evans, C.S. Purification and characterisation of oxalate decarboxylase from *Coriolus versicolor*. *FEMS Microbiol. Lett.* **1994**, *116*, 321–325. [CrossRef]
36. Wu, H.; Lao, X.F.; Wang, Q.W.; Lu, R.R.; Shen, C.; Zhang, F.; Liu, M.; Jia, L. The shiraiachromes: Novel fungal perylenequinone pigments from shiraia bambusicola. *J. Nat. Prod.* **1989**, *52*, 948–951. [CrossRef]
37. Golinski, P.; Krick, T.P.; Blanchette, R.A.; Mirocha, C.J. Chemical characterization of a red pigment (5,8-dihydroxy-2,7-dimethoxy-1,4-naphthalenedione) produced by arthrographis cuboidea in pink stained wood. *Holzforschung* **1995**, *49*, 407–410. [CrossRef]
38. Morito, S.; Tanaka, H.; Konishi, R.; Furuhara, T.; Maki, T. The morphology and crystallography of lath martensite in Fe-C alloys. *Acta Mater.* **2003**, *51*, 1789–1799. [CrossRef]
39. Woolfson, M. *X-ray Crystallography*; Cambridge University Press: Cambridge, UK, 1970.
40. Saikawa, Y.; Watanabe, T.; Hashimoto, K.; Nakata, A. Absolute configuration and tautomeric structure of xylindein, a blue-green pigment of *Chlorociboria* species. *Phytochemistry* **2000**, *55*, 237–240. [CrossRef]
41. Gümbel, W. Über xylindein. *Berichte der Deutschen Chemischen Gesellschaft* **1858**, *41*, 113–115.
42. Lieberman, C. Über xylindein. *Berichte der Deutschen Chemischen Gesellschaft* **1874**, *7*, 1102–1103.

4

Basidiomycetes Associated with *Alnus glutinosa* Habitats in Andros Island (Cyclades, Greece)

Elias Polemis, Vassiliki Fryssouli, Vassileios Daskalopoulos and Georgios I. Zervakis *

Laboratory of General and Agricultural Microbiology, Agricultural University of Athens, 11855 Athens, Greece; teonanac_rec1@hotmail.com (E.P.); vfrisouli@gmail.com (V.F.); vassilismks@gmail.com (V.D.)
* Correspondence: zervakis@aua.gr.

Abstract: Alluvial forests dominated by black alder (*Alnus glutinosa*) are widespread in Europe along river banks and watercourses forming a habitat of renowned ecological/conservation importance. Despite the considerable interest this habitat has attracted in terms of the associated fungal diversity, very few pertinent data are available from the eastern Mediterranean. Andros island (Aegean Sea, Greece) hosts the southernmost population of *A. glutinosa* in the Balkan Peninsula; such stands have been systematically inventoried for several years in respect to macrofungi. In total, 187 specimens were collected and studied by examining morphoanatomic features and by evaluating (when necessary) the outcome of sequencing the internal transcribed spacer (ITS) region of nuclear ribosomal DNA (nrDNA) to elucidate their identity and obtain an insight into phylogenetic relationships. As a result, 106 species were recorded, 92 are saprotrophic and 14 form ectomycorrhizae (ECM) with alders. Twenty-one species are first national records, while 68 other species are reported for the first time from this habitat in Greece. Several findings of particular interest due to their rarity, ecological preferences and/or taxonomic status are presented in detail and discussed, e.g., six *Alnicola* taxa, *Cortinarius americanus*, *Lactarius obscuratus*, *Paxillus olivellus* and *Russula pumila* (among the ECMs), and the saprotrophs *Entoloma uranochroum*, *Gymnopilus arenophilus*, *Hyphoderma nemorale*, *Lepiota ochraceofulva*, *Phanerochaete livescens* and *Psathyrella hellebosensis*.

Keywords: macrofungi; Basidiomycota; mushroom diversity; ectomycorrhiza; saprotroph; alder; Aegean Sea; Mediterranean; *Alnicola*

---

1. Introduction

Alluvial forests with *Alnus glutinosa* Gaertn. and *Fraxinus excelsior* L. (priority habitat 91E0*; Annex I, Directive 92/43/EEC) are distributed throughout Europe, but they are generally rare and threatened since only remnants exist, mainly in central and northern Europe [1]. Alder stands are considerably less frequent in the Mediterranean region, where the repercussions of changes in the hydrological cycle caused by global warming and climate destabilization are much more evident [2]. The southernmost limit of the priority habitat 91E0* in the Balkan Peninsula is located in Andros island (Figure 1), i.e., the northernmost in the Cyclades and situated at a transition zone between continental Greece and other islands of the Aegean Archipelago. From the geomorphological point of view, it is characterized by a remarkably intense relief and by many rivulets and streams of constant flow, which are unique among most of Central and South Aegean islands. *A. glutinosa* trees demonstrate a patchy distribution in Andros, predominantly occurring along the main streams within the Site of Community Importance (SCI) GR4220001 and in altitudes ranging from sea level to as high as 850 m above sea level (a.s.l.), very close to the highest peaks of the island. In many cases black alders are mixed with *Platanus orientalis* L., *Fraxinus ornus* L. and/or *Nerium oleander* L. (in lower altitudes), while they also form pure stands, as it is the case at the estuaries of the Vori stream in NE Andros.

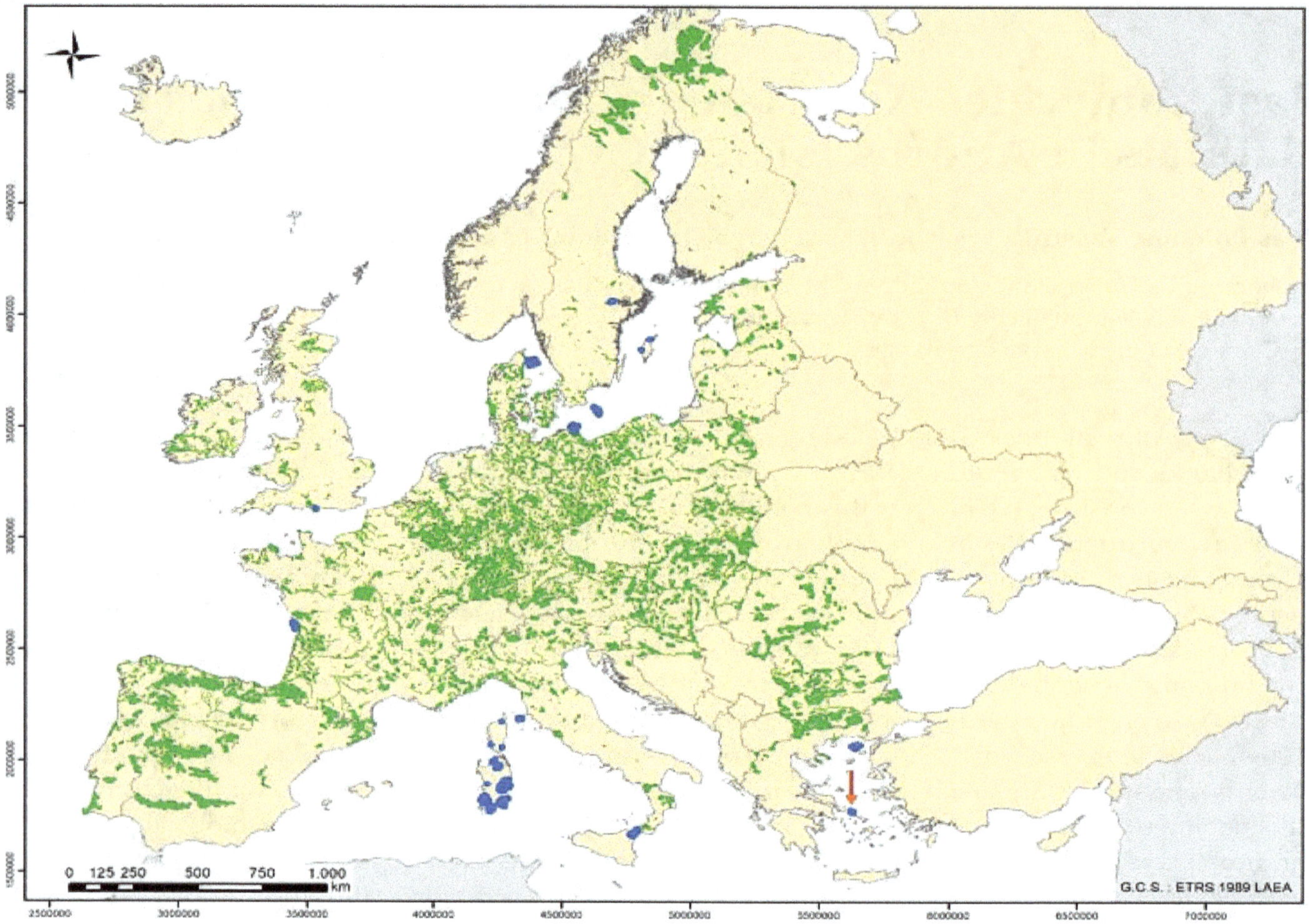

Figure 1. Map presenting Natura 2000 sites, which include the priority habitat 91E0* in continental Europe (in green) and in islands (in blue); Andros island is indicated by the red arrow.

Alder trees are known to form symbiotic relationships with nitrogen-fixing actinomycetes of the genus *Frankia* Brunchorst [3,4], with arbuscular mycorrhizal fungi (AM) of Glomeromycota [5,6] and with various ectomycorrhizal (ECM) fungi of Ascomycota and Basidiomycota [7–9]. European alder stands have been relatively well-studied in terms of both macro- and microfungal communities, and approx. 1000 species of saprotrophic and ECM macrofungi were reported [10–15]. In addition, mycocoenological studies from Europe and North America suggested that ECM fungi of *Alnus* spp. exhibit a remarkably high degree of host specificity compared to other tree species [8,16], while the analysis of both sporophores and ectomycorrhizae evidenced that alders have a low number (<50) of ECM symbionts worldwide [17–19].

Limited knowledge is available on the diversity of fungi associated with alders in Greece, and only preliminary data are reported in the few pertinent publications [20,21]. On the other hand, Andros is the only island of the Aegean Archipelago where a systematic inventory of macrofungi is in progress for more than 20 years. Biotopes characterized by river banks, springs and alluvial forests, where *A. glutinosa* is often the dominant tree species, were forayed in the past and 37 mushroom species were reported from this particular habitat in Andros, including ECM symbionts as well as xylotrophic, litter and/or humus saprotrophs [22–24]. Among the latter, *Entoloma alnicola* Noordel. & Polemis was described asnew species for science and it is still known from the type locality only [25].

Since 2017, mycodiversity studies in alder stands of Andros were intensified in the frame of a LIFE Nature project (LIFE16-NAT_GR_000606), which -among others- aims at the conservation and restoration of the priority habitat 91E0* in the island. Hence, during the last few years, new sites with alder stands were repeatedly forayed (in addition to those previously investigated), and a large number of new collections were made. These, together with previously sampled—but still unidentified—specimens, were subjected to detailed morphoanatomical examination in conjunction with sequencing and phylogenetic analyses (where judged necessary) in order to assess their identity. Moreover, in several occasions, past relevant reports on recorded taxa were revised/re-evaluated according to the latest respective taxonomic and phylogenetic concepts. Hence, this work presents an updated compilation of available data on the diversity of macrofungi in a habitat of significant interest occurring at the limits of its distribution in Europe.

2. Materials and Methods

2.1. Sampling of Biological Material

Data presented in this inventory are based on specimens collected from 10 sampling sites covering almost the entire area of *A. glutinosa* distribution in Andros island, which appears mainly within (or marginally out) the SCI GR4220001, extending from sea-level to an altitude of ca. 850 m a.s.l. (Figure 2; Table S1).The biological material examined for the purpose of this work was sampled in 38 forays performed during the last 25 years from late October to April; more than half of those (#23) were conducted in the period from 2017 to 2020. In total, 187 specimens found exclusively under alder trees or directly on their wood, woody residues or leaf-litter were collected, and voucher specimens are deposited in the Fungarium of the Laboratory of General and Agricultural University of Athens (ACAM).

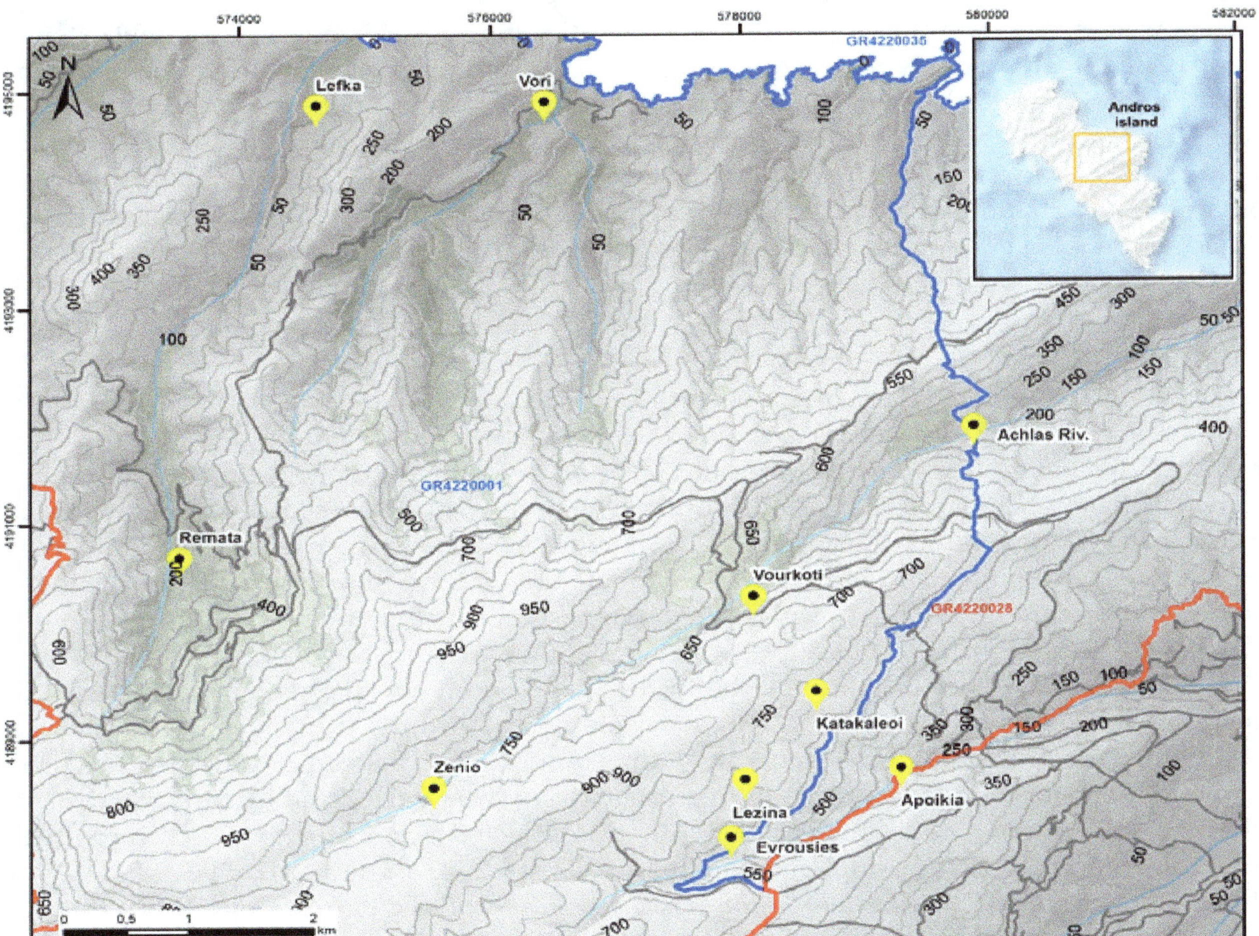

Figure 2. Sampling sites (in yellow marking) in the *Alnus glutinosa* habitat and relative position/size of the area under investigation within Andros island (map in upper right corner).

2.2. *Morpho-Anatomical Features in Basidiomes*

The morphological study included in situ recording of macroscopic features of taxonomic interest, while ex-situ examination involved observations of morphoanatomical characters in dried specimens. Sections were mounted and observed in KOH 3–5% (*w/v*), in Melzer's reagent, in cotton-blue, in cresyl-blue and in sulfovaniline solution. Observations were performed with the use of a Zeiss AxioImager A2 microscope under bright field and differential interference contrast (DIC); microphotographs were taken with the aid of a mounted digital camera (Axiocam). For all examined specimens a minimum of 30 mature basidiospores were measured and the resulting measurements as well as additional observations of other essential microscopical features (hymenial cystidia, pileipellis etc.) were used for determination of the species examined in accordance to pertinent identification keys and monographs (e.g., [26–36]).

2.3. *DNA Extraction, Amplification and Sequencing*

When deemed necessary, DNA sequencing and phylogenetic analyses were performed. Total genomic DNA was obtained from dried basidiomes and DNA extraction was performed through the use of the Nucleospin Plant II DNA kit (Macherey and Nagel, Düren, Germany) by following the manufacturer's protocol. The internal transcribed spacer (ITS; ITS1, 5.8S, ITS2) region within the nuclear ribosomal RNA gene cluster was examined by using the primers ITS1/ITS4 [37]. Polymerase chain reactions (PCR) were performed in 50 μL containing 50 ng DNA template, 0.25 μM of each primer, 0.2 mM of each dNTP, 1 × HiFi Buffer (Takara BIO INC., Shiga, Japan) and 1 U HiFi Taq DNA polymerase (Takara BIO INC., Shiga, Japan). PCR reactions were performed as follows: 94 °C for 5 min, followed by 35 cycles of 94 °C for 30 s, 50 °C for 30 s and 72 °C for 1 min, and a final extension at 72 °C for 10 min. PCR products were run in 1% agarose gels and purified using Invitrogen PureLink kit (Thermo Fisher Scientific, Seoul, S. Korea), and were submitted for sequencing to CeMIA SA (Larissa, Greece). The same PCR primers were used for sequencing. Chromatograms were checked with the aid of BioEdit v. 7.2.5 software [38]. Then sequences were examined against GenBank built-in search tools for obtaining information which could confer at identifying the material under study. A total of 61 validated sequences generated in this work were deposited in GenBank and the accession numbers MT458502 to MT458562 were obtained.

2.4. *Phylogenetic Analysis of Sequence Data*

A total of 42, 29 and 22 ITS sequences corresponding to selected species of the genera *Alnicola* Kühner (and *Naucoria* (Fr.) P. Kumm.), *Lactarius* Pers. and *Paxillus* Fr. (including 12, 5 and 4 sequences generated in this work), respectively, were subjected to phylogenetic analysis. In addition, species of the same or other genera were used as outgroups in each case. Multiple sequence alignment of each ITS rDNA dataset was conducted using the Q-INS-I algorithm as implemented in the online version of MAFFT v. 7 [39]. Alignments were reviewed, manually adjusted at misaligned sites and trimmed at the same position through MEGA X [40] before being used for further analysis.

Phylogenetic relationships of taxa for each alignment were inferred by using maximum likelihood (ML) and Bayesian inference (BI) through the CIPRES web portal (www.phylo.org; Miller et al. 2010). ML analyses were conducted by RAxML BlackBox online server (http://phylobench.vital-it.ch/raxml-bb/) [41] using default parameters and calculating bootstrap statistics according to the program recommendations for the best-scoring ML tree. BI analyses were performed by MrBayes v. 3.2.1 [42]. The best-fit substitution model for each dataset was selected according to the corrected Akaike information criterion (cAIC), as implemented in jModeltest v.2 [43]. The TPM2uf+G, TPM1uf+G and SYM+G models were selected for the *Alnicola*, *Lactarius* and *Paxillus* datasets, respectively. To estimate posterior probabilities, Markov chain Monte Carlo (MCMC) simulation was implemented in two parallel independent runs of four chains, one cold and three heated, with trees sampled every 1000 generations until the standard deviation of split frequencies is below 0.05; the first 25% of trees were omitted as burn-in. A 50% majority rule consensus tree was built and visualized with iTOL [44]. Clades with ML bootstrap support (MLB) ≥ 65% and Bayesian posterior probability (BPP) ≥ 95% were considered as significantly supported.

3. Results and Discussion

The study of 187 specimens of macrofungi associated with the *A. glutinosa* priority habitat in Andros led to the identification of 106 species (74 genera) of basidiomycetes. Among them, 14 (13%) are ECM species (Table 1) strictly associated with alders [18,19]. The other 92 (87%) are saprotrophic; 70 (66%) saproxylic and 22 (21%) saprotrophic on soil, humus or leaf-litter (Table 2). Interestingly, 10 ECM and 11 saprotrophic species are first national records, while other 68 are reported for the first time from this habitat in Greece. Identification of specimens to species was performed by examining their morphoanatomic features and by evaluating (when necessary) the outcome of ITS sequencing and phylogenetic analysis; in the latter case, the respective GenBank accession numbers are provided (Tables 1 and 2). Selected findings of particular interest are presented (and discussed) by providing brief descriptions and comments on characters of potentially diagnostic value.

Table 1. Ectomycorrhizal (ECM) fungi identified during the study: species name, specimen code/collection date, locality and GenBank accession numbers for ITS sequences generated. First national records for Greece are indicated by an asterisk (*) before the species name.

a/a	Species Name	Specimen Code/ Collection Date	Locality	GenBank Accession No.
1	**Alnicola escharoides* (Fr.) Romagn.	EP.17-A1344/11-Nov-2017 EP.17-A1420/24-Nov-2017 EP.18-A1548/22-Feb-2018 EP.18-A1561/1-Nov-2018 EP.18-A1571/2-Nov-2018 EP.19-A1636/16 Nov 2019	Zenio Vori Vori Zenio Vourkoti Katakalaioi	 MT458538 MT458539 MT458540
2	**Alnicola inculta* (Peck) Singer	EP.17-A1346/11 Nov 2017	Zenio	MT458541
3	**Alnicola luteolofibrillosa* Kühner	EP.17-A1430/24-Nov-2017	Vori	MT458542
4	**Alnicola subconspersa* (Kühner ex P.D. Orton) Bon	EP.17-A1421/24-Nov-2017 EP.19-A1637/16-Nov-2019	Vori Katakalaioi	MT458543 MT458544
5	*Alnicola striatula* (P.D. Orton) Romagn.	EP.04-A679/15-Nov-2004 EP.19-A1614/14-Nov-2019	Evrousies Evrousies	 MT458545
6	**Alnicola umbrina* (R. Maire) Kühner	EP.04-A678/15-Nov-2004 EP.17-A1377/2-Nov-2017 EP.18-A1572/2-Nov-2018 EP.19-A1607/12 Nov 2019 EP.19-A1638/16-Nov-2019 EP.19-A1646/17-Nov-2019 EP.19-A1666/2-Dec-2019	Evrousies Lezina Vourkoti Zenio Katakalaioi Achlas riv. Remata	 MT458546 MT458547 MT458548 MT458549
7	**Cortinarius americanus* A.H. Sm.	EP.19-A1622/15-Nov-2019	Vourkoti	
8	*Gyrodon lividus* (Bull.) Sacc.	EP.14-A1263/1-Nov-2014 EP.17-A1428/24-Nov-2017	Vori Vori	
9	**Inocybe calospora* Quél.	EP.18-A1570/2-Nov-2018	Vourkoti	MT458550
10	**Lactarius obscuratus* (Lasch) Fr.	EP.17-A1347/11-Oct-2017 EP.17-A1566/1-Nov-2018 EP.17-A1576/2-Nov-2018 EP.19-A1645/17-Nov-2019 EP.19-A1664/30-Nov-2019	Zenio Zenio Vourkoti Achlas riv. Remata	MT458551 MT458552 MT458553 MT458554 MT458555

Table 1. *Cont.*

a/a	Species Name	Specimen Code/ Collection Date	Locality	GenBank Accession No.
11	**Paxillus olivellus* P.-A. Moreau, J.-P. Chaumeton, Gryta & Jarge	EP.95-A028/13-Nov-1995 EP.02-A353/22-Sep-2002 EP.04-A670/23-Oct-2004 EP.04-A673/24-Oct-2004 EP.14-A1266/1-Nov-2014 EP.17-A1348/11-Nov-2017 EP.17-A1396/23-Nov-2017 EP.17-A1426/24-Nov-2017 EP.18-A1552/22-Feb-2018 EP.18-A1583/2-Nov-2018 EP.19-A1628/16-Nov-2019	Achlas riv. Evrousies Remata Achlas riv. Vori Zenio Evrousies Vori Vori Vourkoti Katakalaioi	MT458556 MT458557 MT458558 MT458559
12	**Russula pumila* Rouzeau & F. Massart	EP.18-A1575/2-Nov-2018	Vourkoti	MT458560
13	*Tomentella stuposa* (Link) Stalpers	EP.02-A327/29-Apr-2002	Vori	MT458561
14	*Tomentella sublilacina* (Ellis & Holw.) Wakef.	EP.02-A452/11-Oct-2002 EP.17-A1437/24-Nov-2017	Achlas riv. Vori	MT458562

Table 2. Saprotrophic basidiomycetes identified during the study: species name, specimen code/collection date, locality, type of substrate and GenBank accession numbers for ITS sequences generated. First national records for Greece are indicated by an asterisk (*) before the species name.

a/a	Species Name	Specimen Code/Collection Date	Locality	Substrate Type	GenBank Accession No.
1	*Abortiporus biennis* (Bull.) Singer	EP.18-A1582/02-Nov-2018	Vourkoti	fallen trunk	
2	*Agaricus moelleri* Wasser	EP.19-A1613/14-Nov-2019	Katakalaioi	leaf-litter	MT458502
3	*Amaropostia stiptica* (Pers.) B.K. Cui, L.L. Shen & Y.C. Dai	EP.17-A1423/24-Nov-2017	Vori	dead stump	MT458503
4	*Armillaria gallica* Marxm. & Romagn.	EP.17-A1443/24-Nov-2017	Vori	dead stump	
5	*Armillaria mellea* (Vahl) P. Kumm.	EP.95-A021/12-Nov-1995 EP.18-A1584/02-Nov-2018 EP.19-A1651/29-Nov-2019	Remata Vourkoti Vori	dead stump standing trunk trunk base	
6	*Auricularia auricula-judae* (Bull.) Quél.	EP.18-A1539/22-Feb-2018 EP.19-A1672/02-Dec-2019	Vori Remata	standing trunk standing trunk	
7	*Bjerkandera adusta* (Willd.) P. Karst.	EP.19-A1670/02-Dec-2019	Remata	fallen trunk	
8	*Botryobasidium candicans* J. Erikss.	EP.01-A275/26-Dec-2001 EP.11-A1023/05-Jan-2011 EP.17-A1434/24-Nov-2017	Vori Vori Vori	fallen trunk fallen trunk fallen trunk	
9	*Brevicellicium olivascens* (Bres.) K.H. Larss. & Hjortstam	EP.17-A1352/11-Nov-2017	Zenio	fallen trunk	MT458504
10	*Calocera cornea* (Batsch) Fr.	EP.17-A1440/25-Nov-2017 EP.19-A1616/14-Nov-2019	Vori Evrousies	dead stump dead stump	
11	*Ceriporia purpurea* (Fr.) Donk	EP.17-A1350/11-Nov-2017 EP.17-A1363/11-Nov-2017	Zenio Zenio	fallen trunk fallen trunk	MT458505
12	*Chondrostereum purpureum* (Pers.) Pouzar	EP.17-A1467/28-Nov-2017 EP.19-A1678/02-Dec-2019	Achlas riv. Remata	standing trunk standing trunk	MT458506
13	*Clavaria fragilis* Holmsk.	EP.19-A1639/16-Nov-2019	Katakalaioi	soil	
14	*Clitocybe nebularis* (Batsch) P. Kumm.	EP.18-A1580/02-Nov-2018	Vourkoti	leaf litter	

Table 2. *Cont.*

a/a	Species Name	Specimen Code/Collection Date	Locality	Substrate Type	GenBank Accession No.
15	*Clitocybe phyllophila* (Pers.) P. Kumm.	EP.18-A1579/02-Nov-2018	Vourkoti	leaf litter	
16	*Clitopilus hobsonii* (Berk.) P.D.Orton	EP.02-A331/29-Apr-2002	Vori	fallen trunk	
17	*Coniophora puteana* (Schumach.) P. Karst.	EP.11-A1017/5-Jan-2011 EP.17-A1411/24-Nov-2017 EP.19-A1679/02-Dec-2019	Vori Lefka Remata	fallen trunk fallen trunk fallen trunk	MT458507 MT458508 MT458509
18	*Coprinellus disseminates* (Pers.) J.E. Lange	EP.01-A260/26-Dec-2001 EP.19-A1654/29-Nov-2019	Vori Vori	rotten wood around stump	
19	*Coprinellus radians* (Fr.) Vilgalys, Hopple & Jacq. Johnson	EP.19-A1655/29-Nov-2019 EP.19-A1668/02-Dec-2019	Vori Remata	woody residues woody residues	MT458510
20	**Coprinopsis melanthina* (Fr.) Örstadius & E. Larss.	EP.17-A1412/24-Nov-2017 EP.19-A1659/29-Nov-2019	Lefka Vori	woody residues woody residues	MT458511
21	*Coriolopsis gallica* (Fr.) Ryvarden	EP.11-A1016/05-Jan-2011	Vori	standing trunk	
22	*Crepidotus luteolus* (Lambotte) Sacc.	EP.01-A268/26-Dec-2001	Vori	woody residues	
23	*Delicatula integrella* (Pers.) Fayod	EP.19-A1634/16-Nov-2019 EP.19-A1640/17-Nov-2019	Katakalaioi Achlas riv.	trunk base bark living tree	
24	*Entoloma alnicola* Noordel. & Polemis	EP.02-A364/22-Sep-2002	Evrousies	soil	
25	*Entoloma incanum* (Fr.) Hesler	EP.02-A362/22-Sep-2002 EP.04-A674/24-Oct-2004 EP.19-A1633/16-Nov-2019	Evrousies Achlas riv. Katakalaioi	soil soil soil	
26	*Entoloma juncinum* (Kühner & Romagn.) Noordel.	EP.02-A362/22-Sep-2002	Evrousies	soil	
27	*Entoloma mougeotii* (Fr.) Hesler	EP.02-A446/11-Oct-2002	Achlas riv.	soil	
28	**Entoloma uranochroum* Hauskn. & Noordel.	EP.19-A1615/14-Nov-2019	Evrousies	soil/leaf-litter	
29	*Exidiopsis galzinii* (L.S. Olive) K. Wells	EP.02-A448/11-Oct-2002	Achlas riv.	fallen trunk	
30	*Fibroporia citrine* (Bernicchia & Ryvarden) Bernicchia & Ryvarden	EP.17-A1356/11-Nov-2017	Zenio	fallen trunk	
31	*Fomes fomentarius* (L.) Fr.	EP.20-A1681/05-Jan-2020	Vori	standing trunk	
32	*Fuscoporia torulosa* (Pers.) T. Wagner	EP.04-A668/15-Oct-2004 EP.04-A672/24-Oct-2004 EP.17-A1447/26-Nov-2017 EP.19-A1677/02-Dec-2019	Evrousies Achlas riv. Achlas riv. Remata	standing trunk standing trunk standing trunk standing trunk	
33	*Ganoderma adspersum* (Schulzer) Donk	EP.14-A1264/01-Nov-2014 EP.17-A1422/24-Nov-2017	Vori Vori	standing trunk standing trunk	
34	*Ganoderma resinaceum* Boud.	EP.19-A1662/30-Nov-2019	Remata	standing trunk	
35	**Gymnopilus arenophilus* A. Ortega & Esteve-Rav.	EP.03-A659/04-Nov-2003 EP.17-A1408/24-Nov-2017 EP.19-A1674/02-Dec-2019	Apoikia Lefka Remata	rotten stump rotten stump rotten stump	 MT458512 MT458513
36	*Gymnopilus junonius* (Fr.) P.D. Orton	EP.04-A667/15-Oct-2004 EP.04-A825/02-Dec-2004 EP.17-A1385/14-Nov-2017 EP.18-A1587/05-Nov-2018	Apoikia Apoikia Evrousies Vori	standing trunk standing trunk standing trunk standing trunk	
37	*Gymnopus brassicolens* (Romagn.) Antonín & Noordel.	EP.17-A1425/24-Nov-2017 EP.19-A1673/02-Dec-2019	Vori Remata	woody residues woody residues	
38	**Hydropus floccipes* (Fr.) Singer	EP.19-A1658/29-Nov-2019	Vori	standing trunk	
39	*Hyphoderma medioburiense* (Burt) Donk	EP.17-A1361/11-Nov-2017 EP.17-A1381/12-Nov-2017	Zenio Katakalaioi	fallen trunk twigs	 MT458514

Table 2. *Cont.*

a/a	Species Name	Specimen Code/Collection Date	Locality	Substrate Type	GenBank Accession No.
40	**Hyphoderma nemorale* K.H. Larss.	EP.02-A326/29-Apr-2002	Vori	fallen trunk	MT458515
41	*Hyphoderma setigerum* (Fr.) Donk	EP.02-A332/29-Apr-2002 EP.17-A1357/11-Nov-2017	Vori Zenio	fallen trunk fallen trunk	 MT458516
42	**Hyphodermella corrugate* (Fr.) J. Erikss. & Ryvarden	EP.17-A1398/23-Nov-2017	Evrousies	fallen branch	MT458517
43	**Lepiota ochraceofulva* P.D. Orton	EP.19-A1627/15-Nov-2019	Vourkoti	leaf-litter	MT458518
44	* *Lepista ovispora* (J.E. Lange) Gulden	EP.19-A1608/12-Nov-2019	Zenio	leaf-litter	
45	*Lepista nuda* (Bull.) Cooke	EP.18-A1564/01-Nov-2018	Zenio	leaf-litter	
46	*Leucoagaricus melanotrichus* (Malençon & Bertault) Trimbach	EP.19-A1604/12-Nov-2019	Zenio	leaf-litter	
47	*Leucopaxillus gentianeus* (Quél.) Kotl.	EP.18-A1577/01-Nov-2018	Zenio	leaf-litter	
48	*Marasmius rotula* (Scop.) Fr.	EP.02-A360/22-Sep-2002	Evrousies	twigs	
49	*Melanoleuca exscissa* (Fr.) Singer	EP.17-A1415/24-Nov-2017	Vori	soil	MT458519
50	*Merulius tremellosus* Schrad.	EP.04-A680/15-Nov-2004 EP.02-A323/29-Apr-2002	Evrousies Vori	fallen trunk fallen trunk	
51	*Mycena galericulata* (Scop.) Gray	EP.17-A1349/11-Nov-2017 EP.19-A1625/15-Nov-2019 EP.19-A1629/16-Nov-2019 EP.19-A1668/02-Dec-2019	Zenio Vourkoti Katakalaioi Remata	around trunk dead stump around trunk dead stump	 MT458520 MT458521
52	*Mycena haematopus* (Pers.) P. Kumm.	EP.01-A262/26-Dec-2001	Vori	rotten trunk	
53	*Mycena pseudocorticola* Kühner	EP.17-A1449/26-Nov-2017 EP.19-A1667/02-Dec-2019	Achlas riv. Remata	bark living tree bark living tree	
54	*Mycena sanguinolenta* (Alb. & Schwein.) P. Kumm	EP.19-A1642/17-Nov-2019	Achlas riv.	twigs	
55	*Mycetinis scorodonius* (Fr.) A.W. Wilson & Desjardin	EP.19-A1644/17-Nov-2019	Achlas riv.	twigs	
56	*Mycoacia aurea* (Fr.) J. Erikss. & Ryvarden	EP.02-A325/29-Apr-2002 EP.11-A1026/05-Jan-2011	Vori Vori	fallen trunk fallen trunk	
57	*Mycoacia uda* (Fr.) Donk	EP.17-A1351/11-Nov-2017	Zenio	fallen trunk	
58	*Mycoaciella bispora* (Stalpers) Erikss. & Ryvarden	EP.01-A272/26-Dec-2001 EP.02-A330/29-Apr-2002	Vori Vori	fallen trunk fallen trunk	
59	*Paralepista flaccida* (Sowerby) Vizzini	EP.18-A1563/01-Nov-2018 EP.18-A1581/02-Nov-2018	Zenio Vourkoti	leaf-litter leaf-litter	
60	*Parasola kuehneri* (Uljé & Bas) Redhead, Vilgalys & Hopple	EP.18-A1546/22-Feb-2018	Vori	soil	
61	*Peniophora tamaricicola* Boidin & Malenç.	EP.02-A329/29-Apr-2002	Vori	fallen trunk	
62	*Peniophorella praetermissa* (P. Karst.) K.H. Larss.	EP.01-A276/26-Dec-2001 EP.17-A1355/11-Nov-2017	Vori Zenio	fallen trunk fallen trunk	 MT458522
63	*Perenniporia ochroleuca* (Berk.) Ryvarden	EP.02-A461/11-Oct-2002	Achlas riv.	dead wood	
64	**Phanerochaete livescens* (P. Karst.) Volobuev & Spirin	EP.01-A273/26-Dec-2001 EP.18-A1541/22-Feb-2018	Vori Vori	fallen trunk fallen trunk	 MT458523
65	*Phellinus lundellii* Niemelä	EP.17-A1342/13-Apr-2017	Zenio	standing trunk	
66	*Phlebia rufa* (Pers.) M.P. Christ.	EP.17-A1354/11-Nov-2017 EP.20-A1682/05-Jan-2020	Zenio Vori	fallen trunk fallen trunk	MT458524 MT458525

Table 2. *Cont.*

a/a	Species Name	Specimen Code/Collection Date	Locality	Substrate Type	GenBank Accession No.
67	*Phlebiopsis ravenelii* (Cooke) Hjortstam	EP.11-A1018/05-Jan-2011	Vori	fallen trunk	
68	*Phloeomana alba* (Bres.) Redhead	EP.19-A1641/17-Nov-2019	Achlas riv.	bark living tree	MT458526
69	*Phloeomana speirea* (Fr.) Redhead	EP.17-A1378/12-Nov-2017	Lezina	twigs	MT458527
70	*Physisporinus vitreus* (Pers.) P. Karst.	EP.20-A1684/05-Jan-2020	Vori	fallen trunk	
71	*Pilatotrama ljubarskyi* (Pilát) Zmitrovich	EP.18-A1551/22-Feb-2018	Vori	fallen trunk	MT458528
72	*Pleurotus ostreatus* (Jacq.) P. Kumm.	EP.01-A261/26-Dec-2001 EP.18-A1586/05-Nov-2018	Vori Vori	standing trunk standing trunk	
73	*Pluteus cervinus* (Schaeff.) P. Kumm.	EP.01-A321/29-Apr-2002 EP.19-A1623/15-Nov-2019 EP.19-A1676/02-Dec-2019	Vori Vourkoti Remata	dead wood dead wood fallen trunk	MT458529 MT458530
74	*Pluteus nanus* (Pers.) P. Kumm.	EP.19-A1653/29-Nov-2019	Vori	woody residues	
75	*Pluteus salicinus* (Pers.) P. Kumm.	EP.19-A1657/29-Nov-2019	Vori	fallen branch	
76	**Pluteus podospileus* Sacc. & Cub.	EP.19-A1643/17-Nov-2019	Achlas riv.	fallen twigs	
77	*Postia balsamea* (Peck) Jülich	EP.19-A1663/30-Nov-2019	Remata	fallen trunk	
78	*Psathyrella candolleana* (Fr.) Maire	EP.02-A336/04-Jun-2002 EP.19-A1624/15-Nov-2019	Achlas riv. Vourkoti	woody residues woody residues	 MT458531
79	*Psathyrella corrugis* (Pers.) Konrad & Maubl.	EP.19-A1601/16-Oct-2019	Vourkoti	soil/buried wood	MT458532
80	**Psathyrella hellebosensis* Deschuyteneer & A. Melzer	EP.17-A1409/24-Nov-2017	Lefka	woody residues	MT458533
81	*Psathyrella microrhiza* (Lasch) Konrad & Maubl.	EP.19-A1603/12-Nov-2019	Vourkoti	woody residues	MT458534
82	*Psathyrella prona* (Fr.) Gill.	EP.02-A363/22-Sep-2002	Evrousies	soil	
83	*Psathyrella vinosofulva* P.D. Orton	EP.17-A1431/24-Nov-2017	Vori	soil	MT458535
84	*Radulomyces confluens* (Fr.) M.P. Christ.	EP.11-A1022/05-Jan-2011	Vori	fallen trunk	
85	*Steccherinum ochraceum* (Pers. ex J.F. Gmel.) Gray	EP.18-A1540/22-Feb-2018	Vori	fallen trunk	
86	*Stereum hirsutum* (Willd.) Pers.	EP.17-A1397/23-Nov-2017 EP.17-A1463/28-Nov-2017 EP.18-A1544/22-Feb-2018 EP.20-A1691/25-Jan-2020	Evrousies Achlas riv. Vori Lefka	trunk/branch trunk/branch trunk branch	
87	*Trechispora nivea* (Pers.) K.H. Larss.	EP.02-A328/29-Apr-2002 EP.20-A1683/05-Jan-2020	Vori Vori	fallen trunk fallen trunk	 MT458536
88	*Trametes versicolor* (L.) Lloyd	EP.11-A1019/05-Jan-2011 EP.18-A1542/22-Feb-2018	Vori Vori	fallen trunk standing trunk	
89	*Tubaria furfuracea* (Pers.) Gillet	EP.19-A1652/29-Nov-2019	Vori	woody residues	
90	*Tulostoma fimbriatum* Fr.	EP.17-A1431/24-Nov-2017	Vori	soil	
91	*Vitreoporus dichrous* (Fr.) Zmitr.	EP.02-A445/11-Oct-2002 EP.17-A1439/25-Nov-2017 EP.18-A1538/22-Feb-2017	Achlas riv. Vori Vori	fallen trunk fallen branch fallen branch	
92	*Xylodon raduloides* Riebesehl & Langer	EP.11-A1025/05-Jan-2011 EP.18-A1543/22-Feb-2018 EP.19-A1671/02-Dec-2019	Vori Vori Remata	fallen trunk fallen trunk fallen trunk	MT458538

3.1. The ECM Element

Among the ECM macrofungi recorded (Table 1), the genus *Alnicola* is represented by six species (Figure 3); five of them form part of the sect. *Alnicola* sensu Moreau [45], which is characterized by urticoid cheilocystidia, and one of the sect. *Submelinoideae* Singer with clavate or capitate cheilocystidia [46].

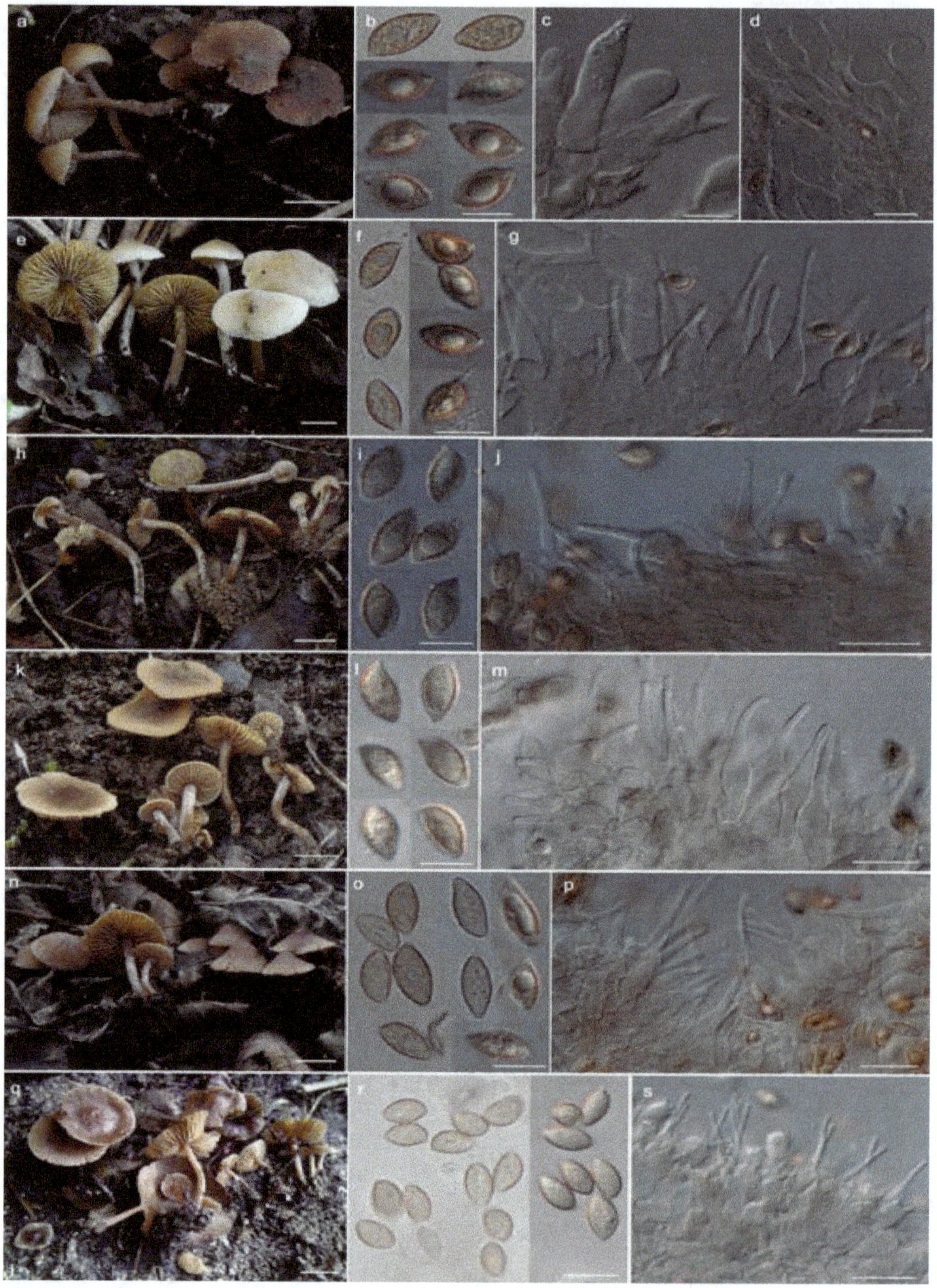

Figure 3. Species of the genus *Alnicola* recorded in Andros alder stands: *A. escharoides* basidiomes, basidiospores and cheilocystidia (**a–c**); *A. umbrina* basidiomes, basidiospores and cheilocystidia (**d–f**); *A. striatula* basidiomes, basidiospores and cheilocystidia (**g–i**); *A. subconspersa* basidiomes, basidiospores and cheilocystidia (**j–l**); *A. luteolofibrillosa* basidiomes, basidiospores and cheilocystidia (**m–o**); *A. inculta* basidiomes, basidiospores, basidia and cheilocystidia (**p–s**). Bars: basidiomes, 1 mm; basidiospores and basidia, 10 μm; cheilocystidia, 20 μm.

Different opinions exist regarding the genus name in pertinent literature since some European authors as well as the Index Fungorum prefer to conserve the name *Naucoria* (Fr.) P. Kumm., whereas Moreau, in his nomenclatural revision, rejected this name in favour of *Alnicola* Kühner [45]; the latter approach is accepted by other European mycologists, the Mycobank, and is also adopted in this work. Moreover, the taxonomy of species of the sect. *Alnicola* remains problematic and, consequently, a phylogenetic analysis was performed to deal with this issue.

The most often found *Alnicola* species in our study were *A. umbrina* (R. Maire) Kühner and *A. escharoides* (Fr.) Romagn., i.e., two of the most common taxa associated with alders in Europe; both constitute new national records for Greece. Particularly *A. escharoides* (syn. *A. citrinella* Moreau & A. de Haan [47]) is distinguished from all other (more or less brownish) species found in Andros by its pale yellowish-buff non striate pileus, the amygdaliform to navicular spores, with prominent ornamentation, measuring 9.9–11.8 × 5.3–5.9 µm, Q = 1.9–2.1 (Figure 3a–c). Following phylogenetic analysis, our specimens are positioned in a distinct group (albeit not adequately supported) together with other sequences from material identified as *A. escharoides* and *A. citrinella* (Figure 4).

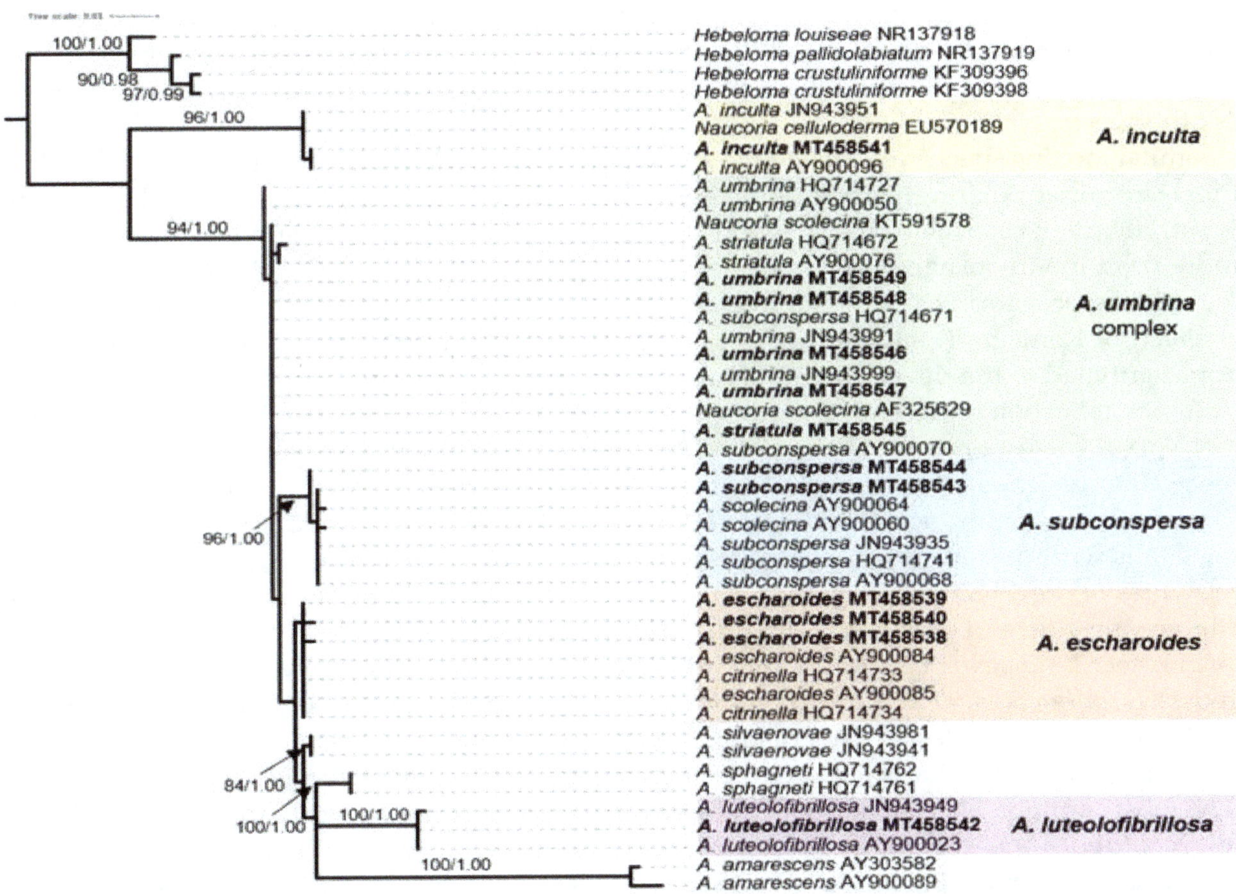

Figure 4. Phylogeny of *Alnicola* species derived from rDNA ITS sequences through ML analysis. Branches are labelled when MLB > 65% and BPP > 0.95. *Hebeloma* species (*H. louiseae, H. pallidolabiatum, H. crustuliniforme*) were used as outgroups. Boxes include sequences from specimens recorded in the *Alnus glutinosa* habitat.

On the other hand, *A. umbrina* (Figure 3d–f) is hereby considered as a species complex following the nomenclatural concept of Moreau [45] and the outcome of the phylogenetic study by Rochet et al. [19]. According to our observations, *A. umbrina* shows a rather large morphological variability with dark brown hygrophanous pilei bearing prominent striations up to their centre when wet, becoming much lighter and indistinctly striate only at margin when dry. Basidiospores are variable in size and shape, often somewhat elongated fusiform, weakly to moderately verrucose, measuring 10.7–13.6 × 5.2–6.1 µm, Q = 1.9–2.4. Sequences generated in this work clustered together with material identified as *A. umbrina*, *N. scolecina* (Fr.) Quél., *A. striatula* (P.D. Orton) Romagn. and *A. subconspersa* (Kühner ex P.D. Orton) Bon into a group that was not adequately supported (Figure 4). However, the morphological features of specimens identified as *N. scolecina* in Europe are very similar to descriptions of *A. umbrina* [22,33,48,49]. Therefore, *N. scolecina* and *A. umbrina* form part of the same complex and the question whether they constitute different entities or not remains open and in need of further research.

One collection representing another closely related taxon, previously reported as *N. striatula* P.D. Orton (Figure 3g–i) from alder stands in Andros [22], derived from the same site during our recent forays. According to Moreau (2005), *A. striatula* might merely correspond to a pale form of *A. umbrina*, but our morphological studies revealed some noteworthy differences when compared to specimens hereby named *A. umbrina*, i.e., pileus always very prominently striate, smooth and shiny, and (most importantly) significantly smaller basidiospores measuring 8.2–10.0 × 4.5–5.6 µm, Q = 1.7–1.9; these features are in accordance to previous descriptions of *N. striatula* [33,48,50]. As evidenced from our phylogenetic analysis (Figure 4), this particular collection forms part of the *A. umbrina* complex (together with the other two *A. striatula* sequences included in the tree) by using ITS alone; however, since it is morphologically distinct and fits to the widely accepted taxonomic concept of *A. striatula*, we provisionally retain it in this inventory as a separate taxon, until a future multigene approach shows otherwise.

A similar looking species to *A. umbrina*—but less common in Andros—is *A. subconspersa* (Figure 3j–l). The most prominent distinguishing features versus our *A. umbrina* specimens are the non (or very faintly) striate pileus as well as the size and shape of spores, being wider, amygdaliform to navicular and more prominently ornamented, measuring 10.9–12.9 × 6.0–6.8 µm, Q = 1.7–2.0. It is noteworthy that *A. subconspersa* forms a well-supported phylogenetic group including sequences labelled as *A. scolecina* (Fr.) Romagn. (Figure 4), which is indicative of the morphological affinity of these taxa that had apparently led to the development of ambiguous species concepts.

Another collection representing a member of the sect. *Alnicola* was recorded in the alluvial littoral forest of Vori; it corresponds to *A. luteolofibrillosa* Kühner and constitutes the first report of this species in Greece (Figure 3m–o). It is morphologically characterized by non-striate, pale buff, fibrillose to tomentose pilei, with abundant whitish veil remnants on stipe and pileal margin; the respective sequence falls within a highly-supported terminal subgroup corresponding to this species (Figure 4). Lastly, *A. inculta* (Peck) Singer (Figure 3p–s) was recorded only at Zenio (i.e., the site with the highest altitude among those of this study, 850 m) and is reported for the first time in Greece. It forms part of the sect. *Submelinoideae*, and, according to Moreau [45] is conspecific to the taxon widely referred as *N. celluloderma* P.D. Orton, as it is also evidenced by our phylogenetic analysis (Figure 4). Morphologically, this species is easily distinguished from all aforementioned taxa thanks to the clavate to capitate cheilocystidia characterizing members of sect. *Submelinoideae* and the 2-spored basidia.

The most common ECM mushroom in alder stands of Andros belongs to the genus *Paxillus*; it was the first recorded *Alnus*-specific symbiont in the island 25 years ago, and was later repeatedly found in this particular habitat (Figure 5a–d). It was initially identified as *P. rubicundulus* P.D. Orton [22]; however, sequencing of recent collections revealed that it forms part of the newly described taxon *P. olivellus* Moreau P-A, Chaumeton J-P, Gryta H, Jargeat P [51]. Although clearly separated by molecular approaches, *P. olivellus* can be hardly distinguished from *P. rubicundulus* and *P. adelphus* Chaumeton JP, Gryta H, Jargeat P, Moreau P-A on the basis of morphology alone, i.e., only by the olivaceous tinges of the young basidiomes and the basidiospores shape, which are ovoid to ellipsoid in *P. olivellus*, cylindrical in *P. rubicundulus* and short cylindrical in *P. adelphus* [51]. Such features were observed in our specimens since olivaceous tints were always evident in young basidiomes, and spores were ovoid to ellipsoid measuring 6.7–8.1× 4.5–5.2 µm, Q = 1.39–1.66. In addition, ITS sequences from our material originating from various sites in the habitat under study were very similar or identical to those corresponding to *P. olivellus* (including the type), and formed a terminal subgroup with high support (Figure 6). Therefore, this particular species seems to be the only representative of the genus *Paxillus* in the black alder stands of Andros island.

Figure 5. Alder-associated ECM fungi recorded in Andros: *Paxillus olivellus* basidiomes (**a**; bar 1 mm), basidiospores (**b**; bar 10 μm), section of lamella (**c**; bar 20 μm), hymenial cystidium and basidia (**d**; bar 20 μm); *Lactarius obscuratus* basidiomes (**e**, bar 1 mm), basidiospores (**f**, bar 10 μm), pileipellis (**g**, bar 20 μm); *Russula pumila* basidiomes (**h**, bar 1 mm), basidiospores (**i**, bar 10 μm), pileipellis (**j**, bar 20 μm); *Cortinarius americanus* basidiomes (**k**, bar 1 mm); *Inocybe calospora* basidiospores, basidium and pleurocystidium (**l**, bar 10 μm).

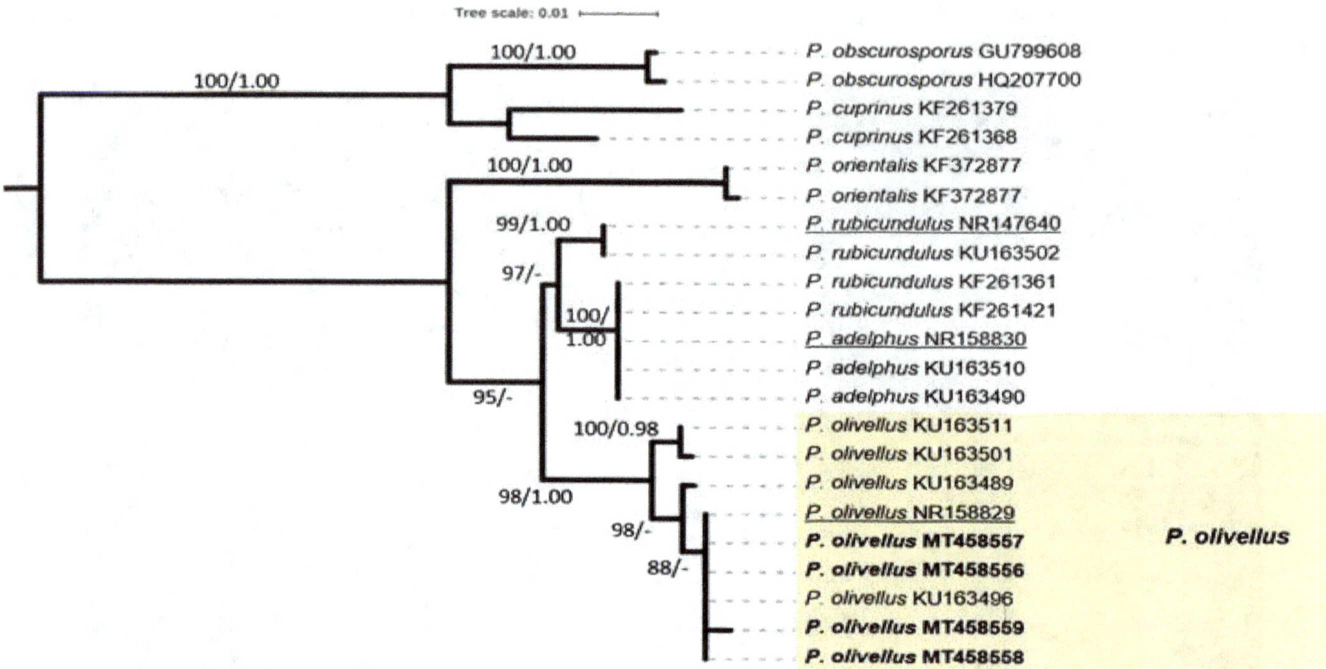

Figure 6. Phylogeny of *Paxillus* species derived from rDNA ITS sequences through ML analysis. Branches are labelled when MLB > 65%, and BPP > 0.95. *P. cuprinus* and *P. obscurosporus* were used as outgroups. The coloured box includes sequences from specimens recorded in the *Alnus glutinosa* habitat.

Lactarius obscuratus (Lasch) Fr. is one the few *Alnus*-specific ECM symbionts of this particular genus; it was found in several inland collection sites dominated by *A. glutinosa*, but not in the alluvial (littoral) forest of Vori (Figure 5e–g). Phylogenetic analysis of our sequences derived from several collections confirmed that they belong to this particular species (Figure 7). However, the respective terminal subgroup in our phylogenetic tree is composed from sequences named either *L. obscuratus* or *L. cyathuliformis* Bon, which is due to the different interpretations existing about this taxon (J. Nuytinck, pers. comm.). In the study of Rochet et al. [19], it is referred as *L. cyathuliformis*, whereas the correct name for the same group is *L. obscuratus* according to Wisitrassameewong et al. [52]. Since the basidiospores average size in our collections (measuring 7.6–8.5 × 6.1–6.3 μm) is in agreement with the concept of *L. obscuratus* (sensu Heilmann-Clausen et al. [31]; according to the same authors, spores of *L. cyathuliformis* have an average size of 8.3–9.9 × 7.0–7.7 μm), we adopt this name for the specimens included in this work. The genus *Russula* Pers. is represented by *R. pumila* Rouzeau & F. Massart (Figure 5h–j) detected in one site only (Vourkoti). Although *R. pumila* is synonymous to *R. alnetorum* Romagn. according to both Index Fungorum and Mycobank, there are different opinions about the synonymy of these two taxa and to the best of our knowledge this issue has not been resolved yet (S. Adamcik, pers. comm.). *R. pumila* is reported to occur mainly in lowlands with *A. glutinosa*, as opposed to *R. alnetorum*, which is mostly recorded in subalpine habitats with *A. viridis* [53,54]; therefore, we adopt the use of the former name.

Tomentella sublilacina (Ellis & Holw.) Wakef. and *T. stuposa* (Link) Stalpers were previously recorded in Andros (in the alluvial alder forest of Vori; [20]) and identified on the basis of their morphology. These names are provisionally retained here due to the absence of precise taxonomic information concerning alder-specific *Tomentella* species. It should be noted that the ITS sequences generated in the frame of this work represent phylogenetically distinct taxa corresponding to entities named "aff. *sublilacina*" and "aff. *stuposa*" in previous studies referring to material originating from alder hosts only [18,55].

It is noteworthy that the first ever report of an alniphilous *Cortinarius* species in Greece derives from a single collection of *C. americanus* A.H. Sm. (Figure 5k), which forms part of a small group of species within the subgenus *Telamonia* (Fr.) Trog known to be associated with *A. glutinosa* and *A. incana* in Europe [33,56]. *C. americanus* is characterized by the minute size, pileus not exceeding 2 cm in diameter, with dark violet to blackish colour, and spores smaller than 10 × 6 μm [33,56]; our collection has spores measuring 7.6–8.4 × 4.9–5.6 μm.

Last, *Inocybe calospora* Quél., recorded for the first time in Greece, was collected in Vourkoti only; it is an easily identified species thanks to its unique star-shaped spiny spores (Figure 5). Among all ECM species included in this inventory it is the only one which is not considered to be exclusively associated with alders, and reported from diverse damp deciduous forests of Europe [33,56].

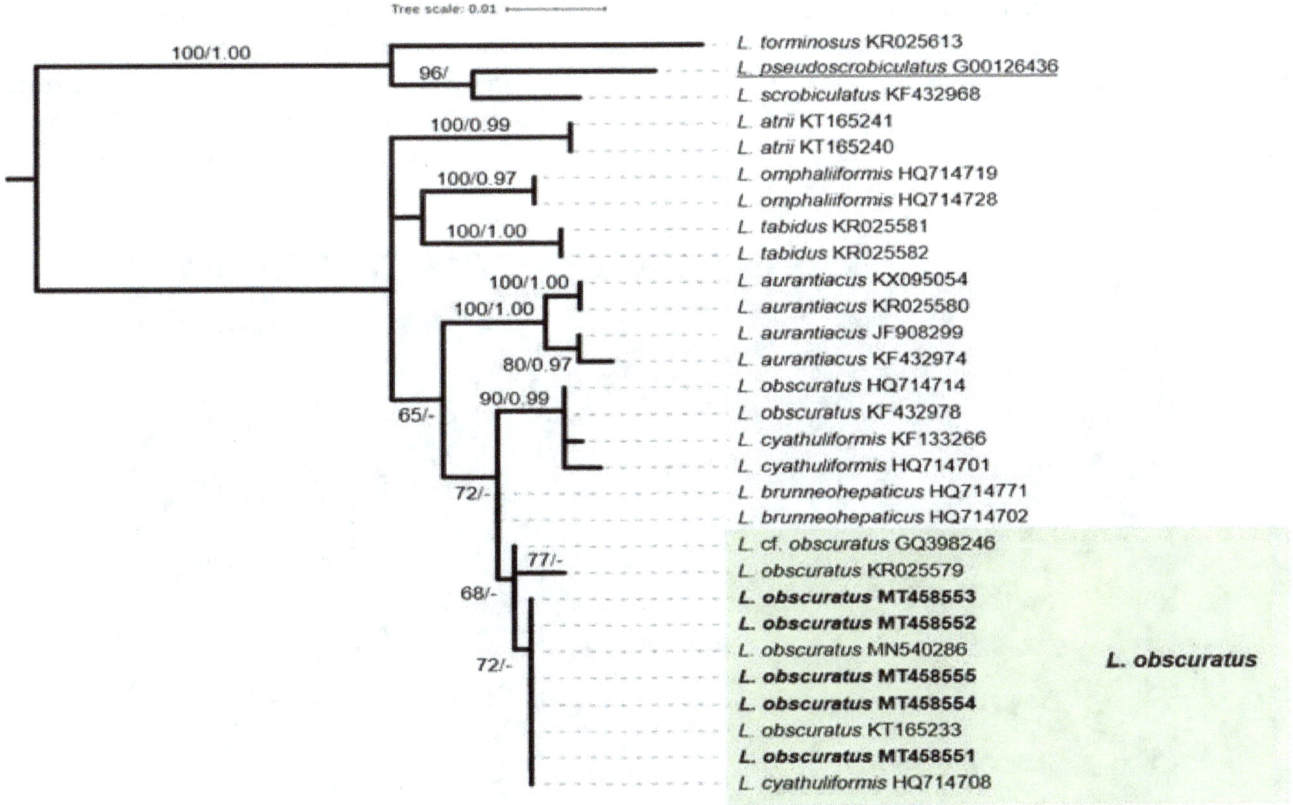

Figure 7. Phylogeny of *Lactarius* species derived from rDNA ITS sequences through ML analysis. Branches are labelled when MLB > 65%, and BPP > 0.95. *L. torminosus*, *L. scrobiculatus* and *L. pseudoscrobiculatus* were used as outgroup. The colored box includes sequences from specimens recorded in the *Alnus glutinosa* habitat.

3.2. The Saproxylic Element

By far the highest number of species recorded in this inventory correspond to white-rot and brown-rot basidiomycetes found on various wood parts of *A. glutinosa* (Table 2). This is quite anticipated since alder trees have a life-span which rarely exceeds 100 years; therefore, they produce large amounts of dead wood. Moreover, in contrast to ECM species, wood-rotting fungi do not show any specificity to alders, with only few exceptions including the common in northern Europe plant-pathogenic polypore *Inonotus radiatus* (Sowerby) P. Karst. [15,57–60], which, however, was not among our findings. Most of the recorded species are wood rotting basidiomycetes that are quite common throughout Europe on deciduous tree species including alders [13,61,62]. The best represented genera of saproxylic fungi were *Hyphoderma* Wallr. (three spp.), *Mycena* (Pers.) Roussel (four spp.), *Pluteus* Fr. (four spp.) and *Psathyrella* (Fr.) Quél. (four spp. on woody residues or buried wood). In total, seven species recorded on dead wood or bark of living alder trees are recorded for the first time in Greece and are presented in more detail below.

Among corticioid basidiomycetes, *Hyphoderma nemorale* K.H. Larss. is a distinct, widely distributed but rare species in Europe [35], which was identified by re-examining an old collection from Vori alluvial forest and further confirmed by ITS sequencing. The presence of thick-walled subicular hyphae and of two types of hymenial cystidia (i.e., short ventricose and subcapitate, and more seldom long tubular with characteristic constrictions, sometimes moniliform, which was the case in our specimen) are the main diagnostic features of this species [63]. *Hyphodermella corrugata* (Fr.) J. Erikss. & Ryvarden (Figure 8a) is fairly common and widespread in Europe [35]. It is easily identified thanks to its characteristic cystidioid hyphal ends appearing in bundles that are heavily incrusted [24]. *Phanerochaete livescens* (P. Karst.) Volobuev & Spirin (Figure 8b) is a species closely related to *Ph. sordida* (P. Karst.) J. Erikss. & Ryvarden which was recently described by using both morphological and phylogenetic criteria [64]. In accordance to the pertinent description, our specimens possessed cystidia with thickened walls to the acute apex, densely covered by crystals, as opposed to the accidentally encrusted, obtuse and thin-walled towards the apex cystidia of *Ph. sordida*. Moreover, the identity of our specimen was confirmed by the respective ITS sequence which was identical to those of *Ph. livescens* as determined by Volobuev et al. [64].

Figure 8. Saproxylic species recorded on Alnus glutinosa wood and litter in Andros: Hyphodermella corrugata (**a**) Phanerochaete livescens (**b**) Pluteus podospileus (**c**) Delicatula integrella (**d**) Coprinopsis melanthina (**e**) Gymnopilus arenophilus (**f**) Hydropus floccipes (**g**) Lepiota ochraceofulva (**h**) Lepista ovispora (**i**) Psathyrella hellebosensis (**j**) Entoloma uranochroum (**k**) Bar: 1 mm.

Among the agaricoid wood-inhabiting fungi, the genus *Pluteus* Fr. is hereby represented by four species, of which *P. podospileus* Sacc. & Cub. (Figure 8c) is reported for the first time in Greece. It belongs to the sect. *Celluloderma* Fay. subsection *Mixtini* Sing. ex Sing, and possesses a pileipellis made up of both fusiform and broadly clavate elements. *P. thomsonii* (Berk. & Broome) Dennis is very similar morphologically but it differs in the absence of pleurocystidia and the shape of cheilocystidia, which are characteristically rostrate [65]. The record of *Delicatula integrella* (Pers.) Rat. (Figure 8d) is worth mentioning since it was reported only once before in Greece, in the content of a regional field-guide [66]. *D. integrella* forms whitish-mycenoid mushrooms of minute size with pileus diameter measuring (in our specimens) 0.4–0.6 cm, reduced, almost vein-like lamellae and non-amyloid, amygdaliform-fusoid spores. It is considered widespread and common in Europe and N. America, and grows on decaying wood and wood debris of deciduous trees [34]. *Coprinopsis melanthina* (Fr.) Örstadius & E. Larss. (Figure 8e) is a striking-looking psathyrelloid species, easily identified due to the relatively large basidiomes with wooly to squamulose pileus (measuring 2–6 cm in diam.) and stipe (up to 6.0 × 0.8 cm), absence of pleurocystidia, almost colourless basidiospores, devoid of germ-pore, measuring 9.8–11.8(13.5) × 5.6–6.5 μm in our collections. This is a rather uncommon European species growing on and around rotten stumps in humid deciduous forests [33]; the only other record of this species in Greece derives from Crete (G. Konstandinidis, pers. comm.).

Gymnopilus arenophilus A. Ortega & Esteve-Rav. (Figure 8f) was described from continental areas of Spain [67] and from maritime dunes in France, under or near Mediterranean pines, on sandy soil by being attached to wood debris or wood, often burnt or buried in the sand [68]. Two of our collections from the alluvial *A. glutinosa* habitat at Lefka were sequenced and found to correspond to this species. Apparently, no native pines exist in Andros while both specimens were growing on rotten alder stumps, a fact that largely expands the so far known ecological and geographical range of this Mediterranean species. Morphologically, our specimens possessed features that fit well to the taxonomic concept of *G. arenophilus* i.e., smooth to fibrillose pileal surface, bitter taste, ellipsoid to subamygdaliform, moderately verrucose spores, measuring 8–10 × 5.5–6.5 μm, lageniform cheilocysidia often with subcapitate apex, 25–45 × 5–8 μm and absence of pleurocystidia. On the other hand, the size of basidiomes was significantly larger, with pilei up to 10 cm in diam. and a sturdy stipe often thicker than 1 cm. One previous collection of ours also found on rotten alder stump, and originally identified as *G. picreus* (Pers.) P. Karst. [22], is now re-assessed as *G. arenophilus*.

Hydropus floccipes (Fr.) Singer (Figure 8g) is a rare mycenoid species generally found to grow on decayed trunks of deciduous trees in damp forests, and is characterized by non-amyloid, subglobose spores, not blackening basidiomes, and typically scabrous stipe with grey-brown spots [33,69]. In addition, our specimens possessed yellowish stipe, previously reported for *H. floccipes* var. *luteipes* Ortega & Zea described from Spain [70]; the latter is otherwise microscopically identical and of unknown phylogenetic status. Two unpublished reports of *H. floccipes* exist from Greece (D. Sofronis and G. Konstandinidis, pers. comm.).

3.3. Litter and Other Terrestrial Decomposers

Apart ECM and saproxylic fungi, several other mushroom species were recorded under black alder trees, and therefore constitute a part of the fungal diversity of the *A. glutinosa* priority habitat in Andros. Needless to say, none of these species is specifically linked to alders; instead they are considered 'generalists' to be found in both deciduous and coniferous forests. Among them, the following four species are recorded for the first time in Greece. *Lepiota ochraceofulva* P.D. Orton (Figure 8h) is a rather rare (but widespread) species in Europe, reported from *Fagus* and other deciduous trees on humus-rich, loamy soil [71,72]; it forms highly toxic mushrooms containing amanitins. Our single collection consisted of few basidiomes growing on a thick layer of leaf-litter under *A. glutinosa*. They are characterized by pilei of up to 7 cm in diam., with orange-brown scales; lamellae forming a distinct collarium and reddish-orange in maturity; spores ellipsoid to oblong, dextrinoid, not metachromatic in cresyl-blue, measuring 5.4–7.5(8.1) × 3.5–4(4.5) μm; basidia (2)4-spored, clamped; cheilocystidia short clavate to cylindrical, rarely papilate, often in chains; pileipellis, a hymeniderm, composed of more or less clavate elements up to 50 μm long.

Lepista ovispora (J.E. Lange) Gulden (Figure 8i) is an uncommon (albeit widespread) European species recorded only once during this study on leaf litter under alders. Typical diagnostic features are the densely caespitose habit, the relatively fleshy basidiomes, the brown hygrophanous pileus with pruinose surface [73]. In addition, our specimens possessed spores ovoid to broadly ellipsoid, finely punctate, 4.7–6.8 × 3.8–4.4 μm, clamped basidia and no cystidia. *Psathyrella hellebosensis* D. Deschuyteneer, A. Melzer (Figure 8j) was recently described from Belgium [74] and was later reported from riparian alder habitats in Italy [75]. The morphological features of our collection are in agreement with the morphology of Belgian and Italian basidiomes, but since it corresponds to a rarely reported species, a detailed description of our material is hereby provided: pileus up to 3 cm in diam., hygrophanous from dark reddish-brown to greyish-beige, with scanty remains of veil; lamellae subdistant with whitish edge; stipe 2–4 × 0.2–0.3 cm, not rooting; spores 7.3–8.7 × 4.5–5.7 μm, Q = 1.43–1.73, ovoid to angular in face-view and not or weakly phaseoliform in side-view, not opaque; lamellae edge sterile composed exclusively of sphaeropendunculate paracystidia (no pleurocystidioid paracystidia were observed); pleurocystidia 34–48 × 11–17 μm, utriform. The material was collected from wet soil by the alluvial stream banks. This species shows high phylogenetic affinity to *P. thujina* A. H. Sm. by using ITS sequences only; however, it is clearly separated when the tef-1α marker is added in the phylogenetic analysis, while it is also distinguished by its distinctly larger and prominently phaseoliform spores [75].

Previous studies on the mycodiversity of Andros island reported the occurrence of four *Entoloma* species, one of them was new to science, i.e., *E. alnicola* Noordel. & Polemis [22,25]. Our recent field work resulted in other interesting collections of *Entoloma* spp. for which the identity and phylogenetic relationships to closely allied taxa are still under investigation. However, by using morphology alone, the presence of a rare European species was confirmed, namely *E. uranochroum* Hauskn. & Noordel. (Figure 8k) recorded for the first time in an alder habitat. This beautiful dark blue-violet mushroom was so far reported from subalpine meadows on calcareous soil in Austria (type locality) and the French Alps. Moreover, its striking microscopical features, e.g., the large fusiform cheilocystidida with granular yellowish-brown content, place it in the distinct section *Ramphocystotae* (Largent) Noordel., together with only one other European representative, namely *E. rhynchocystidiatum* Noordel. & Liiv [76].

4. Conclusions

A long-term study of the diversity of macrofungi in alder stands of Andros resulted in an inventory consisting of 106 species of basidiomycetes, including 21 taxa recorded for the first time in Greece. The majority of findings corresponded to saprotrophs (#92, mainly wood-rotting fungi) and the rest were ECM species. Considering the limited size of the area under study in a small Aegean island, the outcome of this work in terms of the number of taxa and variability is indicative of the wealth of the *A. glutinosa* priority habitat. However, the black alder stands in Andros have suffered considerably from floods in the past (as a consequence of fires that destroyed vegetation in the surrounding mountains which acted as a physical barrier protecting from downhill water runoffs) and their regeneration is hindered due to grazing by feral goats. The importance of fungi in the conservation/restoration of such natural habitats was demonstrated in the past [77,78], and recent activities focus at improving the status of the degenerated alder stands by exploiting indigenous ECM fungi as inoculants to young alder seedlings prior to their transplantation on site. Moreover, new knowledge about mushroom diversity and the ecological role of this group of organisms seems to enhance considerably people's perception and awareness, and hence facilitates implementation of conservations actions which are currently under way in selected alder stands of Andros.

Author Contributions: Conceptualization, E.P. and G.Z.; methodology, E.P., G.I.Z., V.D. and V.F.; validation, E.P., V.D. and V.F.; formal analysis, E.P., G.I.Z. and V.F.; investigation, E.P., G.I.Z., V.D. and V.F.; data curation, E.P., G.I.Z. and V.F.; writing—original draft preparation, E.P.; writing—final draft, G.I.Z.; review and editing—final draft, E.P., G.I.Z., V.D. and V.F.; supervision, G.I.Z.; project administration, G.I.Z.; and funding acquisition, G.I.Z. All authors have read and agreed to the published version of the manuscript.

Acknowledgments: We would like to thank V. Goritsas for the preparation of the map figures included in this work, and S. Adamcik, B. Dima, M. Noordeloos and J. Nuytinck for helpful discussions on some of the findings of this study.

References

1. Kajba, D.; Gračan, J. *EuFORGEN Technical Guidelines for Genetic Conservation and Use for Black Alder (Alnus glutinosa)*; Bioversity International: Rome, Italy, 2003; pp. 1–6.
2. Karl, T.R.; Trenberth, K.E. Modern Global Climate Change. *Science* **2003**, *302*, 1719–1723. [CrossRef] [PubMed]
3. McEwan, N.R.; Wilkinson, T.; Girdwood, S.E.; Snelling, T.J.; Collins, T.; Dougal, K.; Jones, D.L.; Godbold, D.L. Evaluation of the microbiome of decaying alder nodules by next generation sequencing. *Endocyt. Cell Res.* **2017**, *28*, 14–19.
4. Roy, M.; Pozzi, A.C.; Gareil, R.; Nagati, M.; Manzi, S.; Nouioui, I.; Sharikadze, N.; Jargeat, P.; Gryta, H.; Moreau, P.-A.; et al. Alder and the Golden Fleece: High diversity of *Frankia* and ectomycorrhizal fungi revealed from *Alnus glutinosa* subsp. *barbata* roots close to a Tertiary and glacial refugium. *PeerJ* **2017**, *5*, e3479. [CrossRef] [PubMed]
5. Orfanoudakis, M.Z.; Hooker, J.E.; Wheeler-Jones, C.T. Early interactions between arbuscular mycorrhizal fungi and *Frankia* during colonisation and root nodulation of *Alnus glutinosa*. *Symbiosis* **2004**, *36*, 69–82.
6. Põlme, S.; Öpik, M.; Moora, M.; Zobel, M.; Kohout, P.; Oja, J.; Kõljalg, U.; Tedersoo, L. Arbuscular mycorrhizal fungi associating with roots of *Alnus* and *Rubus* in Europe and the Middle East. *Fungal Ecol.* **2016**, *24*, 27–34. [CrossRef]
7. Harley, J.L.; Smith, S.E. *Mycorrhizal Symbiosis*; Academic Press: London, UK, 1983; pp. 1–483.
8. Pritsch, K.; Munch, J.C.; Buscot, F. Characterization and identification of black alder ectomycorrhizas by PCR/RFLP analyses of the rDNA internal described spacer (ITS). *New Phytol.* **1997**, *137*, 357–369. [CrossRef]
9. Pritsch, K.; Munch, J.C.; Buscot, F. Morphological and anatomical characterisation of black alder *Alnus glutinosa* (L.) Gaertn. ectomycorrhizas. *Mycorrhiza* **1997**, *7*, 201–216. [CrossRef]
10. Boyle, H. Aspekte der Macromycetenflora dreier Erlenbrücher Norddeutschlands und vergleichende PCR/RFLP- Analyse ausgewählter ectomycorrhizaler Mycobionten. *EcoSys Suppl.* **1996**, *10*, 1–106.
11. Brunner, I.; Horak, E. Mycoecological analysis of *Alnus* associated macrofungi in the region of the Swiss National Park as recorded by J. Favre (1960). *Mycol. Helv.* **1990**, *4*, 111–139.
12. Bujakiewicz, A.M. Macrofungi in the alder and alluvial forests in various parts of Europe and North America. *Opera Bot.* **1989**, *100*, 29–41.
13. Kunttu, P.; Kotiranta, H.; Kulju, M.; Pasanen, H.; Kouki, J. Occurrence patterns, diversity and ecology of aphyllophoroid fungi on the black alder (*Alnus glutinosa*) in an archipelago in the Baltic Sea. *Ann. Bot. Fenn.* **2016**, *53*, 173–193. [CrossRef]
14. Senn-Irlet, B.; Mürner, R.; Martini, E.; Küffer, N.; de Marchi, R.; Bieri, G. Saprobic fungi on wood and litter of *Alnus alnobetula* in the Swiss Alps. *Mycotaxon* **2012**, *120*, 506.
15. Strid, Å. Wood-inhabiting fungi of alder forests in north-central Scandinavia 1. Aphylloporales (Basidiomycetes). Taxonomy, ecology and distribution. *Wahlenbergia* **1975**, *1*, 1–237.
16. Griesser, B. Mykosoziologie der Grauerlen-und Sanddorn-Auen (Alnetum incanae, Hippophaëtum) am Hinterrhein (Domleschg, Graubünden, Schweiz). *Ver. Geobot. Inst. ETH* **1992**, *109*, 1–235.
17. Molina, R. Ectomycorrhizal specificity in the genus *Alnus*. *Can. J. Bot.* **1981**, *59*, 325–334. [CrossRef]
18. Tedersoo, L.; Suvi, T.; Jairus, T.; Ostonen, I.; Põlme, S. Revisiting ectomycorrhizal fungi of the genus *Alnus*: Differential host specificity, diversity and determinants of the fungal community. *New Phytol.* **2009**, *182*, 727–735. [CrossRef]
19. Rochet, J.; Moreau, P.-A.; Manzi, S.; Gardes, M. Comparative phylogenies and host specialization in the alder ectomycorrhizal fungi *Alnicola*, *Alpova* and *Lactarius* (Basidiomycota) in Europe. *BMC Evol. Biol.* **2011**, *11*, 40. [CrossRef]

20. Dimou, D.M.; Polemis, E.; Zervakis, G.I. Macromycetes associated with *Alnus glutinosa* in Greece. *Phytopathol. Mediterr.* **2006**, *45*, 78.
21. Dimou, D.M.; Zervakis, G.I.; Polemis, E. Mycodiversity studies in selected ecosystems of Greece: IV. Macrofungi from *Abies cephalonica* forests and other intermixed tree species. (Oxya mountain, central Greece). *Mycotaxon* **2008**, *104*, 39–42.
22. Polemis, E.; Dimou, D.M.; Tzanoudakis, D.; Zervakis, G.I. Diversity of Basidiomycota (subclass Agaricomycetidae) in the island of Andros (Cyclades, Greece). *Nova Hedwig.* **2012**, *95*, 25–58. [CrossRef]
23. Polemis, E.; Dimou, D.; Zervakis, G.I. The family *Hymenochaetaceae* (Agaricomycetes, Basidiomycota) in the islands of the Aegean Archipelago (Greece). *Plant Biosyst.* **2013**, *147*, 306–314. [CrossRef]
24. Polemis, E.; Roberts, P.; Dimou, D.M.; Zervakis, G.I. Heterobasidiomycetous fungi from Aegean Islands (Greece): New annotated records for a neglected group. *Plant Biosyst.* **2016**, *150*, 295–303. [CrossRef]
25. Noordeloos, M.; Polemis, E. Studies in the genus *Entoloma* (Basidiomycetes, Agaricales) from the Kiklades (C. Aegean, Greece). *Mycotaxon* **2008**, *105*, 301–312.
26. Eriksson, J.; Ryvarden, L. *The Corticiaceae of North Europe, Vol. 4: Hyphodermella—Mycoacia*; Fungiflora: Oslo, Norway, 1976; pp. 549–886.
27. Bas, C.; Kuyper, T.W.; Noordeloos, M.E.; Vellinga, E.C. (Eds.) *Flora Agaricina Neerlandica*; A. A. Balkema: Rotterdam, The Netherlands, 1990; Volume 2, pp. 1–137.
28. Bas, C.; Kuyper, T.W.; Noordeloos, M.E.; Vellinga, E.C. (Eds.) *Flora Agaricina Neerlandica*; A. A. Balkema: Rotterdam, The Netherlands, 1995; Volume 3, pp. 1–183.
29. Bas, C.; Kuyper, T.W.; Noordeloos, M.E.; Vellinga, E.C. (Eds.) *Flora Agaricina Neerlandica*; A. A. Balkema: Rotterdam, The Netherlands, 1999; Volume 4, pp. 1–191.
30. Noordeloos, M.E.; Kuyper, T.W.; Vellinga, E.C. (Eds.) *Flora Agaricina Neerlandica*; A. A. Balkema: Rotterdam, The Netherlands, 2001; Volume 5, pp. 1–170.
31. Heilmann-Clausen, J.; Verbeken, A.; Vesterholt, J. *The Genus Lactarius. Fungi of Northern Europe*; Danish Mycological Society, Svampetryk: Copenhagen, Denmark, 1998; Volume 2, pp. 1–287.
32. Bernicchia, A. *Polyporaceae S.L.*; Candusso: Alassio, Italy, 2005; pp. 1–808.
33. Knudsen, H.; Vesterholt, J. (Eds.) *Funga Nordica. Agaricoid, Boletoid and Cypheloid Genera*; Nordsvamp: Copenhagen, Denmark, 2008; pp. 1–965.
34. Antonín, V.; Noordeloos, M.E. *A Monograph of Hemimycena, Delicatula, Fayodia, Gamundia, Myxomphalia, Resinomycena, Richenella and Pseudomphalina (Tribus Mycenae Sensu Singer, Mycena Excluded)*; IHW Verlag: Eching, Germany, 2004; pp. 1–279.
35. Bernicchia, A.; Gorjón, S.P. *Corticiaceae S.L.*; Candusso: Alassio, Italy, 2010; pp. 1–1008.
36. Aronsen, A.; Læssøe, T. *The Genus Mycena. Fungi of Northern Europe*; Svampetryk: Copenhagen, Denmark, 2016; Volume 5, pp. 1–373.
37. White, T.J.; Bruns, T.; Lee, S.; Taylor, J.W. Amplification and direct sequencing of fungal ribosomal RNA genes for phylogenetics. In *PCR Protocols: A Guide to Methods and Applications*; Innis, M.A., Gelfand, D.H., Sninsky, J.J., White, T.J., Eds.; Academic Press Inc.: New York, NY, USA, 1990; pp. 315–322.
38. Hall, T.A. BioEdit: A User-Friendly Biological Sequence Alignment Editor and Analysis Program for Windows 95/98/NT. *Nucleic Acids Symp. Ser.* **1999**, *41*, 95–98.
39. Katoh, K.; Rozewicki, J.; Yamada, K.D. MAFFT online service: Multiple sequence alignment, interactive sequence choice and visualization. *Brief Bioinform.* **2019**, *20*, 1160–1166. [CrossRef]
40. Kumar, S.; Stecher, G.; Li, M.; Knyaz, C.; Tamura, K. MEGA X: Molecular Evolutionary Genetics Analysis across computing platforms. *Mol. Biol. Evol.* **2018**, *35*, 1547–1549. [CrossRef] [PubMed]
41. Stamatakis, A.; Hoover, P.; Rougemont, J. A rapid bootstrap algorithm for the RAxML Web servers. *Syst. Biol.* **2008**, *57*, 758–771. [CrossRef]
42. Ronquist, F.; Teslenko, M.; van der Mark, P.; Ayres, D.L.; Darling, A.; Höhna, S.; Larget, B.; Liu, L.; Suchard, M.A.; Huelsenbeck, J.P. MrBayes 3.2: Efficient Bayesian phylogenetic inference and model choice across a large model space. *Syst. Biol.* **2012**, *61*, 539–542. [CrossRef]
43. Darriba, D.; Taboada, G.L.; Doallo, R.; Posada, D. jModelTest 2: More models, new heuristics and parallel computing. *Nat. Methods* **2012** *9*, 772. [CrossRef] [PubMed]
44. Letunic, I.; Bork, P. Interactive tree of life (iTOL) v3: An online tool for the display and annotation of phylogenetic and other trees. *Nucleic Acids Res.* **2016**, *44*, W242–W245. [CrossRef] [PubMed]

45. Moreau, P.-A. A nomenclatural revision of the genus *Alnicola* (Cortinariaceae). *Fungal Divers.* **2005**, *20*, 121–155.
46. Moreau, P.-A.; Peintner, U.; Gardes, M. Phylogeny of the ectomycorrhizal mushroom genus *Alnicola* (Basidiomycota, Cortinariaceae) based on rDNA sequences with special emphasis on host specificity and morphological characters. *Mol. Phylogen. Evol.* **2006**, *38*, 794–807. [CrossRef]
47. De Haan, A.; Moreau, P.-A. Waarnemingen in het genus *Alnicola* (Zompzwam) in Vlaanderen (3). *Steerbeckia* **2012**, *31*, 3–15.
48. Moser, M. *Keys to Agarics and Boleti*; R. Phillips: Tonbridge, UK, 1983; pp. 1–535.
49. Horak, E. *Röhrlinge und Blätterpilze in Europa—Unter der Mitarbeit von Anton Hausknecht (Bolbitiaceae) und P.A. Moreau (Alnicola)*; Elsevier Spektrum Akademischer: Heidelberg, Germany, 2005; pp. 1–557.
50. Henrici, A. Keys to *Naucoria* in Britain. *Field Mycol.* **2009**, *9*, 55–62. [CrossRef]
51. Jargeat, P.; Moreau, P.-A.; Gryta, H.; Chaumeton, J.P.; Gardes, M. *Paxillus rubicundulus* (Boletales, Paxillaceae) and two new alder-specific ectomycorrhizal species, *Paxillus olivellus* and *Paxillus adelphus*, from Europe and North Africa. *Fungal Biol.* **2016**, *120*, 711–728. [CrossRef]
52. Wisitrassameewong, K.; Looney, B.P.; Le, H.T.; De Crop, E.; Das, K.; Van de Putte, K.; Eberhardt, U.; Jiayu, G.; Stubbe, D.; Hyde, K.D.; et al. *Lactarius* subgenus *Russularia* (Basidiomycota, Russulales): Novel Asian species, worldwide phylogeny and evolutionary relationships. *Fungal Biol.* **2016**, *120*, 1554–1581. [CrossRef] [PubMed]
53. Galli, R. *Le Russule. Atlante Pratico—Monografico per la Determinazione delle Russule*; Dalla Natura: Milano, Italy, 2003; pp. 1–480.
54. Floriani, M.; Partacini, G. Sull'identità di *Russula puellaris* var. *leprosa*. *Boll. Gruppo Micol. G. Bres.* **1998**, *40*, 213–218.
55. Roy, M.; Rochet, J.; Manzi, S.; Jargeat, P.; Gryta, H.; Moreau, P.-A.; Gardes, M. What determines *Alnus*-associated ectomycorrhizal community diversity and specificity? A comparison of host and habitat effects at a regional scale. *New Phytol.* **2013**, *198*, 1228–1238. [CrossRef] [PubMed]
56. Breitenbach, J.; Kranzlin, F. *Fungi of Switzerland*; Verlag Mykologia: Lucerne, Switzerland, 2000; Volumes 1–5, pp. 1–342.
57. Yurchenko, E.O. Natural substrata for corticioid fungi. *Acta Mycol.* **2006**, *41*, 113–124. [CrossRef]
58. Ainsworth, M. Some British alder-associated wood- inhabiting fungi. *Field Mycol.* **2010**, *11*, 10–15. [CrossRef]
59. Kotiranta, H.; Saarenoksa, R.; Kytövuori, I. Aphyllophoroid fungi of Finland. A check-list with ecology, distribution, and threat categories. *Norrlinia* **2009**, *19*, 1–223.
60. Niemelä, T.; Kotiranta, H. Polypore survey of Finland 3. The genera *Coltricia*, *Inonotopsis*, *Inonotus* and *Onnia*. *Karstenia* **1983**, *23*, 15–25. [CrossRef]
61. Küffer, N.; Senn-Irlet, B. Diversity and ecology of corticioid basidiomycetes in green alder stands in Switzerland. *Nova Hedwig.* **2000**, *71*, 131–143.
62. Piętka, J.; Grzywacz, A. Grzyby wielkoowocnikowe stwierdzone na olszy czarnej *Alnus glutinosa* (L.) Gaertn. w drzewostanach olszowych wykazujących objawy zamierania. *Sylwan* **2018**, *162*, 22–31.
63. Larsson, K.-H. Two new species in *Hyphoderma*. *Nord. J. Bot.* **1998**, *18*, 121–127. [CrossRef]
64. Volobuev, S.; Okun, M.; Ordynets, A.; Spirin, V. The *Phanerochaete sordida* group (Polyporales, Basidiomycota) in temperate Eurasia, with a note on *Phanerochaete pallida*. *Mycol. Prog.* **2015**, *14*, 1–13. [CrossRef]
65. Vellinga, E.C. Pluteaceae Kolt. & P. In *Flora Agaricina Neerlandica*; Bas, C., Noordeloos, M.E., Kuyper, T.W., Vellinga, E.C., Eds.; A. A. Balkema: Rotterdam, The Netherlands, 1990; Volume 2, pp. 31–64.
66. Konstantinidis, G. *1000 Mushrooms of Western Makedonia*; The Mushroom Friends Society of Western Makedonia: Kastoria, Greece, 2006; pp. 1–523. (In Greek)
67. Ortega, A.; Esteve-Raventós, F. A new species of *Gymnopilus* (Cortinariaceae) from sandy soils in *Pinus* forests. *Persoonia* **2005**, *18*, 505–510.
68. Guzman-Davalos, L.; Ortega, A.; Contu, M.; Vizzini, A.; Rodriguez, A.; Villalobos-Arambula, A.R.; Santerre, A. *Gymnopilus maritimus* (Basidiomycota, Agaricales), a new species from coastal psammophilous plant communities of northern Sardinia, Italy, and notes on *G. arenophilus*. *Mycol. Prog.* **2009**, *8*, 195–205. [CrossRef]
69. Bas, C. *Hydropus* (Kühner) ex Singer. In *Flora Agaricina Neerlandica*; Bas, C., Kuyper, T.W., Noordeloos, M.E., Vellinga, E.C., Eds.; A. A. Balkema: Rotterdam, The Netherlands, 1999; Volume 4, pp. 166–172.
70. Ortega, A.; Zea, M. *Hydropus floccipes* var. *luteipes* Ortega & Zea var. nov. en España meridional. *Bol. Soc. Micol. Madr.* **1991**, *15*, 189–191.

71. Vellinga, E.C. *Lepiota* (Pers.: Fr.) S.F. Gray. In *Flora Agaricina Neerlandica*; Noordeloos, M.E., Kuyper, T.W., Vellinga, E.C., Eds.; A. A. Balkema: Rotterdam, The Netherlands, 2001; Volume 5, pp. 109–151.
72. Gierczyk, B.; Kujawa, A.; Szczepkowski, A.; Chachuła, P. Rare species of *Lepiota* and related genera. *Acta Mycol.* **2011**, *46*, 137–178. [CrossRef]
73. Noordeloos, M.E.; Bas, C. *Flora Agaricina Neerlandica*; Bas, C., Kuyper, T.W., Noordeloos, M.E., Vellinga, E.C., Eds.; A. A. Balkema: Rotterdam, The Netherlands, 1995; Volume 3, pp. 67–75.
74. Deschuyteneer, D.; Melzer, A. *Psathyrella hellebosensis*, a new species from Belgium. *Bull. AMFB* **2017**, *10*, 3–10.
75. Voto, P.; Dovana, F.; Garbelotto, M. A revision of the genus *Psathyrella*, with a focus on subsection *Spadiceogriseae*. *FUSE* **2019**, *4*, 97–170. [CrossRef] [PubMed]
76. Noordeloos, M.E. *Entoloma s.l. Supplemento. Fungi Europei Vol. 5a*; Candusso: Alassio, Italy, 2004; pp. 1–760.
77. Glassman, S.; Levine, C.; DiRocco, A.; Battles, J.J.; Bruns, T.D. Ectomycorrhizal fungal spore bank recovery after a severe forest fire: Some like it hot. *ISME J.* **2016**, *10*, 1228–1239. [CrossRef] [PubMed]
78. Zotti, M.; Persiani, A.M.; Ambrosio, E.; Vizzini, A.; Venturella, G.; Donnini, D.; Angelini, P.; Di Piazza, S.; Pavarino, M.; Lunghini, D.; et al. Macrofungi as ecosystem resources: Conservation versus exploitation. *Plant Biosyst.* **2013**, *147*, 219–225. [CrossRef]

5

Utilization of High Performance Liquid Chromatography Coupled to Tandem Mass Spectrometry for Characterization of 8-*O*-methylbostrycoidin Production by Species of the Fungus *Fusarium*

Mark Busman

Mycotoxin Prevention and Applied Microbiology Research Unit, National Center for Agricultural Utilization Research, Agricultural Research Service, United States Department of Agriculture, Peoria, IL 61604, USA; Mark.Busman@ars.usda.gov.

Abstract: The pigment 8-*O*-methylbostrycoidin is a polyketide metabolite produced by multiple species of the fungus *Fusarium* that infects plant crops, including maize. A technique was developed for the analysis of 8-*O*-methylbostrycoidin by high performance liquid chromatography coupled to electrospray ionization tandem mass spectrometry. The quantitative nature of the LC-MS/MS experiment was demonstrated over a range of concentrations in maize. Limits of detection for the method (10 ng/g from 8-*O*-methylbostrycoidin spiked into ground maize) were shown, and susceptibility of the method to matrix effects from maize was also evaluated. The method was applied to evaluate the ability of the maize pathogen *Fusarium verticillioides* to produce 8-*O*-methylbostrycoidin in developing maize ears grown in an agricultural field.

Keywords: maize; liquid chromatography-mass spectrometry; 8-*O*-methylbostrycoidin; pigment

---

1. Introduction

The pigment, 8-*O*-methylbostrycoidin (Figure 1), is an azaanthraquinone secondary metabolite that is widely produced by species of the filamentous fungus *Fusarium*. A complete structural characterization was accomplished by Steyn [1]. Total synthesis of the compound was demonstrated by Cameron [2,3]. The compound was first observed from *Fusarium* isolates taken from maize [1] and citrus [4]. Deng has recently noted production from a marine *Aspergillus* isolate [5].

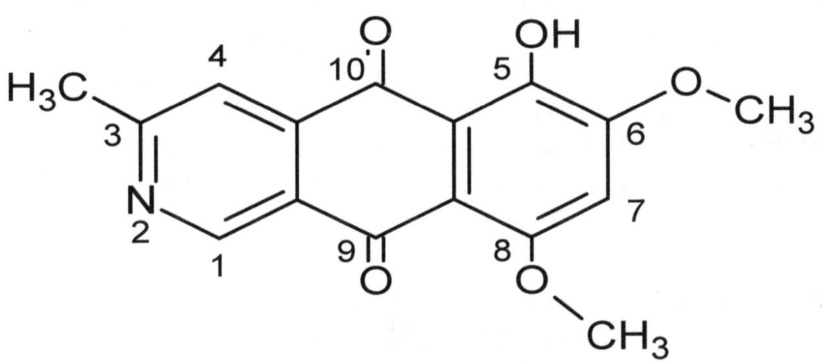

Figure 1. Structure of 8-*O*-methylbostrycoidin.

Early studies described the isolation of 8-O-methylbostrycoidin and showed its cytotoxicity against several mammalian cell lines, including rat hepatoma, Madin Darby canine kidney and Chinese hamster ovary [6]. Later, Hashimoto again observed the low cytotoxicity against the rat hepatoma cell line, but observed a strong ability to inhibit hepatic glucose production [7]. More recently, 8-O-methylbostrycoidin has been evaluated for cytotoxicity to a number of human cancer cell lines and has been shown to strongly inhibit α-acetylcholine esterase [5].

It is thought that 8-O-methylbostrycoidin biosynthesis involves a polyketide synthase that catalyzes the synthesis of a heptaketide, the first intermediate in 8-O-methylbostrycoidin biosynthesis. It has been suggested that a common polyketide synthase is responsible for the initial stages in the synthesis of several anthraquinones produced by *Fusaria* [8–10].

Chemical characterization of 8-O-methylbostrycoidin frequently is based upon its deep red color. For example, 8-O-methylbostrycoidin production in fungi grown on culture plates can be monitored visually, based on the color of the substrate. Levels of 8-O-methylbostrycoidin have also been monitored by absorbance at 510–520 nm either in HPLC detection or in spectrophotometric analysis of *Fusarium* extracts. Such UV-VIS based detection of related compounds has been implicated as an interfering actor in the detection of another Fusarium metabolite, zearalanone [11].

Here, we demonstrate a convenient technique for the analysis of 8-O-methylbostrycoidin by high performance liquid chromatography coupled to electrospray ionization tandem mass spectrometry. The technique is applied to the screening of *Fusarium* species for 8-O-methylbostrycoidin production on cracked maize substrate and to the evaluation of field production by *F. verticillioides* in maize.

2. Materials and Methods

2.1. Materials

Unless otherwise noted, all solvents were of HPLC grade from Sigma Chemical Co. (St. Louis, MO, USA). Water was from a Millipore (Billerica, MA, USA) water purification system. Purified 8-O-methylbostrycoidin was obtained from cultures of *F. verticillioides* as previously described [6]. Mature cultures of *F. verticillioides* MRC-826 on cracked maize were extracted with 60% aqueous methanol and purified on with successive Amberlite XAD-2 (Sigma, St. Louis, MO, USA) and normal phase (silica gel) liquid chromatographies. Eluted 8-O-methylbostrycoidin was monitored by ESI-MS. Purified 8-O-methylbostrycoidin was evaluated by ^{1}H and ^{13}C NMR. A stock solution of 1 mg/mL 8-O-methylbostrycoidin in acetonitrile was used for constructing calibration solutions and for spike solutions.

2.2. Fungal Strains and Preparation of Maize Materia

Two strains of *F. verticillioides* isolates were obtained from field-grown maize kernels as previously described [12]. Briefly, maize kernels were surface sterilized by soaking in a 10% chlorine bleach solution for 1 min. After rinsing three times in sterile water, the kernels were placed on Nash agar medium [12] and incubated for 3–5 days in the dark at room temperature. Single-spore derived isolates of the resulting fungal growth were obtained as described previously [13]. The resulting colonies were identified as *F. verticillioides* first by morphology in comparison with morphological descriptions of the fungus [13] and then by DNA sequence analysis of the gene encoding the translation elongation factor 1-α and comparison of the resulting sequence data to the Fusarium ID Database [14]. Cracked maize kernel cultures of *Fusarium* were prepared as previously described [15].

2.3. Extraction of 8-O-methylbostrycoidin from Maize

Twenty grams of cracked or ground maize were accurately weighed and extracted with 50 mL of acetonitrile for 2 h on a Gyratory Model G2 Shaker (New Brunswick Scientific, Edison, NJ, USA). Slurries of sample with solvent were centrifuged for 5 min at ~1500× *g*. One milliliter aliquots of the centrifugate were decanted by pipette into sample vials for LC-MS analysis.

2.4. ESI-MS and MS/MS

Unless otherwise noted, all experiments were conducted utilizing a ThermoFinnigan LCQ-DECA (Thermo Scientific, San Jose, CA, USA) ion trap mass spectrometer equipped with an electrospray ionization source. Flow injection experiments were accomplished with by injecting 10 μL plugs of analyte solution into a 300 μL/min flow of 1/1 methanol-water with 0.3% acetic acid. Injection of the analyte plugs was done by use of a Rheodyne Model 7125 injector (IDEX, Oak Harbor, WA, USA) valve fitted with a 10 μL injection loop.

2.5. HPLC-MS/MS

Ten microliters of extract analyzed with a MetaChem (Torrance, CA, USA) Inertsil C18 (150 mm length, 3 mm diameter, 5 μm diameter particle, 100 Å pore size) column. Chromatography was performed utilizing a ThermoSpectraPhysics (Thermo Scientific, San Jose, CA, USA) high performance liquid chromatography system consisting of an AS4000 autosampler coupled to a P2000 gradient pump. Elution of analyte was achieved with a 300 μL/min gradient flow of methanol and water (0.3% acetic acid was added to the mobile phase.) The solvent program used a 35–95% gradient over 25 min. The mass spectrometer was operated in positive mode electrospray ionization mode with an ionization voltage of 4.5 kV. The inlet capillary temperature was 255 °C, the inlet capillary voltage was 46 V and the tube lens offset was −5 V. The collision energy was optimized to 33%. Operation of the chromatography and mass spectrometry instrument and quantitation of the eluting 8-*O*-methylbostrycoidin was done utilizing ThermoFinnigan Xcalibur software (Version 1.4, Thermo Scientific, San Jose, CA, USA). Quantitative data was processed using Xcalibur QualBrowser (ThermoFinnigan) software. Uncertainties in mean determinations are expressed in terms of standard deviation.

Calibration curves were plotted using pure standards as well as maize matrix-assisted standards (acetonitrile extract). Maize matrix assisted standards were prepared as follows: Maize extract (considered blank for 8-*O*-methylbostrycoidin), which was obtained as described above, was spiked with 8-*O*-methylbostrycoidin dissolved at an appropriate concentration in acetonitrile.

2.6. Spike/Recovery Experiments and Method Validation

Spike recovery experiments were carried out in order to validate the method. Spiking solutions were prepared in acetonitrile and appropriate amounts added to cracked maize 1 h before extraction. Maize samples were spiked with 8-*O*-methylbostrycoidin by adding 250 μL aliquots (0.5, 5 and 50 μg/mL) of 8-*O*-methylbostrycoidin. Spiked samples were extracted with the same methods used for other samples. Spike/recovery samples were analyzed in triplicate on three successive days to evaluate method stability. Quantitations were based on integration of chromatographic peak area corresponding to the detection of characteristic fragment ions arising from the collisionally induced dissociation of the parent pseudomolecular ion $[M + H]^+$ of the analyte. Analyte responses were compared to responses obtained from solubilized (acetonitrile) 8-*O*-methylbostrycoidin. 8-*O*-methylbostrycoidin was quantified using an external calibration curve, which was constructed by plotting the concentration against the signal area.

2.7. Field Test

Maize line B73 was grown in field plots at the USDA, ARS, NCAUR (Peoria County, IL, USA) following guidelines of the Animal and Plant Health Inspection Service (APHIS) outlined in APHIS permit number P526-09-01825. The resulting maize ears were infected with the *F. verticillioides* by injection of 2 mL of a solution of 1×10^6 *F. verticillioides* spores per mL water into the silk channel of each ear at 4 to 6 days after silk emergence [16]. Control ears were similarly injected with water. For each treatment 10 ears were inoculated. Upon maturity, ears were harvested and visually evaluated for extent of fungal infestation as described by Desjardins et al. [17]. The disease evaluation method provides a 0–7 scoring of the damage of the kernels of the ear where a higher score indicates greater damage and a lower score indicates lesser damage. The ears were shelled and the resulting kernels were ground with a Stein Model M2 Laboratory Mill (Steinlite Corporation, Atchison, KS, USA). Ground kernels for each ear were extracted for LC-MS/MS analysis.

3. Results

3.1. MS and MS/MS by Flow Injection

The ESI-MS spectrum for the flow injection of a solution of 8-*O*-methylbostrycoidin is shown in Figure 2a. Operating the ion-trap mass analyzer in the MS/MS mode has potential to increase the selectivity of the method, minimize background signal and improve method performance. To further assess 8-*O*-methylbostrycoidin mass spectrometric behavior, tandem mass spectra for the collisional induced dissociation of the $[M + H]^+$ peak were acquired. When a certain threshold collision energy level was exceeded, the molecule broke up into several fragments. Features of the spectra did not change substantially despite use of higher collision energies. The resulting spectra were extremely simple. A spectrum for a CID-MS/MS experiment is shown in Figure 2b (33% collision energy). The instrument was operated to allow selection of a parent ion (m/z 300) and scanning of product ions. The dominant fragment ions at m/z 241, 256 and 271 are likely the result of facile losses of acetic acid, carbon dioxide and methanol. Other minor fragments are likely the result of extensive rearrangements. The molecular ion shows a great deal of stability despite use of relatively high collision energy. Lower mass fragments resulting from a more major decomposition of the parent ion would likely fall beneath the lower mass limit of the MS detector during the MS/MS experiment.

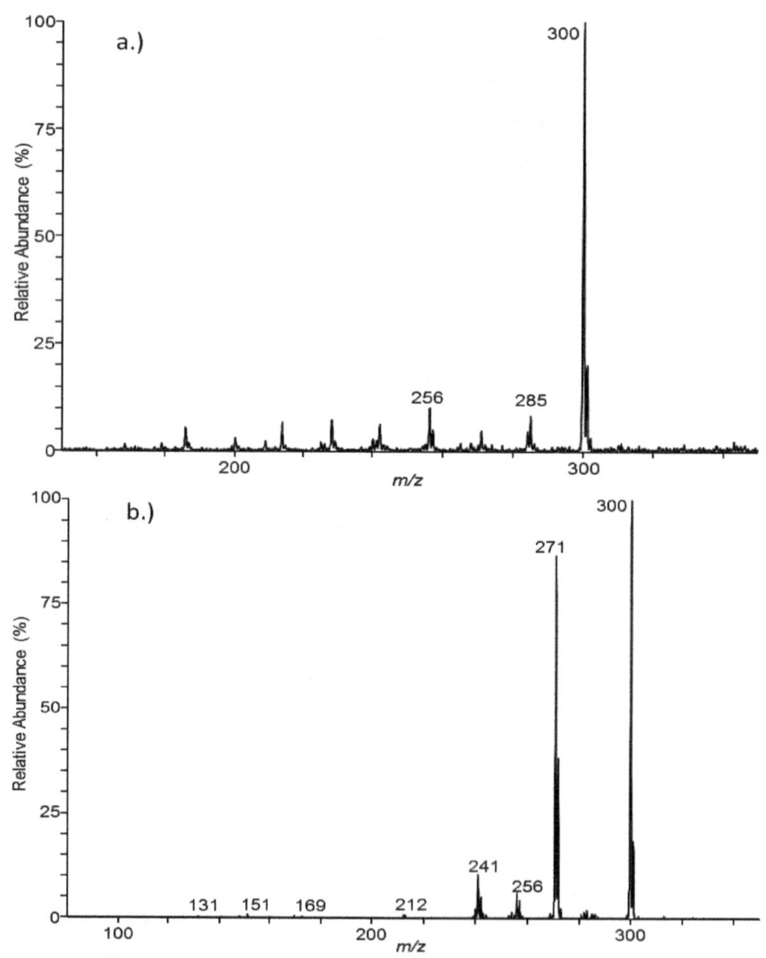

Figure 2. (**a**) MS and (**b**) MS/MS spectra for 8-*O*-methylbostrycoidin.

3.2. *HPLC Coupled to MS*

Injections of 8-*O*-methylbostrycoidin (10 μL, 50 μg/mL) were made into reversed phase gradient flows of methanol and water on the HPLC column. Good chromatographic behavior was observed with 10 μL injections of concentrations up to 1000 μg/mL. Figure 3 shows a LC-MS/MS profile resulting from of a 50 μg/mL solvent standard injection using multiple reaction monitoring (MRM) reflecting detection of the three fragments and intact parent $[M + H]^+$ ion.

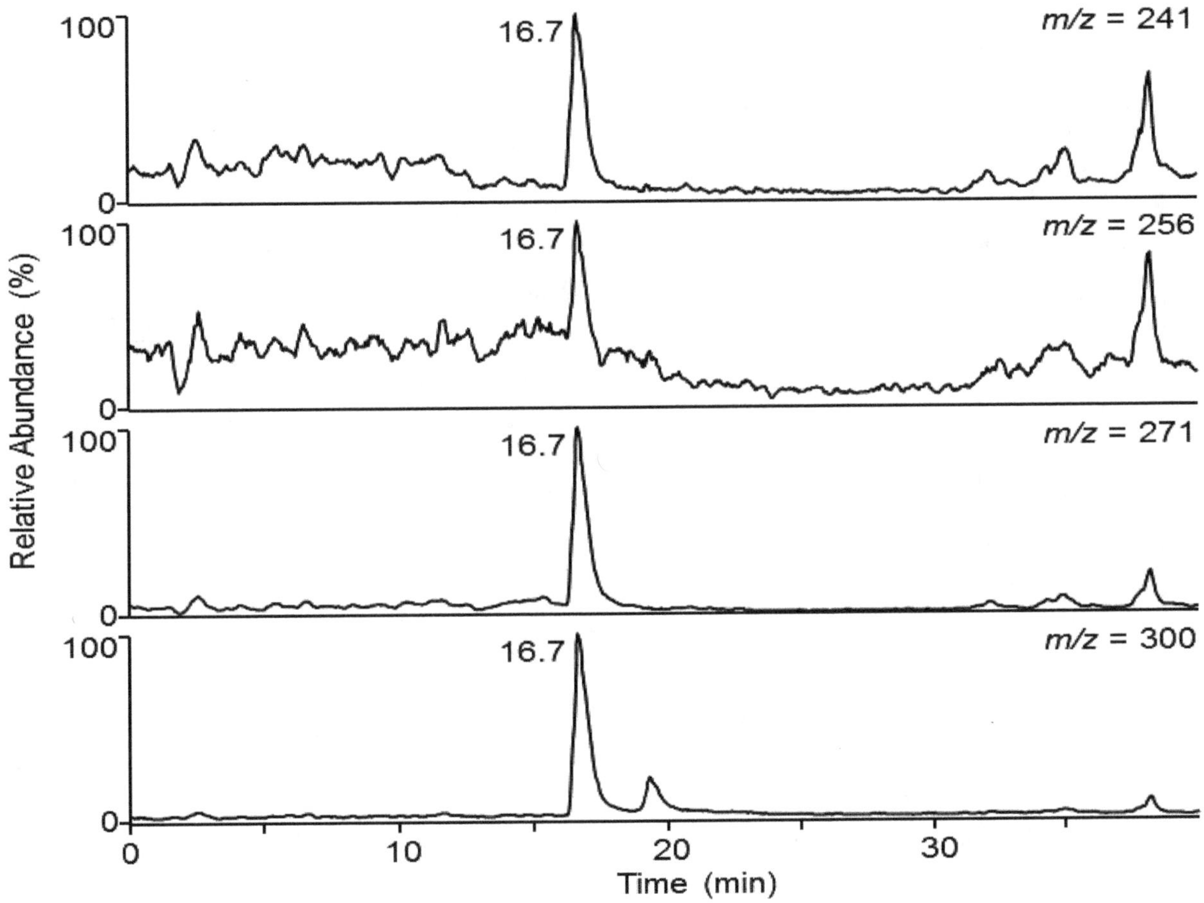

Figure 3. Chromatogram for elution of a 50 μg/mL solvent standard of 8-*O*-methylbostrycoidin.

3.3. *LC-MS/MS Quantitation*

For quantitative analyses, abundant fragments from the $[M + H]^+$ ion of the analyte, 8-*O*-methylbostrycoidin, were monitored in the MRM mode. Signals from the fragments were maximized by optimizing collision energy. To assess quantitative behavior of the LC-MS system, 10 μL injections over a wide range of 8-*O*-methylbostrycoidin concentrations were made. Response curves were obtained over a range of 8-*O*-methylbostrycoidin concentrations. Based on the observed response, it was determined that the monitoring of the m/z 271 fragment ion provided a good basis for quantitation of 8-*O*-methylbostrycoidin. Figure 4 shows an example plot of integrated response against concentration for the m/z 271 fragment ion from solvent and matrix standards. Good linearity, up to 1000 μg/mL, is observed. Based on the observation of the transition to m/z 271 ion, the absolute minimum detection limit (based on the signal-to-noise ratio of 3) was better than 3 ng/g from 8-*O*-methylbostrycoidin spiked into ground maize. Further, the limit of quantitation was 10 ng/g from 8-*O*-methylbostrycoidin spiked into ground maize.

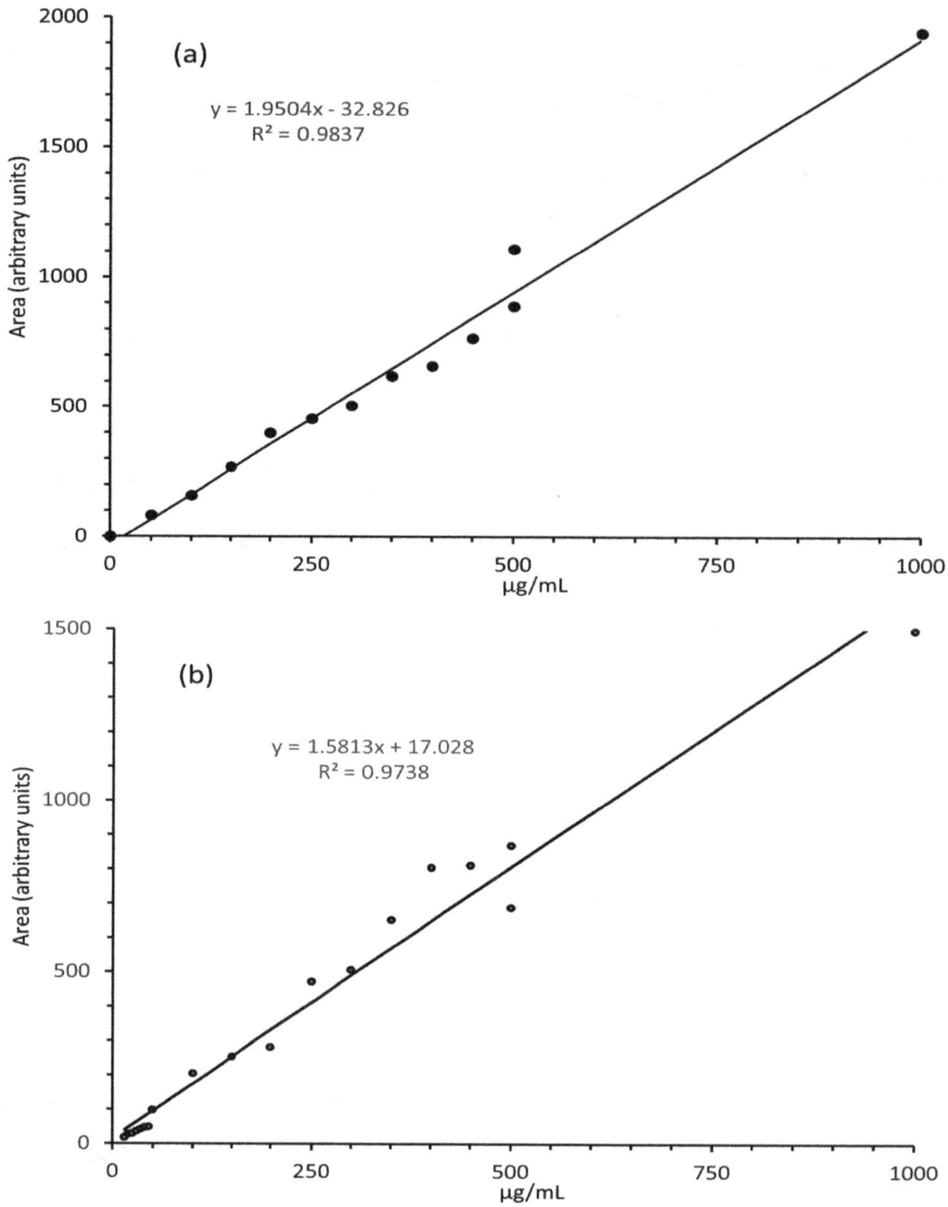

Figure 4. Calibration curves for 8-*O*-methylbostrycoidin (**a**) solvent and (**b**) matrix standards.

3.4. Spike Recovery Studies

The goal of this work was to have a convenient method for determination of 8-*O*-methylbostrycoidin from maize substrate. To assess the ability of the LC-MS/MS technique to serve as the basis for such a method, spike-recovery experiments were conducted. A series of portions of cracked maize were spiked with 8-*O*-methylbostrycoidin (acetonitrile) in order to achieve levels corresponding to 0.0125, 0.125 and 1.25 µg/g 8-*O*-methylbostrycoidin. The samples were then extracted and processed according to the above described method. The analysis of the spiked samples ($n = 9$ for each spike level) yielded recoveries (72.8 $\pm$ 35.6, 0.0125 µg/g; 41.3 $\pm$ 17.5, 0.125 µg/g; 59.9 $\pm$ 32.3, 1.25 µg/g) of the 8-*O*-methylbostrycoidin upon quantitation utilizing the m/z 271 fragment ion.

3.5. Matrix Interference

Experiments to evaluate matrix effects were in conducted according to guidelines described by Matuszewski et al. [18]. LC-MS/MS areas of standards formulated in solvent (acetonitrile) were compared with those measured in a blank cracked maize extract spiked, after extraction, with the same analyte amounts. Tests were conducted on spiked maize extract samples (8-*O*-methylbostrycoidin concentration of 0.5 μg/mL). The level of 8-*O*-methylbostrycoidin in the maize extract used for the spiking experiments was assumed to be zero. The average of measured matrix effect (%) from solvent and extract samples run in pairs ($n = 8$) was 154 ± 12%. In the case of maize extract, the matrix actually provides an enhancement to the response of the LC-MS/MS detection.

3.6. Determination of 8-O-methylbostrycoidin from Field Grown Maize

Field tests were conducted to evaluate the ability of *F. verticillioides* to produce 8-*O*-methylbostrycoidin in maize kernels under agriculturally relevant conditions. Injection of suspensions of spores from two different strains of *F. verticillioides* into developing maize ears resulted in relatively high levels of maize ear rot symptoms at harvest (Table 1). Injection of the different strains of *F. verticillioides* resulted in markedly different levels of 8-*O*-methylbostrycoidin in the resulting kernels (Table 1). Control injection of ears with water resulted in low levels of disease symptoms, as previous reported [17], and 8-*O*-methylbostrycoidin was not detected in kernels harvested from control-injected ears.

Table 1. Results for field evaluation of disease level and 8-*O*-methylbostrycoidin production in maize ears upon inoculation with *F. verticillioides* isolates.

Strain	Disease Score [1]	Level of 8-*O*-methylbostrycoidin (μg/g) [1]
AMR 5	4.9 (1.9)	0.5 (0.9)
AMR 10	4.4 (1.6)	2.6 (4.8)
Control inoculation	1.3 (0.9)	0 (–)

[1] Uncertainties in determinations in terms of standard deviations are indicated in parentheses. Each maize ear was considered a repetition of the inoculation experiment $n = 10$. No standard deviation was calculated for the levels of the analyte detected in the control inoculation kernels. This is indicated by the (–).

4. Discussion

Currently, common methods for the evaluation of 8-*O*-methylbostrycoidin production by fungi are based upon UV absorbance. Potentially, such methods would be susceptible to interferences from extracted absorbing compounds produced by maize, as well as invading fungi. The method described here offers the potential of advantages in sensitivity and selectivity associated with LC-MS/MS. Our examination of the performance of the method provides us with an indication of its sensitivity, as well as its tolerance of matrix effects from the maize. The developed method allows the convenient determination of 8-*O*-methylbostrycoidin without a substantial effort for sample clean-up. As far as we are aware, this is the first report of production of 8-*O*-methylbostrycoidin in field-grown maize.

Acknowledgments: The author wishes to acknowledge the technical contributions by Debbie Shane.

Author Contributions: Mark Busman conceived and designed the experiments, performed the experiments, analyzed the data and wrote the paper.

References

1. Steyn, P.S.; Wessels, P.L.; Marasas, W.F.O. Pigments from *Fusarium moniliforme* sheldon. Structure and ^{13}C nuclear magnetic resonance assignments of an azaanthraquinone and three naphthoquinones. *Tetrahedron* **1979**, *35*, 1551–1555. [CrossRef]
2. Cameron, D.W.; Deutscher, K.R.; Feutrill, G.I. Synthesis of bostrycoidin and 8-*O*-methlylbostrycoidin. *Tetrahedron Lett.* **1980**, *21*, 5089–5090. [CrossRef]

3. Cameron, D.W.; Deutscher, K.R.; Feutrill, G.I. Nucleophilic alkenes. Ix. Addition of 1,1-dimethoxyethene to azanaphthoquinones: Synthesis of bostrycoidin and 8-*O*-methylbostrycoidin. *Aust. J. Chem.* **1982**, *35*, 1439. [CrossRef]
4. Tatum, J.H.; Baker, R.A.; Berry, R.E. Naphthoquinones produced by *Fusarium oxysporum* isolated from citrus. *Phytochemistry* **1985**, *24*, 457–459. [CrossRef]
5. Deng, C.M.; Liu, S.X.; Huang, C.H.; Pang, J.Y.; Lin, Y.C. Secondary metabolites of a mangrove endophytic fungus *Aspergillus terreus* (no. Gx7-3b) from the South China Sea. *Mar. Drugs* **2013**, *11*, 2616–2624. [CrossRef] [PubMed]
6. Vesonder, R.F.; Gasdorf, H.; Peterson, R.E. Comparison of the cytotoxicities of *Fusarium* metabolites and *Alternaria* metabolite AAL-toxin to cultured mammalian cell lines. *Arch. Environ. Contam. Toxicol.* **1993**, *24*, 473–477. [CrossRef] [PubMed]
7. Hashimoto, J.; Motohashi, K.; Sakamoto, K.; Hashimoto, S.; Yamanouchi, M.; Tanaka, H.; Takahashi, T.; Takagi, M.; Shin-ya, K. Screening and evaluation of new inhibitors of hepatic glucose production. *J. Antibiot. (Tokyo)* **2009**, *62*, 625–629. [CrossRef] [PubMed]
8. Awakawa, T.; Kaji, T.; Wakimoto, T.; Abe, I. A heptaketide naphthaldehyde produced by a polyketide synthase from *Nectria haematococca*. *Bioorg. Med. Chem. Lett.* **2012**, *22*, 4338–4340. [CrossRef] [PubMed]
9. Brown, D.W.; Butchko, R.A.E.; Busman, M.; Proctor, R.H. Identification of gene clusters associated with fusaric acid, fusarin, and perithecial pigment production in *Fusarium verticillioides*. *Fungal Genet. Biol.* **2012**, *49*, 521–532. [CrossRef] [PubMed]
10. Studt, L.; Wiemann, P.; Kleigrewe, K.; Humpf, H.U.; Tudzynski, B. Biosynthesis of fusarubins accounts for pigmentation of *Fusarium fujikuroi* perithecia. *Appl. Environ. Microbiol.* **2012**, *78*, 4468–4480. [CrossRef] [PubMed]
11. Smith, J.S.; Fotso, J.; Leslie, J.F.; Wu, Y.; VanderVelde, D.; Thakur, R.A. Characterization of bostrycoidin: An analytical analog of zearalenone. *J. Food Sci.* **2004**, *69*, C227–C232. [CrossRef]
12. Nelson, P.E.; Desjardins, A.E.; Plattner, R.D. Fumonisins, mycotoxins produced by *Fusarium* species: Biology, chemistry, and significance. *Annu. Rev. Phytopathol.* **1993**, *31*, 233–252. [CrossRef] [PubMed]
13. Leslie, J.F.; Summerell, B.A. *The Fusarium Laboratory Manual*; Blackwell Publishing: Ames, IA, USA, 2006.
14. Geiser, D.M.; Jimenez-Gasco, M.D.M.; Kang, S.; Makalowska, I.; Veeraraghavan, N.; Ward, T.J.; Zhang, N.; Kuldau, G.A.; O'Donnell, K. Fusarium-ID v. 1.0: A DNA sequence database for identifying *Fusarium*. *Eur. J. Plant Pathol.* **2004**, *110*, 473–480. [CrossRef]
15. Seo, J.A.; Proctor, R.H.; Plattner, R.D. Characterization of four clustered and coregulated genes associated with fumonisin biosynthesis in *Fusarium verticillioides*. *Fungal Genet. Biol.* **2001**, *34*, 155–165. [CrossRef] [PubMed]
16. Reid, L.; Hamilton, R.; Mather, D. Screening maize for resistance to *gibberella* ear rot. In *Technical Bulletin 1996-5e*; Agriculture & Agri-food Canada: Ottawa, ON, Canada, 1996; pp. 1–40.
17. Desjardins, A.E.; Plattner, R.D. Fumonisin B1-nonproducing strains of *Fusarium verticillioides* cause maize (*Zea mays*) ear infection and ear rot. *J. Agric. Food Chem.* **2000**, *48*, 5773–5780. [CrossRef] [PubMed]
18. Matuszewski, B.K.; Constanzer, M.L.; Chavez-Eng, C.M. Strategies for the assessment of matrix effect in quantitative bioanalytical methods based on HPLC-MS/MS. *Anal. Chem.* **2003**, *75*, 3019–3030. [CrossRef] [PubMed]

6

Ecology, Phylogeny, and Potential Nutritional and Medicinal Value of a Rare White "Maitake" Collected in a Mediterranean Forest

Maria Letizia Gargano [1], Georgios I. Zervakis [2], Omoanghe S. Isikhuemhen [3], Giuseppe Venturella [4], Roberta Calvo [4,*], Anna Giammanco [5], Teresa Fasciana [5] and Valeria Ferraro [4]

1 Department of Agricultural and Environmental Science, University of Bari Aldo Moro, Via Amendola 165/A, I-70126 Bari, Italy; marialetizia.gargano@uniba.it

2 Laboratory of General and Agricultural Microbiology, Agricultural University of Athens, 11855 Athens, Greece; zervakis@aua.gr

3 Mushroom Biology and Fungal Biotechnology Laboratory, Department of Natural Resource & Environmental Design, North Carolina A&T State University, Greensboro, NC 27411, USA; omon.isi@fungibiotech.com

4 Department of Agricultural, Food and Forest Sciences, University of Palermo, Viale delle Scienze, Bldg. 5, I-90128 Palermo, Italy; giuseppe.venturella@unipa.it (G.V.); valeria.ferraro@unipa.it (V.F.)

5 Department of Health Promotion, Mother and Child Care, Internal Medicine and Medical Specialities, University of Palermo, 90127 Palermo, Italy; anna.giammanco@unipa.it (A.G.); teresa.fasciana@unipa.it (T.F.)

* Correspondence: roberta.calvo@unipa.it.

Abstract: Albino *Grifola frondosa* (Dicks.) Gray "maitake" mushrooms (described as *G. albicans* Imazeki and then placed in synonymy with *G. frondosa*) are particularly rare, and the few pertinent records are not treated in scientific publications. A field investigation carried out in Sicily (Italy) led to the collection of an unusual white *Grifola* specimen at the base of a living tree of *Quercus pubescens* Willd. s.l. The outcome of sequencing the internal transcribed spacer (ITS) region of nuclear ribosomal DNA (nrDNA) indicated that it belongs to *G. frondosa* and provided an insight to the phylogenetic relationships within the genus. The results of nutritional composition analysis showed that the albino basidioma possesses relatively high contents of Ca, Fe, K, and Cu and is rather low in Na when compared with literature data on edible mushrooms. Vitamin (B_1, B_2, B_3, B_5, B_9, and D_2) contents ranged from 0.15 to 3.89 mg per 100 g of mushroom dry weight. The cold-water extract of this specimen was effective at inhibiting the growth of *Staphylococcus epidermidis* ATCC 12228 and *Pseudomonas aeruginosa* ATCC 15442 at the maximum screening concentration of 50% *v/v*. In addition, the extract slowed down the ability of *Staphylococcus aureus* ATCC 43300 to form biofilms. According to data hereby reported, the albino *G. frondosa* is a culinary-medicinal mushroom with a promising exploitation potential.

Keywords: *Grifola frondosa*; fungal diversity; Mediterranean forest; medicinal mushroom; bioprospecting; ITS rDNA; phylogenetics; basidiomycete; polypore fungus; *Quercus pubescens*

---

1. Introduction

The family Grifolaceae Jülich (Polyporales, Basidiomycota) comprises only the genus *Grifola* Gray [1], which consists of nine species according to Index Fungorum. *Grifola frondosa* (Dicks) Gray includes several forms and varieties, such as *G. frondosa* f. *frondosa* (Dicks.) Gray, *G. frondosa* f. *intybacea* (Fr.) Pilát, *G. frondosa* var. *frondosa* (Dicks.) Gray, and *G. frondosa* var. *intybacea* (Fr.) Cetto, while *G. intybacea* (Fr.) Imazeki is considered as a synonym. The culinary-medicinal mushroom

G. frondosa, widely known as "maitake", is a white-rot polypore associated primarily with deciduous trees of different genera (*Quercus* L., *Acer* L., *Carpinus* L., *Castanea* Mill., *Fagus* L., *Ulmus* L.) while it appears more rarely on conifers. This species is distributed in temperate regions of North America, Europe, and Asia (Japan and China) [2]. Albino maitake mushrooms (described as *G. albicans* by Imazeki in 1943 [3] and then placed in synonymy with *G. frondosa*) are quite rare and the few available pertinent records do not appear in scientific publications. Only recently, Kawaguchi et al. [4,5] reported on melanin biosynthesis in *G. frondosa* and suggested that the albino mutation is caused by a single base deletion in the coding region of the tyrosinase 2 (*tyr2*) gene. It is noteworthy that white strains of *G. frondosa* are much sought after since they can be used for culinary purposes without the unwanted dark brown pigment resulting from processing the common form of this mushroom [4].

Edible mushrooms are widely acknowledged for their nutritional and medicinal properties [6]; however, their potential market is still far from being fully developed considering that there is a high consumer demand. Consequently, the private sector is eager to generate new relevant products. In the frame of a research project aiming at collecting mushrooms with bioprospecting potential, a field investigation carried out in Sicily (Italy) led to the collection of an unusual *Grifola* basidioma of white color and considerable weight (ca. 7 kg). The objective of the present work was to provide morphological, molecular/phylogenetic, and ecological data about this rare albino *Grifola* specimen. Information on the nutritional value and antimicrobial activity is also reported.

2. Materials and Methods

2.1. Sample Collection, Habitat Details, and Evaluation of Morphological Characters

Research carried out in forest ecosystems in Sicily (Italy) led to the collection of a single albino *Grifola* basidioma found at the base of a monumental downy oak tree (*Quercus pubescens* Willd. s.l.), 10 January 2016, Castelbuono, Madonie Regional Park (province of Palermo, Sicily), 800 m a.s.l., 37°54′48″ N, 14°04′41″ E. In Castelbuono, summers are rather brief, hot, and dry, with clear skies and condensation often appearing in the late afternoon, while winters are long, cold, rainy, and windy. Annual temperature ranges from 7 to 29 °C (rarely below 4 °C or above 32 °C) while the average value is 15.8 °C. The average annual rainfall is 515 mm. In January, the month when the specimen under study was harvested, the average temperature is 8.9 °C, and the respective maximum and minimum values are 11.5 and 6.3 °C; the rainfall is 67 mm. The forest is characterized by evergreen and deciduous oak vegetation with a prevalence of *Quercus ilex* L. and *Q. pubescens* Willd. s.l. In the upper part of the mountain, the vegetation is characterized by mixed woods of oaks, *Fagus sylvatica* L. and *Ilex aquifolium* L. The specimen was transferred to the Department of Agricultural, Food, and Forest Sciences (SAAF) of the University of Palermo, and stored at <4 °C for up to 24 h prior to morphological examination, which was carried out according to Bernicchia [7]. Observations on macromorphological characters (pileus, cuticle, pores, stipe, context, etc.) were performed on fresh material, while microscopic characters (hyphal system, generative hyphae, basidia, sterigmata, cystidia, cystidioles, and basidiospores) were studied by using 3% potassium hydroxide and ammoniacal Red Congo. In addition, the reaction of the flesh to iodine-potassium iodide, potassium hydroxide, and iron salts was also evaluated.

2.2. Establishment of Pure Cultures and Mushroom Cultivation Trials

The collected basidioma was dried and then deposited in the Herbarium of the Department of Agricultural, Food, and Forest Sciences (SAAF 450). Prior to this, a piece of pseudo-tissue was removed from the fresh specimen, placed on potato dextrose agar (PDA) in Petri dishes under aseptic conditions, and incubated for 15 days at 25 ± 2 °C. The established pure culture was stored in the Mycotheca of the Herbarium SAF (SAF 323), and it was subsequently used for the inoculation of mushroom cultivation substrates composed of either holm oak (*Quercus ilex* L.) or chestnut (*Castanea sativa* Miller) wood residues. The former (holm oak wood shavings) was filled in plastic bottles (volume: 1 L each), while the latter (chestnut wood chips of ca. 2–3 cm and sawdust) was filled in plastic bags (volume: 6 L each) after their moisture content was adjusted to 50–60%; in both cases, non-hydrophilic cotton filters were used to allow air exchange. Sterilization was performed twice at 120 °C, 1.1 Atm for 20 min with a 24-h interval between each cycle. After the final cooling, substrates were inoculated using actively growing mycelium and incubation was carried out at 25 °C in the dark.

2.3. DNA Extraction, PCR Amplification, and Sequencing

Total genomic DNA was extracted from the herbarium specimen using the Nucleospin Plant II DNA kit (Macherey and Nagel, Düren, Germany) following the manufacturer's protocol. The internal transcribed spacer (ITS) region of nuclear ribosomal DNA (nrDNA) was amplified using the primer combination ITS1/ITS4 [8]. Polymerase chain reactions (PCRs) were performed in 50 μL containing 50 ng DNA template, 0.25 μM of each primer, 0.2 mM of each dNTP, 1 × HiFi Buffer (Takara BIO INC., Shiga, Japan), and 1 U HiFi Taq DNA polymerase (Takara BIO INC., Shiga, Japan). Conditions for PCR amplification were as follows: 94 °C for 5 min, followed by 35 cycles of 94 °C for 30 s, 50 °C for 30 s and 72 °C for 1 min, and a final extension at 72 °C for 10 min.

PCR products were purified using an Invitrogen PureLink kit (Thermo Fisher Scientific, Seoul, Korea) and were submitted for sequencing to CeMIA S.A. (Larissa, Greece). The resulting chromatograms were proofread, and the sequence generated was deposited in GenBank under the accession number MN944407.

2.4. Phylogenetic Analysis of Sequence Data

In addition to the biological material examined in this study, 53 additional ITS sequences were included in the phylogenetic analysis; 38 of them representing other *Grifola* Spp., i.e., all available sequences of *G. colensoi* (Berk.) G. Cunn., *G. gargal* Singer, and *G. sordulenta* (Mont.) Singer, and the rest corresponding to two other species used as outgroups, i.e., *Polyporus umbellatus* (Pers.) Fr. (syn. *Grifola umbellata* (Pers.) Pilát) and *Rhodonia placenta* (Fr.) Niemelä, K.H. Larss. & Schigel (syn. *Oligoporus placentus* (Fr.) Gilb. & Ryvarden). Sequence alignment was performed through the aid of the Clustal Omega software (https://www.ebi.ac.uk/Tools/msa/clustalo/); alignments were inspected and manually adjusted at misaligned sites using MEGA X (Pennsylvania State University, State College, PA, USA) [9].

Phylogenetic relationships were inferred by using the maximum likelihood (ML) method based on the Kimura 2-parameter model [10]. Initial tree(s) for the heuristic search were obtained automatically by applying the Neighbor-Join and BioNJ algorithms to a matrix of pairwise distances estimated using the maximum composite likelihood (MCL) approach, and then selecting the topology with the superior log likelihood value. Phylogenetic analysis was conducted in MEGA X [9].

2.5. Evaluation of Nutritional Value and Determination of Antimicrobial Activity

The powder resulting after freeze-drying part of the collected basidioma was also subjected to analysis for the evaluation of the nutritional value (proximate composition, and content in elements and vitamins) according to standard methodologies reported by Palazzolo et al. [11], AOAC [12], Thompson et al. [13], and Loewus [14].

With regard to the antibacterial properties, the study initially included the evaluation of the activity of the lyophilized powder deriving from the basidioma against four Gram-positive and Gram-negative bacteria strains: *Staphylococcus aureus* ATCC 25923, *S. epidermidis* ATCC 12228, *Pseudomonas aeruginosa* ATCC 15442, and *Escherichia coli* ATCC 25922. Tryptic soy broth (Sigma-Aldrich, Darmstadt, Germany) containing glucose (2% *w/v*) or Mueller Hinton (Sigma-Aldrich, Darmstadt, Germany) was used as growth media. The lyophilized sample (3 g) was placed in a beaker with 200 mL of demineralized water, and then stored at −20 °C.

The minimum inhibitory concentration (MIC) was determined in a Minisart syringe with a 2.5-μm cellulose acetate filter by preparing solutions in vitro at increasing concentrations, incubating them with separate lots of cultivated bacteria, and measuring the results using agar dilutions. In addition, the effect of extract obtained from the basidioma powder was examined in respect to the biofilm produced by strains of *S. aureus* ATCC 43300, *Enterococcus faecalis* ATCC 29212, *P. aeruginosa* ATCC 27853, *E. coli* ATCC 25922, and *Klebsiella pneumoniae* ATCC 700603.

The bacterial strains were inoculated in Trypticase™ Soy Agar (BBL™; Becton, Dickinson and Company, Sparks, MD, USA) media and incubated at 37 °C for 24 h. In total, 100 µL of overnight grown culture (0.5 McFarland in tryptose broth, BT) was added in a sterile 96-well flat bottom microtiter plate (Biosigma S.r.l. Dominique Dutscher Group, Brumath, France). The negative control was composed by BT medium only; to demonstrate biofilm production, excess bacterial suspension was removed. Subsequently, the extract from the basidioma powder (25% *v/v*) was re-suspended in BT, 100 µL were added in the wells, and the cultures were incubated at 37 °C for 24 h. Finally, the biofilm was stained with crystal violet dissolved in ethanol (0.5% *w/v*). The optical density was measured at 540 nm (Spectrophotometer Multiskan Go; Thermo Fisher Scientific, Waltham, MA, USA). The experiment was conducted in triplicate [15,16].

3. Results and Discussion

3.1. Morphological Description of the *G. frondosa* Specimen and Cultivation Tests

Basidioma is composed of multiple pilei in the form of a rosette, sharing a branched stem-like structure (Figure 1a). Individual pilei were found to be more or less fan-shaped or deltoid, entirely white and yellowing with age, and finely velvety or bald, with wavy margins, 4 to 10 cm wide and 5 to 10 mm thick. The pore surface was shown to be white, staining yellowish when ripe, 2 to 3 mm deep, and irregular, and varying in shape from round to elongated (Figure 1b); it is slightly decurrent to the stipe. The stipe structure was found to be branched, whitish, tough, and often off-center. The flesh was firm, white, and unchanging when sliced. The odor and taste were found to be mild and pleasant, although less so when decaying or acrid. The chemical reaction of flesh to iodine-potassium iodide and potassium hydroxide was negative. The iron salts reaction on the pileus and flesh was negative. Basidiospores, slightly reduced when compared to the brown maitake (5–7 × 3.2 µm), were found to be broadly ellipsoidal, smooth, 5.0–7.0 × 3.2–4.0 µm, slightly narrower than those of the brown maitake, and inamyloid. Basidia were 25–30 × 6–8 µm; clavate; and 4-sterigmata (Figure 2a). Clamp connections were found to be present on generative hyphae but are rare or absent on skeletal hyphae. Hymenial cystidia and cystidioles were absent. The hyphal system is dimitic, and the spore print white. Sterigmata were found to be up to 6 µm long, and robust, with a base diameter up to 1.6 µm (Table 1). Relevant information referring to the description of common brown form [7,17] is included in Table 1. It should be mentioned that the length of the sterigmata and the diameter of their base were not reported by Ryvarden [17], while Bernicchia [7] stated that sterigmata appear "thin and very divaricate".

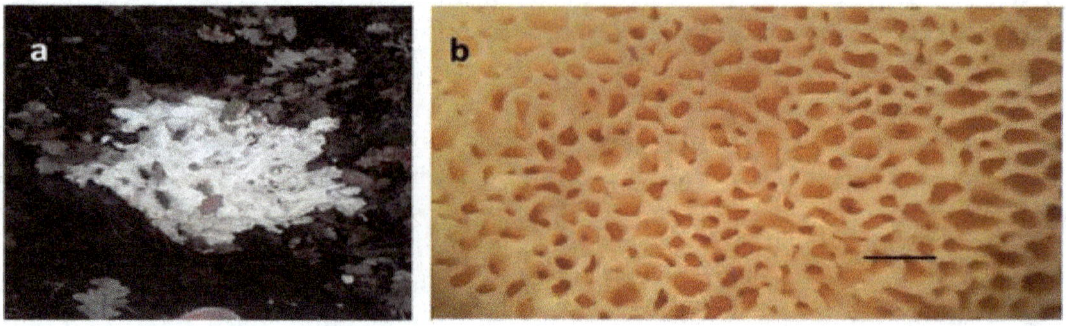

Figure 1. The albino maitake specimen: (**a**) Basidioma in situ; (**b**) Pore surface under the stereomicroscope (scale bar = 1 mm).

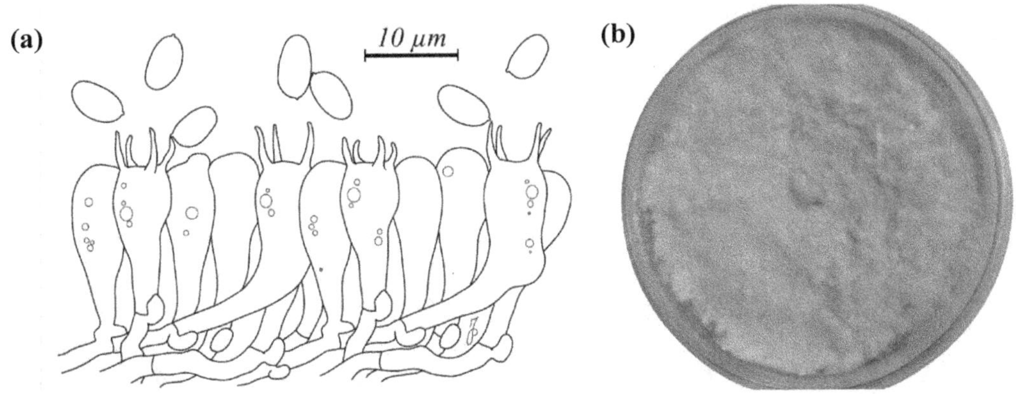

Figure 2. (**a**) Microscopic features (basidia and spores) of albino maitake. (**b**) Pure culture isolated from fresh basidioma.

Table 1. Main macro- and micromorphological features of brown [7,17] and albino (this study) forms of maitake.

Descriptive Characters	Brown Form	Albino Form
Habit	Cluster of pilei	Cluster of pilei
Pileus	Irregular, wrinkled, flat and uneven	Fan-shaped or deltoid
Stipe	Whitish	Whitish
Tubules	Short, thick, whitish	Short, thick, whitish
Pores	Very small, then angular and wider	Round to elongated, slightly decurrent to the stipe
Stem	Whitish and fleshy, with numerous flattened branches and bifurcations that end in a fan-shaped pileus	Structure branched, whitish, tough, often off-center
Flesh	White, immutable, tenacious especially towards the base, fragile towards the pileus	Firm, white, unchanging when sliced
Smell	Penetrating, intense then nauseating	Mild and pleasant, unpleasant or acrid when decaying
Taste	Pleasant and delicate	Mild and pleasant, unpleasant or acrid when decaying
Basidiospores	Ellipsoidal, 5.0–6.5 × 3.5–5.0 μm	Broadly ellipsoidal, smooth, 5.0–7.0 × 3.2–4.0 μm
Basidia	23–30 × 5–8 μm; clavate; 4-sterigmata	25–30 × 6–8 μm; clavate; 4-sterigmata
Hyphal system	Dimitic with clamp connections	Dimitic with clamp connections
Habitat	At the base of hardwoods, particularly chestnut and oak trees	At the base of oak trees
Period of fructification	Late August to October	December to January

Mycelium growth was very slow in the pure cultures established, and the surface of the Petri dish (diameter 90 mm) was completely overgrown in 30 days. The colony was found to be slightly floccose and flat, with dense white hyphae without any zonation and with a filiform margin (Figure 2b). Regarding the cultivation tests performed either in holm oak or in chestnut-based substrates, colonization was slow and ca. two months elapsed before they were completely overgrown by mycelium. At this time, a few primordia were formed on the substrate surface, which, however, failed to produce fully developed basidiomata.

3.2. Phylogenetic Analysis

The main objective of the molecular analysis was to identify the biological material under examination (albino maitake). Following BLAST, the sequence (MN944407) from our specimen showed high identity (>99%) to several other sequences deposited as *G. frondosa* in GenBank. The generated ML trees had similar topologies, and only the tree with the highest log likelihood (−7662.14) was selected to be presented (Figure 3).

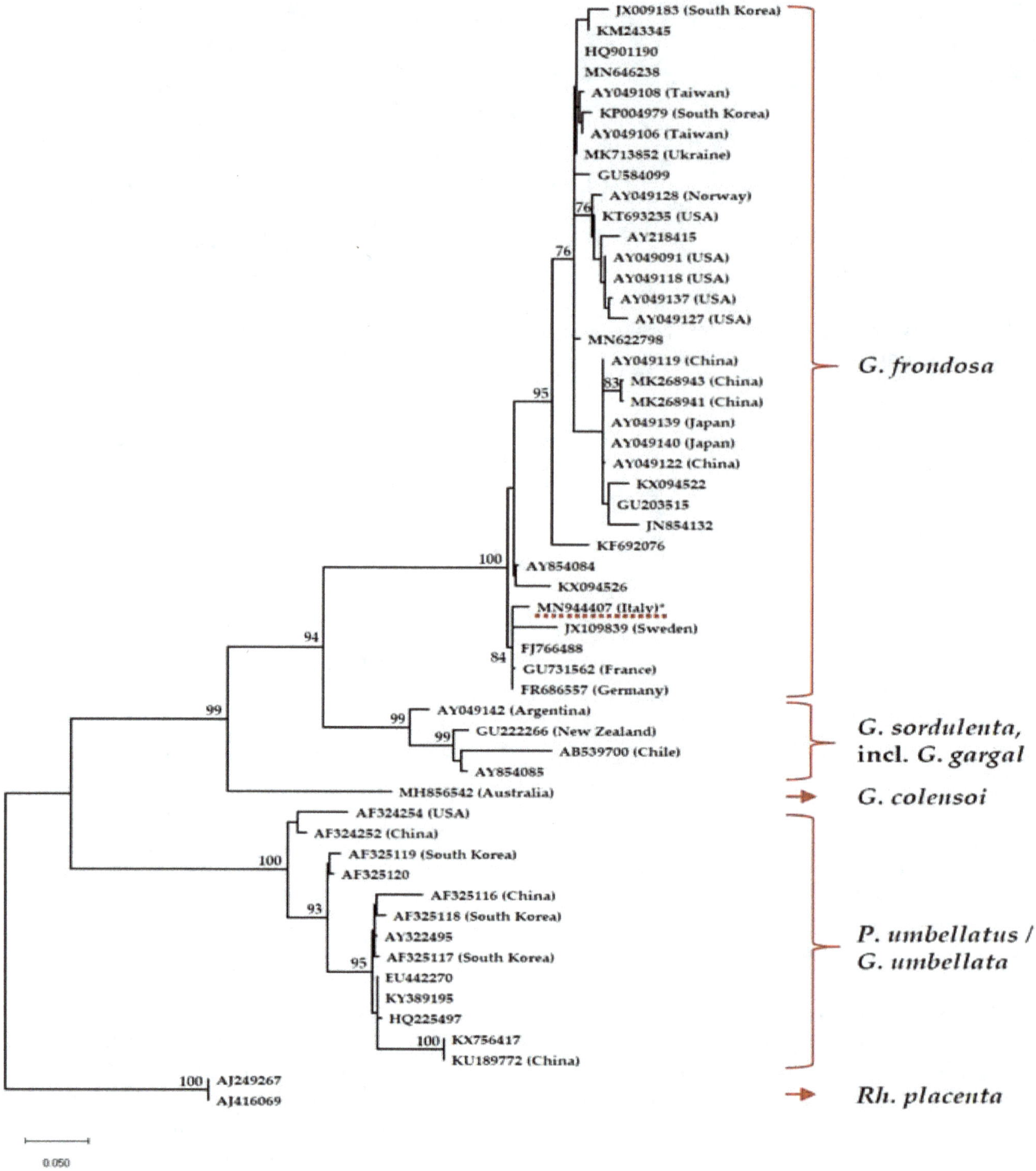

Figure 3. Maximum likelihood tree produced on the basis of ITS (internal transcribed spacer) sequence analysis presenting the phylogenetic position of the albino *G. frondosa* specimen (underlined in red and marked with an asterisk) among members of the genus *Grifola;* GenBank accession numbers are quoted together with information of their geographic origin (when available). Sequences of *Polyporus umbellatus* (syn. *G. umbellata*) and *Rhodonia placenta* were used as outgroups. The percentage of trees in which the associated taxa clustered together is shown next to the branches when ≥70% (ML bootstrap support values derived from a total of 1000 replicates). The tree is drawn to scale, with branch lengths measured as the number of substitutions per site.

To the best of our knowledge, the sequence obtained in this study is the only one available from an albino *Grifola* specimen. The outcome of the phylogenetic analysis revealed that the albino maitake clusters within the *G. frondosa* terminal clade, which receives high bootstrap support (100%), and forms a subgroup (84%) together with other sequences from material collected in Europe; the rest of the *G. frondosa* sequences derive almost exclusively from east Asia (China, Japan, South Korea, Taiwan) and the USA.

Although this subgroup could represent a distinct taxonomic entity, the information generated from our work (through the examination of one basidioma and the use of a single marker only) does not allow such inferences, at least before additional specimens are included and analyzed in the frame of a multigene approach. However, it is noteworthy that *G. frondosa* sequences from Asian and eastern North American isolates were separated through the use of a partial beta-tubulin gene and ITS [18]. *G. frondosa* forms a sister group to *G. sordulenta* (94%).

The latter is highly supported (99%) and consists of three sequences under this name and one representing *G. gargal* (AB539700). Hence, the distinct taxonomic position of the latter is dubious by examining the ITS marker alone. Of interest was also the position of the sole *G. colensoi* sequence available (MH856542; this species is known so far from Australia and New Zealand only), which appears to be quite distant from the rest of the material placed in the genus *Grifola*. Lastly, sequences deposited either as *G. umbellata* or *P. umbelatus* form a distinct highly supported group (100%), which is well separated from the clade corresponding to *Grifola* Spp. To the best of our knowledge, the sequence generated in the frame of this work is the first one deriving from an albino *G. frondosa* specimen.

3.3. Mushroom Proximate Composition and Content in Elements and Vitamins

In the frame of this work, it was not possible to analyze (and hence directly compare to our albino specimen) the composition of wild basidiomata of brown maitake since they were not available during the period of the study, this species being very rare in Sicily. On the other hand, only limited literature data are available referring to the chemical composition and vitamin contents in cultivated *G. frondosa* [19,20]; the respective values reported—apart from a few exceptions—are considerably lower than those found in the white maitake. However, such information should be evaluated with care since it is known that the nature of the cultivation substrate could considerably affect the mushroom content [21,22]. Similarly, metal accumulation in wild edible mushrooms was found to be influenced by various factors, including the species, the available concentration of elements in soil substrates, the substrates' properties (e.g., pH, organic matter content), the antagonistic and/or synergistic effects among elements, and the concentration of other elements in mushrooms [23–26].

The results of the proximate composition analysis of the albino maitake showed rather low values of protein, fat, and carbohydrate contents (Table 2), while the calorie content was 35.84 kcal·100 g^{-1}. In addition, element analysis showed that this specimen presents relatively high contents in Ca, Fe, K, and Cu, and is rather low in Na (Table 2). The Ca content is higher than in some baby foods, such as oatmeal and whole milk, as well as ham, cream, spread cheese, white chocolate, and others [27]. The content in Fe is considerably higher than in most other foods; it is noteworthy that dried thyme (i.e., one of the richest foods in Fe) contains about 50 mg less Fe per 100 g when compared to the albino *G. frondosa*. The K content in albino maitake is lower only to that of cremor tartar and much higher than that of dried spirulina algae [27], while the Na content is lower than that of miso soup and baking powder [27]. The Cu content of albino maitake is higher than most foods, including dark chocolate, oysters, liver, and lobster [27].

Table 2. Outcome of proximate composition analysis (% for nitrogen, proteins, fats, carbohydrates, and ash), and content in elements and vitamins (mg·100 g^{-1}) of the albino maitake specimen. Values are expressed as means ± standard error, $n = 3$.

Composition/Content	Albino Maitake
Nitrogen	2.18 ± 0.01
Proteins	13.65 ± 0.00
Fats	1.02 ± 0.04
Carbohydrates	6.78 ± 0.11
Ash	1.06 ± 0.10
Ca	245.30 ± 0.17
Fe	178.40 ± 0.15
Mg	455.90 ± 0.19
K	11,785.60 ± 0.36
Na	3433.40 ± 0.34
P_2O_5	2399.61 ± 0.02
Cu	2.81 ± 0.02
Mn	2.41 ± 0.03
Zn	10.33 ± 0.02
Se	4.00 ± 0.00
Pb	2.10 ± 0.00
Thiamine (B_1)	0.15 ± 0.03
Riboflavin (B_2)	3.89 ± 0.06
Niacin (B_3)	0.36 ± 0.03
Pantothenic Acid (B_5)	0.68 ± 0.02
Folic acid (B_9)	0.38 ± 0.01
D_2	0.41 ± 0.02

The content in vitamins was: B_1, 0.15 mg 100 g^{-1}; B_2, 0.36 mg·100 g^{-1}; B_3, 3.89 mg·100 g^{-1}; B_5, 0.68 mg 100 g^{-1}; B_9, 0.38 mg 100 g^{-1}; and D_2, 0.41 mg 100 g^{-1} (Table 2). Noteworthy is the value of vitamin B_9, or folic acid, since the daily requirement is about 0.2 mg, which doubles for pregnant women [28].

3.4. Antibacterial and Antibiofilm Activity

The minimal inhibitory concentration (MIC) of the extract deriving from the albino maitake was determined by a micro-method, i.e., by diluting the protein extract in the range from 25% to 0.025% *v/v* as previously described [29].

The antibacterial activity test carried out on albino maitake showed that the cold water extract was effective in inhibiting the growth of two bacterial strains, i.e., *S. epidermidis* ATCC 12228 and *P. aeruginosa* ATCC 15442, at the maximum screening concentration of 50% *v/v*. *S. epidermidis* is a Gram-positive bacterium typically present on the skin, which could be particularly dangerous in surgical procedures, while *P. aeruginosa*, is a ubiquitous bacterium and opportunistic pathogen in humans. As regards the effect of the water extract from the albino *G. frondosa* on biofilm formation, the results revealed that it decreased the biofilm produced by *S. aureus* ATCC43300 [30], whereas it slightly enhanced biofilm formation by *E. faecalis* ATCC 29212, *E. coli* ATCC 25922, *K. pneumoniae* ATCC 700603, and *P. aeruginosa* ATCC 15442 (Figure 4).

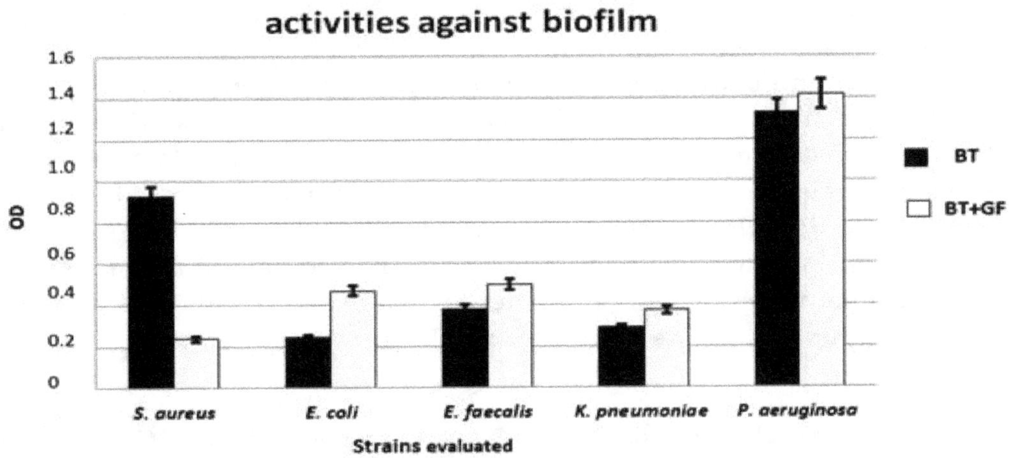

Figure 4. Effect of the cold water extract from the albino *Grifola frondosa* basidioma on biofilm production (expressed as optical density, OD) by five bacterial species. BT: biofilm produced in the absence of the fungal extract; BT+GF: biofilm produced in the presence of the fungal extract. The experiment was conducted in triplicate.

3.5. Ecological and Conservation Issues

The characterization of mushrooms growing in the Mediterranean area is one of the main targets of studies carried out by our research group. The principal objective is to provide reliable/robust scientific data amid the considerable confusion existing in the commercial exploitation of pertinent material in Italy and elsewhere, since accurate identification and/or origin of products placed on the market are often missing or of ambiguous validity. Previous results have highlighted the importance of mushrooms of the genus *Pleurotus* in human nutrition and in the feeding of farm animals [31–35]. Particular attention is also drawn to edible mushroom species that are infrequent/rare or linked to particular habitats (e.g., *Pleurotus nebrodensis* and *P. opuntiae*) [36,37], and for which only few or no studies exist regarding their nutritional and medicinal properties.

Although *G. frondosa* is widely distributed in the northern hemisphere, it is considered very rare in Italy. Particularly in Sicily, the only reports available concern the brown maitake found in the northeastern part of the island and date back to the end of the 19th century [38]; more recent informal records by amateur mycological groups derive from the same wider region (however, no pertinent samples are available for further study). Hence, the occurrence of an albino maitake is of apparent importance in a biogeographical, ecological, and applied context. In addition, the presence of this fungus on a monumental *Q. pubescens* tree of high significance points out the potential risks for host plants in the area because *G. frondosa* demonstrates a parasitic action resulting in root and butt decay [39,40]. At the same time, it is necessary to ensure the necessary conditions for its ex situ conservation due to the exploitation potential it exhibits as food and source of bioactive compounds as it was reported for wild mushrooms in general [41]. We have therefore adopted well-established protocols for its long-term maintenance [42] since our investigation showed that fructification of the albino maitake is not constant over time and the ability of the strain to be preserved under laboratory conditions—without significant degeneration of its properties—is lower than in other mushroom species.

4. Conclusions

An albino maitake specimen was reported for the first time in Italy. The examined basidioma differs morphologically from the common brown/gray form, with the most significant differences being the color and shape of the pileus, the type of pores, the shape of the stipe, and the period of fructification.

Phylogenetic analysis evidenced that the specimen under study groups among other European isolates within a larger terminal clade corresponding to *G. frondosa*. Moreover, the nutritional composition and the potential of the albino maitake to serve as source of functional compounds is of interest, and further investigation is in progress on the characterization of polysaccharides and at evaluating the antitumor activity against several types of cancer cells. The outcome of tests performed using the mushroom water extract against human pathogenic bacteria demonstrated its efficacy at inhibiting the growth of *S. epidermidis* and *P. aeruginosa* strains, while it also hindered biofilm formation by *S. aureus*, i.e., the causal agent of serious infections in immunocompromised patients. The antibacterial agents present in the albino maitake are promising candidates for dealing with pathogens that show resistance to a large spectrum of antibiotics and are widespread in hospitals.

Author Contributions: Conceptualization, G.V. and M.L.G.; methodology, G.V., G.I.Z., M.L.G. and O.S.I.; validation, R.C. and V.F.; formal analysis, G.I.Z., M.L.G. and O.S.I.; investigation, G.V., G.I.Z., M.L.G. and O.S.I.; data curation, A.G. and T.F.; biofilm analysis, R.C. and V.F.; writing—original draft preparation, G.V. and M.L.G.; writing—review and editing, G.V. and G.I.Z.; visualization, G.V.; supervision, G.V. All authors have read and agreed to the published version of the manuscript.

References

1. Justo, A.; Miettinen, O.; Floudas, D.; Ortiz-Santana, B.; Sjökvist, E.; Lindner, D.L.; Nakasone, K.; Niemelä, T.; Larsson, K.-H.; Ryvarden, L.; et al. A revised family-level classification of the Polyporales (*Basidiomycota*). *Fungal Biol.* **2017**, *121*, 798–824. [CrossRef] [PubMed]
2. Chen, A.W.; Stamets, P.; Cooper, R.B.; Huang, N.L.; Han, S.-H. Ecology, morphology, and morphogenesis in nature of edible and medicinal mushroom *Grifola frondosa* (Dicks.: Fr.) S.F. Gray—Maitake (*Aphyllophoromycetideae*). *Int. J. Med. Mushrooms* **2000**, *2*, 8. [CrossRef]
3. Imazeki, R. Genera of *Polyporaceae* of Nippon. *Bull. Nat. Sci. Mus. Tokyo* **1943**, *6*, 1–111.
4. Kawaguchi, N.; Hayashi, M.; Chen, F.-C.; Shimomura, N.; Yamaguchi, T.; Aimi, T. Genetic analyses of causal genes of albinism (white fruiting body) in *Grifola frondosa*. *J. Wood Sci.* **2019**, *65*, 32. [CrossRef]
5. Kawaguchi, N.; Hayashi, M.; Nakano, S.; Shimomura, N.; Yamaguchi, T.; Aimi, T. Expression of tyrosinase genes associated with fruiting body formation and pigmentation in *Grifola frondosa*. *Mycoscience* **2019**, *60*, 262–269. [CrossRef]
6. Gargano, M.L.; Van Griensven, L.J.L.D.; Isikhuemhen, O.S.; Lindequist, U.; Venturella, G.; Wasser, S.P.; Zervakis, G.I. Medicinal mushrooms: Valuable biological resources of high exploitation potential. *Plant Biosyst.* **2017**, *151*, 548–565. [CrossRef]
7. Bernicchia, A. *Polyporaceae s.l.*; Edizioni Candusso: Alassio, Italy, 2005; p. 808.
8. White, T.J.; Bruns, T.; Lee, S.; Taylor, J. *Amplification and Direct Sequencing of Fungal Ribosomal RNA Genes for Phylogenetics*; Academic Press: Cambridge, MA, USA, 1990; pp. 315–322.
9. Kumar, S.; Stecher, G.; Li, M.; Knyaz, C.; Tamura, K. MEGA X: Molecular evolutionary genetics analysis across computing platforms. *Mol. Biol. Evol.* **2018**, *35*, 1547–1549. [CrossRef]
10. Kimura, M. A simple method for estimating evolutionary rates of base substitutions through comparative studies of nucleotide sequences. *J. Mol. Evol.* **1980**, *16*, 111–120. [CrossRef]
11. Palazzolo, E.; Gargano, M.L.; Venturella, G. The nutritional composition of selected wild edible mushrooms from Sicily (southern Italy). *Int. J. Food Sci. Nutr.* **2011**, *63*, 79–83. [CrossRef]
12. AOAC. Official Methods of Analysis of the AOAC. *Anal. Chem.* **1980**, *52*, 148A. [CrossRef]
13. Thompson, T.E.; Senter, S.D.; Grauke, L. Lipid content and fatty acids of pecan pollen. *HortScience* **1993**, *28*, 1191–1193. [CrossRef]
14. Loewus, F.A. Improvement in anthrone method for determination of carbohydrates. *Anal. Chem.* **1952**, *24*, 219. [CrossRef]
15. Calà, C.; Amodio, E.; Di Carlo, E.; Virruso, R.; Fasciana, T.; Giammanco, A. Biofilm production in *Staphylococcus* epidermidis strains, isolated from the skin of hospitalized patients: Genetic and phenotypic characteristics. *New Microbiol.* **2015**, *38*, 521–529. [PubMed]

16. Fasciana, T.; Gentile, B.; Aquilina, M.; Ciammaruconi, A.; Mascarella, C.; Anselmo, A.; Fortunato, A.; Fillo, S.; Petralito, G.; Lista, F.; et al. Co-existence of virulence factors and antibiotic resistance in new *Klebsiella pneumoniae* clones emerging in south of Italy. *BMC Infect. Dis.* **2019**, *19*, 928. [CrossRef]
17. Ryvarden, L.; Gilbertson, R.L. European Polypores. Part 1. *Synopsis Fungorum* **1993**, *6*, 1–387.
18. Shen, Q.; Geiser, D.M.; Royse, D.J. Molecular Phylogenetic Analysis of *Grifola frondosa* (Maitake) reveals a species partition separating Eastern North American and Asian Isolates. *Mycologia* **2002**, *94*, 472–482. [CrossRef]
19. Stamets, P. Notes on Nutritional properties of culinary-medicinal mushrooms. *Int. J. Med. Mushrooms* **2005**, *7*, 103–110. [CrossRef]
20. Niedzielski, P.; Mleczek, M.; Budka, A.; Rzymski, P.; Siwulski, M.; Jasińska, A.; Gąsecka, M.; Budzyńska, S. A screening study of elemental composition in 12 marketable mushroom species accessible in Poland. *Eur. Food Res. Technol.* **2017**, *243*, 1759–1771. [CrossRef]
21. Koutrotsios, G.; Mountzouris, K.; Chatzipavlidis, I.; Zervakis, G.I. Bioconversion of lignocellulosic residues by *Agrocybe cylindracea* and *Pleurotus ostreatus* mushroom fungi—Assessment of their effect on the final product and spent substrate properties. *Food Chem.* **2014**, *161*, 127–135. [CrossRef]
22. Sakellari, A.; Karavoltsos, S.; Tagkouli, D.; Rizou, C.; Sinanoglou, V.J.; Zoumpoulakis, P.; Koutrotsios, G.; Zervakis, G.I.; Kalogeropoulos, N. Trace elements in *Pleurotus ostreatus*, *P. eryngii*, and *P. nebrodensis* Mushrooms Cultivated on various agricultural by-products. *Anal. Lett.* **2019**, *52*, 2692–2709. [CrossRef]
23. Garcia, E.M.; Siegert, I.G.; Suarez, P. Toxicity assays and naphthalene utilization by natural bacteria selected in marine environments. *Bull. Environ. Contam. Toxicol.* **1998**, *61*, 370–377. [CrossRef] [PubMed]
24. Mendil, D.; Özgür, D.U.; Tuzen, M.; Hasdemir, E.; Sarı, H.; Sari, H. Trace metal levels in mushroom samples from Ordu, Turkey. *Food Chem.* **2005**, *91*, 463–467. [CrossRef]
25. Venturella, G.; Gargano, M.L.; Compagno, R.; Saitta, A.; Alaimo, M.G. The mineral contents of some boletaceae species from Sicily (Southern Italy). *J. AOAC Int.* **2014**, *97*, 612–623. [CrossRef]
26. Kokkoris, V.; Massas, I.; Polemis, E.; Koutrotsios, G.; Zervakis, G.I. Accumulation of heavy metals by wild edible mushrooms with respect to soil substrates in the Athens metropolitan area (Greece). *Sci. Total. Environ.* **2019**, *685*, 280–296. [CrossRef] [PubMed]
27. Fidanza, F. Alimenti e Tabelle di Composizione. In *Nutrizione Umana*; Fidanza, F., Liguori, G., Eds.; Idelson: Napoli, Italy, 1984; pp. 221–238.
28. Barua, S.; Kuizon, S.; Junaid, M.A. Folic acid supplementation in pregnancy and implications in health and disease. *J. Biomed. Sci.* **2014**, *21*, 77. [CrossRef] [PubMed]
29. Schillaci, D.; Petruso, S.; Sciortino, V. 3,4,5,3′,5′-Pentabromo-2-(2′-hydroxybenzoyl) pyrrole: A potential lead compound as anti-Gram-positive and anti-biofilm agent. *Int. J. Antimicrob. Agents* **2005**, *25*, 338–340. [CrossRef] [PubMed]
30. Giammanco, A.; Galia, E.; Fasciana, T.; Sciortino, M.; Cannella, S.; Arrigo, I.; Tricoli, M.R.; Di Paola, L.; Venturella, G.; Gargano, M.L. The Anti-Biofilm Potential of Pleurotus eryngii var. elaeoselini and White form of Grifola frondosa Extracts Against Staphylococcus aureus methicillin Resistance. In Proceedings of the 10th International Medicinal Mushrooms Conference, Nantong, China, 19–22 September 2019; International Society for Medicinal Mushrooms & Jiangsu Provincial Department of Agriculture and Rural Affairs: Nantong, China, 2019; pp. 184–185.
31. La Guardia, M.; Venturella, G.; Venturella, F. On the chemical composition and nutritional value of *Pleurotus taxa* growing on Umbelliferous plants (*Apiaceae*). *J. Agric. Food Chem.* **2005**, *53*, 5997–6002. [CrossRef] [PubMed]
32. Koutrotsios, G.; Kalogeropoulos, N.; Stathopoulos, P.; Kaliora, A.C.; Zervakis, G.I. Bioactive compounds and antioxidant activity exhibit high intraspecific variability in *Pleurotus ostreatus* mushrooms and correlate well with cultivation performance parameters. *World J. Microbiol. Biotechnol.* **2017**, *33*, 98. [CrossRef]
33. Koutrotsios, G.; Kalogeropoulos, N.; Kaliora, A.C.; Zervakis, G.I. Toward an Increased functionality in oyster (*Pleurotus*) mushrooms produced on Grape Marc or Olive Mill Wastes Serving as sources of bioactive compounds. *J. Agric. Food Chem.* **2018**, *66*, 5971–5983. [CrossRef]
34. Gaglio, R.; Guarcello, R.; Venturella, G.; Palazzolo, E.; Francesca, N.; Moschetti, G.; Settanni, L.; Saporita, P.; Gargano, M.L. Microbiological, chemical and sensory aspects of bread supplemented with different percentages of the culinary mushroom *Pleurotus eryngii* in powder form. *Int. J. Food Sci. Technol.* **2018**, *54*, 1197–1205. [CrossRef]

35. Bonanno, A.; Di Grigoli, A.; Vitale, F.; Di Miceli, G.; Todaro, M.; Alabiso, M.; Gargano, M.L.; Venturella, G.; Anike, F.N.; Isikhuemhen, O.S. Effects of diets supplemented with medicinal mushroom myceliated grains on some production, health, and oxidation traits of dairy ewes. *Int. J. Med. Mushrooms* **2019**, *21*, 89–103. [CrossRef]
36. Zervakis, G.I.; Ntougias, S.; Gargano, M.L.; Besi, M.I.; Polemis, E.; Typas, M.A.; Venturella, G. A reappraisal of the *Pleurotus eryngii* complex—New species and taxonomic combinations based on the application of a polyphasic approach, and an identification key to *Pleurotus taxa* associated with *Apiaceae plants. Fungal Biol.* **2014**, *118*, 814–834. [CrossRef] [PubMed]
37. Zervakis, G.I.; Venturella, G.; Fryssouli, V.; Inglese, P.; Polemis, E.; Gargano, M.L. *Pleurotus opuntiae* revisited—An insight to the phylogeny of dimitic *Pleurotus* species with emphasis on the P. djamor complex. *Fungal Biol.* **2019**, *123*, 188–199. [CrossRef] [PubMed]
38. Venturella, G. A check-list of Sicilian fungi. *Bocconea* **1991**, *2*, 1–221.
39. Antonio, A.; Valeria, B.; Marcello, D.; Caterina Patrizia Di, F.; Di Gesaro, M.; Emanuele, S. Monumental trees and their existence value: The case study of an Italian natural park. *J. For. Sci.* **2016**, *61*, 55–61. [CrossRef]
40. Schwarze, F.W.M.R.; Engels, J.; Mattheck, C. *Fungal Strategies of Wood Decay in Trees*; Springer: Berlin/Heidelberg, Germany, 2000.
41. Zotti, M.; Persiani, A.M.; Ambrosio, E.; Vizzini, A.; Venturella, G.; Donnini, D.; Angelini, P.; Di Piazza, S.; Pavarino, M.; Lunghini, D.; et al. Macrofungi as ecosystem resources: Conservation versus exploitation. *Plant Biosyst.* **2013**, *147*, 219–225. [CrossRef]
42. Moore, D.; Nauta, M.M.; Evans, S.E.; Rotheroe, M. *Fungal Conservation. Issues and Solutions*; Cambridge University Press: Cambridge, UK, 2001; p. 262.

7

Biodiversity of Pigmented Fungi Isolated from Marine Environment in La Réunion Island, Indian Ocean: New Resources for Colored Metabolites

Mireille Fouillaud [1,2,*], Mekala Venkatachalam [1], Melissa Llorente [1], Helene Magalon [3], Pascale Cuet [3] and Laurent Dufossé [1,2]

1 Laboratoire de Chimie des Substances Naturelles et des Sciences des Aliments—LCSNSA EA 2212, Université de La Réunion, 15 Avenue René Cassin, CS 92003, F-97744 Saint-Denis CEDEX 9, Ile de La Réunion, France; mekalavenkat@gmail.com (M.V.); melissa.llorente@gmail.com (M.L.); laurent.dufosse@univ-reunion.fr (L.D.)

2 Ecole Supérieure d'Ingénieurs Réunion Océan Indien—ESIROI, 2 Rue Joseph Wetzell, F-97490 Sainte-Clotilde, Ile de La Réunion, France

3 UMR ENTROPIE and LabEx CORAIL, Université de La Réunion, 15 Avenue René Cassin, CS 92003, F-97744 Saint-Denis CEDEX 9, Ile de La Réunion, France; helene.magalon@univ-reunion.fr (H.M.); Pascale.cuet@univ-reunion.fr (P.C.)

* Correspondence: mireille.fouillaud@univ-reunion.fr.

Abstract: Marine ecosystems cover about 70% of the planet surface and are still an underexploited source of useful metabolites. Among microbes, filamentous fungi are captivating organisms used for the production of many chemical classes of secondary metabolites bound to be used in various fields of industrial application. The present study was focused on the collection, isolation, screening and genotyping of pigmented filamentous fungi isolated from tropical marine environments around La Réunion Island, Indian Ocean. About 150 micromycetes were revived and isolated from 14 marine samples (sediments, living corals, coral rubble, sea water and hard substrates) collected in four different locations. Forty-two colored fungal isolates belonging to 16 families, 25 genera and 31 species were further studied depending on their ability to produce pigments and thus subjected to molecular identification. From gene sequence analysis, the most frequently identified colored fungi belong to the widespread *Penicillium, Talaromyces* and *Aspergillus* genera in the family Trichocomaceae (11 species), then followed by the family Hypocreaceae (three species). This study demonstrates that marine biotopes in La Réunion Island, Indian Ocean, from coral reefs to underwater slopes of this volcanic island, shelter numerous species of micromycetes, from common or uncommon genera. This unstudied biodiversity comes along with the ability for some fungal marine inhabitants, to produce a range of pigments and hues.

Keywords: fungi; biodiversity; Indian Ocean; Marine; coral reef; genotyping; pigment production

1. Introduction

With the growing demand for natural compounds in the industrial sector, marine derived fungi appear to present many interests. Filamentous fungi are ubiquitous in nature due to their huge capacity of adaptation and their ability to produce an assortment of new secondary metabolites. Literature now abundantly reports the significant involvement of fungi in the industry, through the production of various useful substances, such as antibiotics, immunosuppressants, anti-cancer drugs, plant hormones, enzymes, acids and also natural pigments [1–5]. Both the pigments and enzymes equally find their usages in food and beverages, animal feeds, pharmaceuticals, cosmetics, textile, leather, pulp and paper industries, biofuel production, and environment bioremediation [6].

Nevertheless, the distribution of the marine-derived fungal species and their contribution to marine biotopes are still in infancy, and more has to be explored [7–11]. The highest diversity of marine fungi seems to appear in tropical regions, mainly in tropical mangroves, which are extensively studied because of their high richness in organic matters and especially lignocellulosic materials, favorable to the development of a wide range of heterotrophic microorganisms [11–14]. Anyway, many marine ecological niches are still unexplored and it seems plausible that unique features of marine environments can be the inducers of unique substances, biosynthesized by marine or marine-derived microorganisms [15,16].

Considering the immense genetic and biochemical diversity of these fungi, partially derived from the specificity of the biotopes they are facing, marine-derived fungi are regarded as a potential bright source of new molecules with likely application in pigment production [17,18]. Many genera producing pigments have then been isolated either from water, sediments, and decaying organic residues, or from living organisms such as invertebrates, plants or algae. Fungi belonging to genera such as *Aspergillus, Penicillium, Paecilomyces, Eurotium, Alternaria, Fusarium, Halorosellinia, Monodictys* and *Microsphaerospsis* have already been identified from marine biotopes [19–21]. They are therefore able to exhibit bright colors, from yellow to black, mainly belonging to polyketides. Indeed, polyketides pigments and particularly azaphilones and anthraquinones seem to dominate marine natural products of fungal origin [22]. Colored compounds, usually described as secondary metabolites, do not seem to be directly involved in the primary growth of the fungus in which they occur [23]. However they may play some important roles in the resistance to a variety of adverse environmental factors (desiccation, exposure at extreme temperatures, irradiations and photo-oxidation) or in ecological interactions with other organisms (macroorganisms such as sponges, corals or other microbial communities) [24]. For this reason, many fungal secondary metabolites exhibit useful biological activities and are of interest to the pharmaceutical, food, and agrochemical industries [16,25].

This study initiated the search for filamentous fungi in some tropical marine biotopes of coral reefs and underwater slopes of the volcano from La Réunion Island. Fungal isolates from samples of sediments, seawater, hard substrates, coral rubbles or living coral individuals (*Pocillopora* sp.) were characterized both by phenotypic and molecular ways. The production of pigments of quinoid-type produced from the mycelia cultured in liquid media was used as a first approach to screen for the pigment production. This work reveals a part of the mycofloral biodiversity in La Réunion Island tropical marine environment and its potentiality to propose new pigment sources to expand in an industrial setting.

2. Materials and Methods

2.1. Samples Collection

La Réunion Island lies in the Indian Ocean and is located 800 km east of Madagascar (21°06′54.5″ S and 55°32′11.0″ E) (Figure 1a). This tropical island arose two million years ago from a volcanic hot spot (Piton de La Fournaise) and is known for its rainforested interior and its fringing reefs holding most of the marine wealth.

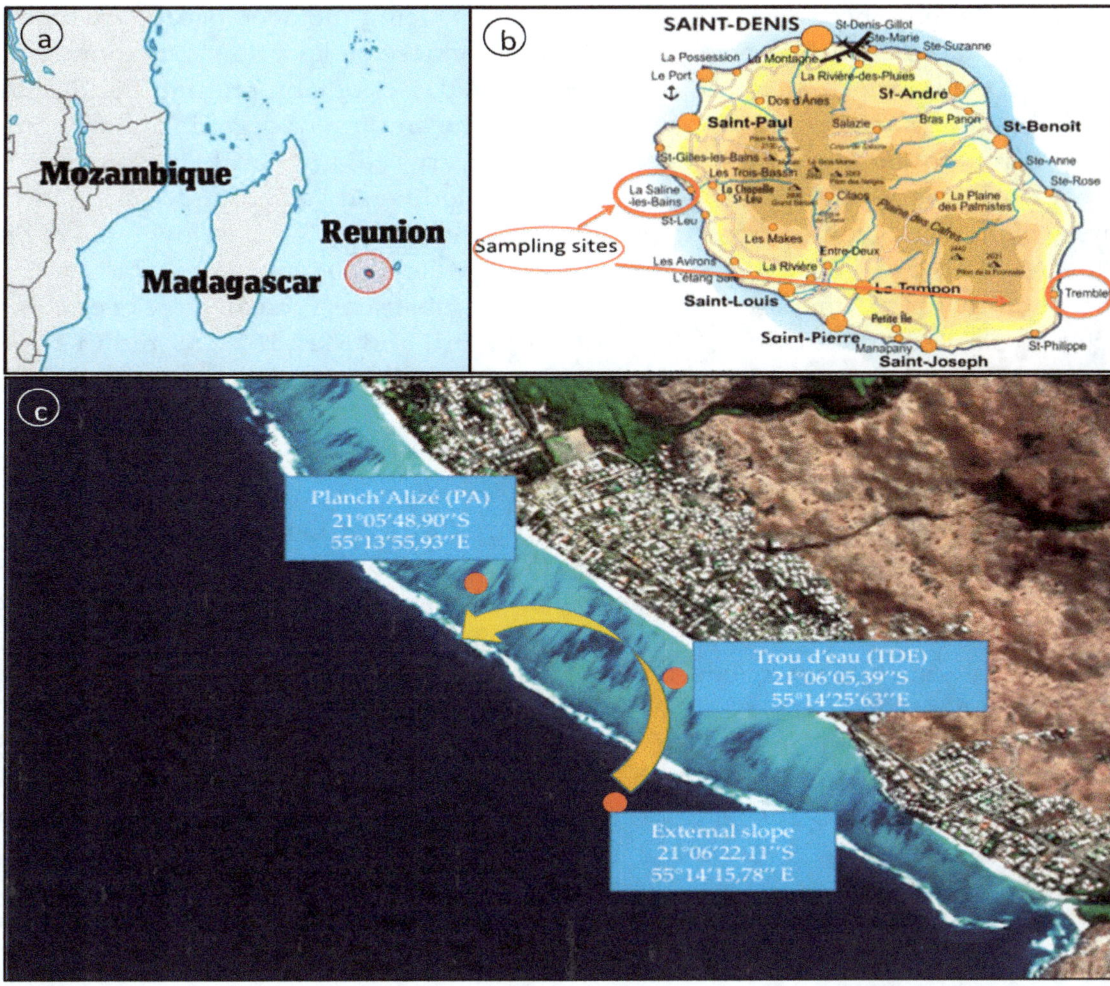

Figure 1. (**a**) La Réunion island location (Indian Ocean, 21°06′54.5′′ S and 55°32′11.0′′ E); (**b**) geolocation of sampling sites around La Réunion Island (West: La Saline; and East: Sainte Rose and Tremblet); and (**c**) geolocation of the three sampling spots at La Saline fringing reef: Trou d'eau (TDE inner reef and TDE outer slope) and Planch'Alizé (PA) (back arrow represents the main water flow).

A first set of samples was collected on the fringing reef from La Saline, which lies on the dry west coast of the island. It is more than 9 km long and ranges in width from 50 m in its northern part to 600 m in the south [26]. For the purpose of research, samples were collected from three sampling spots on the west coast that cover the sites of Trou d'Eau (TDE) (inner reef flat at −1 m depth and outer slope at −17 m) and Planch' Alizé (PA) (inner reef flat, −1 m) (Figure 1b,c). Planch' Alizé is considered as a sheltered site, located downstream of seawater flowing over the Trou d'Eau (Figure 1c). The outer slope is found at the outer edge of the reef, closest to the open ocean, and is characterized by spurs and grooves extending downward to the sand bottom, while the inner reef flat displays wide transversal strips of branched coral colonies alternating with narrow detrital channels perpendicular to the reef flat [27–31]. Low water flow and high solar radiation contribute to heating the reef water during the day, inducing important daily sea surface temperature variations. This area is also heavily laden with organic and mineral matter coming from nearby human activity (seaside area).

A second set of samples was collected in Sainte Rose area (south-east) on the submerged lava flows (Figure 1b). Indeed, the Piton de la Fournaise is one of the most active effusive volcanoes in the world with 27 eruptions between 1998 and 2007 and a mean frequency, over a century, of an eruptive phase every nine months. Submerged lava flows appear on the south-east part of the island when, during eruptions, the pool of lava overflows the active volcano mouth and pours down on the slopes of the volcano, to the sea. These costal marine ecosystems facing the deep ocean, are then regularly subjected to natural hazards such as being covered by incandescent lava flows, temporary changes in physicochemical conditions of water bodies and exceptional rises of temperature. Besides, this area is poorly inhabited and urbanized and, as a consequence, the amount of organic matter poured in the sea is reduced compared to other coastal ecosystems. It provides a natural laboratory to study the colonization of a blank substrate and the evolution of the biodiversity all around, during the following years. Samples were then obtained from sediments extracted from 1977 lava flow (−25 m depth) and 2004 lava flow (−70 m), as well as from surrounding free water at −70 m.

Seawater, sediments, parts of living corals and hard substrates (volcanic rocks or coral rubbles) were collected in sterile bottles, during the months of April and May 2012, stored in a cooling box (4 °C), brought to the laboratory, and treated immediately for the fungal isolation.

2.2. *Culture and Purification of Fungi*

To cultivate the revivable fungi from the collected seawater, 100 mL of water was filtered using a 0.45 μm sterile cellulose-nitrate filter (Sartorius Stedim, Göttingen, Germany). The filters were then placed in Petri plates containing malt extract agar (MEA) and Sabouraud agar (BD Difco, Franklin Lakes, NJ, USA) prepared with natural seawater collected near La Saline, and beforehand sterilized at 121 °C, 15 min.

The other samples such as sediments, hard substrates and parts of living/dead coral were treated separately. The samples were first washed with 70% alcohol and rinsed in sterile seawater. Then they were ground using sterile pestle and mortar. Ground material (5 g) was taken from each sample and added to 15 mL of sterile diluent (1.6 g of tryptone (Sigma- Aldrich, T-9410, Saint Louis, MO, USA), 0.05 g of tween 20, 1 L of sterile seawater of pH = 7.5). After stirring for 20 min at 150 rpm on a shaking table (Edmunt Bühler GmbH, VKS 75 Control, Hechningen, Germany), the suspension was diluted by employing serial decimal dilution method up to 10^{-3} [32]. Each diluted sample (1 mL) was poured on Petri plates containing MEA and Sabouraud agar prepared with natural seawater.

All the platings were performed in triplicates and incubated at 25 °C for 21 days. During this period, the plates were checked each day for the appearance of new colonies. Each new colony was individually isolated and cultured on new MEA solid medium. During the growth period, the production of colors was observed.

All the isolated fungi were cultured using monospore technique for future experiments and long-term storage. The fungi grown after 5 days were scraped and transferred into a sterile vial containing a cryoprotectant medium composed of 15% skimmed milk and 2% glycerol for long term storage at −80 °C [33,34]. In total, 42 fungal isolates were then selected for pigment production based on the visual appearance of the thalli grown on solid culture media.

2.3. *Fungal Identification*

2.3.1. Fungal DNA Extraction

To extract DNA from the 42 purified isolates, a small amount of mycelium along with spores was cultivated on potato dextrose agar (PDA) at 25 °C under day light exposure. After 5 days of growth, the fungal mycelium was scraped and DNA was extracted using DNeasy Blood and Tissue kit (Qiagen, Hilden, Germany) following the manufacturer's instructions. DNA amount and purity contained in each extract were evaluated by measuring the absorbances at 230, 260 and 280 nm (Nanodrop 2000, Thermo Scientific, Waltham, MA, USA) and calculating the ratio A_{260}/A_{280} and A_{260}/A_{230}. DNAs were stored at −20 °C prior to use for amplification studies [35].

2.3.2. Primers, PCR Amplification and Sequencing

The choice of PCR primers was made based on observed phenotypic characteristics for molecular identification. *Aspergillus* species were amplified for calmodulin gene using primers Cmd5/Cmd6 and *Penicillium* species for β-tubulin using primers T10/Bt2b [36]. To amplify and sequence the DNA from *Trichoderma* and *Hypocreales* species, EF-1H/EF-2T primer pair was used to amplify a fragment of the translation elongation factor 1 alpha gene (*Tef1*) [37]. For uncharacterized fungi, the fragments containing ITS region were amplified using ITS1-F_KYO2/ITS2 or ITS3_KYO2/ITS4, and, when necessary, the large subunit rDNA was also amplified using V9G/LR3 primer pair (Table 1) [36–38].

Table 1. PCR amplification and the sequencing primers used for the identification of fungal isolates.

Primers	Direction	Sequences (5′→ 3′)	Note	Hybrid. T °C	Refs.
ITS1-F_KYO2	Forward	TAGAGGAAGTAAAAGTCGTAA	Small sub-unit, ITS 1, 5.8S, ITS 2, Largest sub unit rDNA	56	[36]
ITS2_KYO2	Reverse	TTYRCTRCGTTCTTCATC		47	
ITS3_KYO2	Forward	GATGAAGAACGYAGYRAA		47	
ITS 1	Forward	TCCGTAGGTGAACCTGCGG		55	[39]
ITS 2	Reverse	GCTGCGTTCTTCATCGATGC		55	
ITS 3	Forward	GCATCGATGAAGAACGCAGC		55	
ITS 4	Reverse	TCCTCCGCTTATTGATATGC		55	
V9G	Forward	TTACGTCCCTGCCCTTTGTA	Large sub unit D1/D2 for basidiomycetous yeast	55	[38]
LR3	Reverse	TGACCATTACGCCAGCATCC		57	
Cmd 5	Forward	CCGAGTACAAGGARGCCTTC	Calmodulin, specific for *Aspergillus*	52	
Cmd 6	Reverse	CCGATRGAGGTCATRACGTGG		52	
T 10	Forward	ACGATAGGTTCACCTCCAGAC	β- tubulin, specific for *Penicillium*	55	[38]
Bt2b	Reverse	ACCCTCAGTGTAGTGACCCTTGGC		55	
EF1-728F	Forward	CATCGAGAAGTTCGAGAAGG	Elongation factor 1 for *Trichoderma*	55	[38]
TEF1-LLErev	Reverse	AACTTGCAGGCAATGTGG		55	

PCR reactions were carried out in a total volume of 30 μL: 1× of MasterMix (Applied Biosystems, Foster city, CA, USA), 0.5 μM of forward and reverse primers and at least 1.3 ng/μL of genomic DNA. Amplifications were carried out on a thermal cycler GeneAmp® PCR System 9700 (Applied Biosystems) according to the following program: 94 °C for 5 min + 35 × (94 °C for 30 s, 55 °C (or 52 °C for the primers of calmodulin: Cmd5/Cmd6) for 60 s, 72 °C for 60 s) + 72 °C for 5 min for final elongation step.

2.3.3. Sequence Analysis

Amplicons were sequenced in both directions (GENOSCREEN, Lille, France). The obtained electropherograms were read and corrected with Chromas software (version 2.13, Technelysium pty Ltd., South Brisbane, Australia). The extracted sequences for each gene were separately used to perform nucleotide searches using online BLAST algorithm, provided by NCBI (http://www.ncbi.nlm.nih.gov/BLAST/). BLAST results were sorted based on the maximum identity to the query sequence and considered as the best hit. Sequence-based identities with a cutoff of 97% or above and query coverage >90% were considered as significant [40,41]. Because of low recovery rates and concordance values, some isolates were amplified and sequenced a second time, with additional sets of primers, mainly among the isolates of the genera *Penicillium* and *Trichoderma*.

2.4. Culture Conditions for Pigment Production and Separation of Biomass from Liquid Medium

2.4.1. Culture Conditions

Erlenmeyer flasks (250 mL) containing 80 mL of potato dextrose broth (PDB) medium were autoclaved at 121 °C for 15 min. Then, 120 mg of mycelia from interesting fungal species grown on PDA Petri plates were transferred into the sterile flasks and incubated at 25 °C under daylight exposure, with an agitation of 150 rpm for 10 days (Multitron Pro, Infors HT, Bottmingen, Switzerland).

2.4.2. Separation of Biomass from Culture Liquid

After the end of the fermentation period, the culture medium containing extracellular pigments was separated from mycelia by vacuum filtration using Whatman filter paper No. 2 (Merck, Darmstadt, Germany). Thus, liquid medium and biomass were treated separately. The wet mycelium was further used for the extraction of pigment content.

2.5. *Production of Pigments*

2.5.1. Determination of Pigments Production in Liquid Cultures

Chromophore is a chemical group that absorbs light of specific frequency and confers color to a molecule. Widespread polyketide pigments such as anthraquinones or azaphilones are often highly substituted aromatic molecules, with fused benzene rings [42]. Thus the majority of the common chromophores from fungi absorb in the UV region (one or several peaks between 200–300 nm), whereas absorbance in the visible region (400–700 nm) highly depends on the nature and the number of substituted groups.

To compare the pigment production of all isolates cultured in PDB medium, the amount of pigments produced in liquids was expressed as mg equivalent (mg eq.) of a chosen commercial standard per liter of culture medium (mg eq. purpurin L^{-1}). Purpurin was chosen as a polyketide pigment in orange-red hue, which absorbs in the UV area (250–270 nm) as many polyketides [43], and also in the visible range 458–520 nm [44]. Thus, the absorbance of an authentic colored standard purpurin (Sigma-Aldrich) was estimated at different concentrations using an UV-visible spectrophotometer (Shimadzu UV-1800 Spectrophotometer). Then, in regard with the diversity of pigments content in the fungal cultures and as a preliminary approach, the absorbance of each sample was measured at 254 nm and the amount of pigments produced was expressed in "mg equivalent purpurin L^{-1}" (Figure S1). In addition, for each isolate, the intracellular (IC) pigments (extracted from the biomass) and extracellular (EC) contents (liquid from culture, separated from the biomass) were scanned between 200 and 600 nm with a UV-1800 spectrophotometer (Shimadzu UV spectrophotometer, Shimadzu Corporation, Kyoto, Japan) in a quartz cell of 10 mm path length.

2.5.2. Extraction of Pigments

IC pigments contained in the wet mycelium were extracted using a methanol: water combination (1:1 v/v) as conventional extraction method. The mixture was immersed in an ultrasonic bath at 45 °C for 30 min. The suspension was allowed to stir overnight at room temperature on a shaking table (VKS 75 Control, Edmunt Bühler GmbH). On the following day, it was filtered through Whatman filter paper No. 2 to recover the solvent containing the pigments extracted from biomass.

To compare the amount of pigments produced within the cells with the one diffused into the extracellular medium, we performed the nonparametric Mann–Whitney–Wilcoxon test as our data did not follow the normal distribution using the R software (R Development Core Team 2016) [45].

3. Results

3.1. *Diversity of Isolated Fungi*

More than 150 isolates were first recovered from the 14 samples collected among four locations. Among them, 42 were selected for identification, according to their capacity to develop colored mycelia or to secrete colored compounds in the media.

After sequencing, the 42 colored isolates were assigned to 16 families, 25 genera and 31 species (accession numbers mentioned in Table 2). The vast majority of the isolates have been identified with more than 98% concordance rate and recovered with high precision at the species level. However, few fungi (*Acremonium* sp., *Periconia* spp. and *Biscogniauxia* sp.) were identified to the genus level only, according to the gene chosen. The genetic characterization with these primers partially failed for two isolates (*Whalleya microplaca B* and *Wallemia sebi*). *Wallemia sebi* was only characterized according to morphological criteria.

Table 2. Fungal isolates from La Réunion Island marine biotopes from different sample types and sampling sites: Trou d'Eau (TDE); Planch' Alizé (PA); Lava flow corresponds to 1977 lava flow in Sainte Rose/Tremblet area.

Family	Fungal Species	Sampling Site	Gene Accession Number
	Water Bodies		
Davidiellaceae	*Cladosporium Cladosporioides*	Lava flow (−70 m)	JF949719.1
Didymellaceae	*Peyronellaea glomerata* (syn: *Phoma glomerata*)	Lava flow (−70 m)	JQ936163.1
Nectriaceae	*Nectria haematococca* A	Lava flow (−70 m)	XM_003053163.1
Pleosporales Incertae Sedis	*Periconia* sp. A	Lava flow (−70 m)	HQ608027.1
	Periconia sp. B	Lava flow (−70 m)	HQ608027.1
Sporidiobolaceae	*Rhodosporidium paludigenum*	Lava flow (−70 m)	AF444493.1
Stachybotryaceae	*Myrothecium atroviride*	Lava flow (−70 m)	AJ302002.1
Teratospheriaceae	*Hortaea werneckii (syn: Cladosporium werneckii)*	Lava flow (−70 m)	JN997372.1
Trichocomaceae	*Aspergillus sydowii* B	Lava flow (−70 m)	KC253961.1
	Emericella qinqixianii	TDE outer slope	AB249008.1
	Penicillium brocae NRRL 32599	TDE outer slope	DQ123642.1
	Penicillium viticola B	TDE inner reef	AB606414.1
	Talaromyces rotundus	TDE inner reef	EU497950.1
	Talaromyces verruculosus	PA inner reef	KC416631.1
Wallemiaceae	*Wallemia sebi*	Lava flow (−70 m)	Morphological Identification
	Living Coral *Pocillopora sp.*		
Hypocreaceae	*Acremonium* sp.	PA inner reef	FJ770373.1
	Hypocrea koningii	TDE inner reef	JX174420.1
Trichocomaceae	*Aspergillus creber* A	TDE inner reef	JN854049.1
	Aspergillus creber B	TDE inner reef	JN854049.1
	Aspergillus sydowii A	PA inner reef	JN854052.1
	Eurotium amstelodami	TDE outer slope	FR727111.1
	Penicillium viticola C	PA inner reef	AB606414.1
	Coral Rubbles		
Chaetomiaceae	*Chaetomium globosum or Chaetomium murorum*	TDE outer slope	JN209898.1
Trichocomaceae	*Penicillium herquei*	TDE outer slope	JN246042.1
	Talaromyces albobiverticillius B	TDE outer slope	JN899313.1
	Talaromyces albobiverticillius C	TDE outer slope	JN899313.1
	Hard Substrate/Rock Substrate		
Nectriaceae	*Fusarium equiseti* A	Lava flow (−25 m)	JQ936153.1
	Fusarium equiseti B	Lava flow (−25 m)	JF311925.1
	Fusarium equiseti C	Lava flow (−25 m)	JQ936153.1
	Nectria haematococca B	Lava flow (−25 m)	XM_003053163.1
Pleosporaceae	*Epicoccum sorghi* (syn: *Phoma sorghina; Peyronellaea stemphylioides*)	Lava flow (−25 m)	KC106717.1
Sordariomycetes	*Nigrospora sphaerica* (or Env. sample from marine air*)*	TDE outer slope	KC505176.1
Sporidiobolaceae	*Rhodotorula mucilaginosa*	TDE outer slope	KC515367.1
Trichocomaceae	*Penicillium citrinum*	Lava flow (−25 m)	EU030332.1
	Penicillium viticola A	Lava flow (−25 m)	AB606414.1
Ustilaginaceae	*Sporisorium exsertum*	TDE outer slope	JN367293.1
Xylariaceae	*Biscogniauxia* sp.	PA inner reef	FJ884075.1
	Whalleya microplaca A	TDE outer slope	JQ760548.1
	Sordariomycete (or *Whalleya microplaca* B)	TDE outer slope	FJ416301.1
	Sediments		
Didymosphaeriaceae	*Paraconiothyrium variabile*	TDE outer slope	JQ936271.1
Hypocreaceae	*Trichoderma atroviride*	TDE outer slope	KC008065.1
Trichocomaceae	*Talaromyces albobiverticillius* A	TDE outer slope	JN899313.1

Pigment producing fungi (42) were isolated from all types of samples: sediments (3), living coral *Pocillopora* sp. (7), unidentified coral rubbles (4), hard substrates (reef basis or volcanic rocks) (13) and seawater (15) (Table 2).

The most represented fungi, in the selection of colored micromycetes, belonged to the family Trichocomaceae with *Penicillium, Talaromyces* and *Aspergillus* genera (11 species); and then came the Hypocreaceae with *Trichoderma, Hypocrea* and *Acremonium.*

A high diversity of pigmented isolates was observed from the so-called "hard substrates" (rocky basis on which the coral colonies recruit, or submerged lava flows). Some *Nigrospora, Sporisorium, Whalleya,* and *Rhodotorula* isolates were collected from the outer slope at Trou d'Eau (TDE), although they are rather rarely isolated from marine environment. In Sainte Rose, *Penicillium* species (*P. citrinum, P. viticola*) as well as *Fusarium equiseti, Epicoccum sorghi, Nectria haematococca,* were successfully revived from lava flow, sampled at −25 m.

From our study, the coral rubbles (dead parts of corals) contained colored *Penicillium* or related species: *P. herquei* and two isolates of *Talaromyces albobiverticillius* (B and C), as well as an isolate of *Chaetomium globosum.* Coral rubbles or hard substrates naturally appear diversely colored underwater. Indeed, they support the colonization by multiple organisms (colored algae or other aquatic organisms), visually detectable when sampling.

The revivable colored fungi sheltered by the living coral *Pocillopora* sp. belonged to the genera *Aspergillii* (*A. creber, A. sydowii* and *Eurotium amstellodami* (the teleomorphic form of *A. amstellodami*), as well as to *Penicillium* (*P. viticola*), *Hypocrea* (*H. koningii*) and *Acremonium.*

Some fungal species were identified from different types of samples in the same area. At TDE outer slope, *Talaromyces albobiverticillius* A came from sediment and *T. albobiverticillius* B and C were revived from coral rubbles. *Nectria haematococca* was found in lava substrate (−25 m) (isolate B) as well as seawater (−70 m near lava flow) in the same area (isolate A).

Some similar species also appeared in separate locations: *Aspergillus sydowii* was found near lava flow on the east coast (seawater, −70 m) (isolate B) and also in living *Pocillopora* colonies (isolate A), from the west coast back reef (PA site). *Penicillium viticola* was isolated from the west coast on living *Pocillopora* sp. coral in PA (isolate C), from seawater in TDE back reef (isolate B), as well as from lava flow hard substrate (−25 m), on the east coast (isolate A).

These fungi found in several samples and/or in different locations may be considered as frequent in this marine environment.

3.2. Pigment Production

3.2.1. In Culture Broth

The majority of the isolates produced pigments after four days of fermentation in PDB. The colors of the broth (biomass plus liquid culture medium) always darkened over time, which indicated their potential for pigment production (Figure 2).

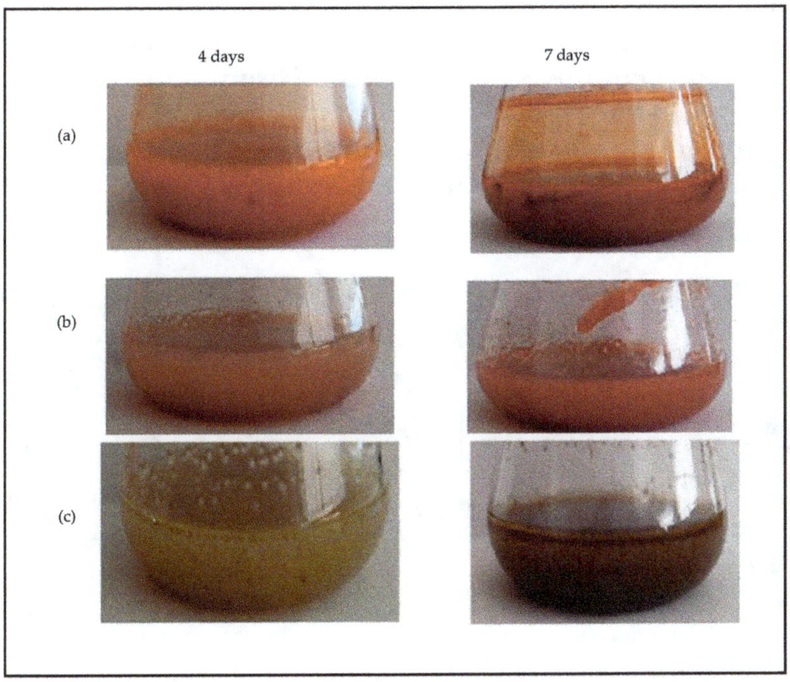

Figure 2. Colors observed in potato dextrose broth cultures from (a) *Talaromyces albobiverticillius* B, (b) *T. albobiverticillius* C, and (c) *Aspergillus creber* B, after four and seven days.

Overall, it was observed that the coloring trend was not directly related to the genus. Even if dominant colors such as yellow, red, brown, purple, orange, pink and green were observed in flasks, the hues were extremely diverse according to the species, even to the isolates (Table 3).

Table 3. Dominant colors of culture broth [1], extracellular (EC) [2] and intracellular (IC) [3] pigments from fungal isolates.

Fungal Isolates	Approximate Hues			Fungal Isolates	Approximate Hues		
Isolates with Intense Hues (Purple/Red/Maroon)				***Isolates with Orange Hues***			
	Broth	***EC***	***IC***		**Broth**	***EC***	***IC***
Acremonium sp.				*Penicillium viticola* A			
Talaromyces albobiverticillius A				*Penicillium viticola* B			
Talaromyces albobiverticillius B				*Epicoccum sorghi*			
Talaromyces albobiverticillius C				*Penicillium brocae* NRRL 32599			
Aspergillus sydowii A				*Penicillium herquei*			
Aspergillus creber A				*Aspergillus sydowii* B			
Aspergillus creber B				*Chaetomium globosum* or *C. murorum*			
Emericella qinqixianii				*Penicillium viticola* C			
Trichoderma atroviride				*Penicillium citrinum*			
Biscogniauxia sp.				*Hypocrea koningii*			
Paraconiothyrium variabile							
Myrothecium atroviride							
Isolates with Yellow Hues				***Isolates with Green/Brown Hues***			
	Broth	***EC***	***IC***		**Broth**	***EC***	***IC***
Peyronellaea glomerata				*Talaromyces verruculosus*			
Eurotium amstelodami				*Talaromyces rotundus*			
Rhodosporidium paludigenum				*Wallemia sebi*			
Periconia sp. A				*Sporisorium exertum*			
Periconia sp. B				*Hortea werneckii*			
Rhodotorula mucilaginosa				*Whalleya microplaca* A			
Fusarium equiseti A				*Whalleya microplaca* B			
Fusarium equiseti B				*Nigrospora sphaerica or Env. sample from marine air*			
Fusarium equiseti C							
Nectria haematococca A				*Cladosporium cladosporioïdes*			
Nectria haematococca B							

[1] Culture broth: mycelium + liquid medium; [2] EC: filtrate from liquid culture medium; [3] IC: intracellular extract of fungal pigments.

Indeed, some similar-looking fungi, identified under a unique accession number (i.e., sharing the same sequence for the considered gene), nevertheless developed different color-phenotypes, while cultured under the same culture conditions. As an example *A. creber* A developed a red hue, clearly different from the green-like color *of A. creber* B ("broth" column in Table 3). Moreover, if the same coloring trend was applicable to all the three isolates of *T. albobiverticillius* (A–C) or *P. viticola* (A–C), different shades of red or yellow-orange hues, respectively, were noticed (Figure 3).

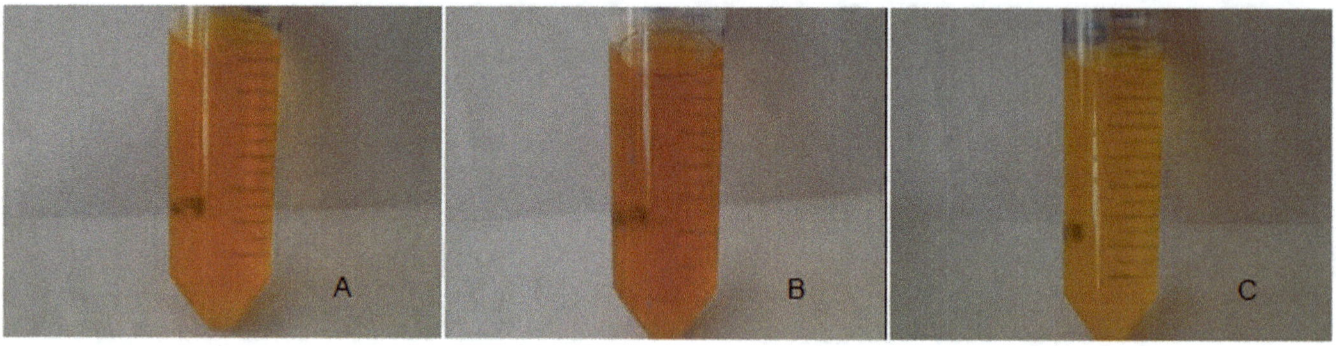

Figure 3. Colors observed from culture filtrates from three isolates *of Penicillium viticola* (**A–C**) (seven-day cultures in potato dextrose broth).

Oppositely, no clear difference could be visually established among the pale pink shades of the three isolates of *F. equisetti* (A–C) or the two *N. haematococca* isolates (A and B).

3.2.2. Pigmented Contents from Mycelium

For all pigment-producing isolates, the intracellular pigments (from mycelium) were extracted from the biomass. The approximate colors visualized after extraction are presented in Table 3. The pigments from most of the extracellular fungal culture filtrates were of dominating red, orange, yellow, green, brown, pink and violet. However, after extraction from biomass, many intracellular samples were uncolored, especially for isolates producing extracellular culture filtrates of pink, yellow and green color. This is probably characteristic of isolates essentially secreting water-soluble colored molecules in the culture media.

Instead, many dark colored cultures, mainly in the shades of red or maroon extracellular pigments, gave dark pigmented intracellular extracts from the biomass, indicating that the pigment was also highly concentrated inside the mycelium. These mainly concerned the isolates included in the group "isolates with intense hues", and in the group "isolates with orange hues" to a lesser extent (Table 3). Thus, isolates appeared with different status and varying capacities, towards pigment production.

3.3. Spectrophotometric Characterization of Pigments

As shown in Figure 4, the absorbance spectra of intra- and extracellular solutions from a single isolate revealed quite similar profiles characterized by a strong absorbance in the UV region and also an area of absorbance in the visible range of wavelengths. The values were principally located in the 400–480 nm area for pale yellow to yellow-orange pigments. The maximal absorbances spread in the 500–550 nm region for red colors.

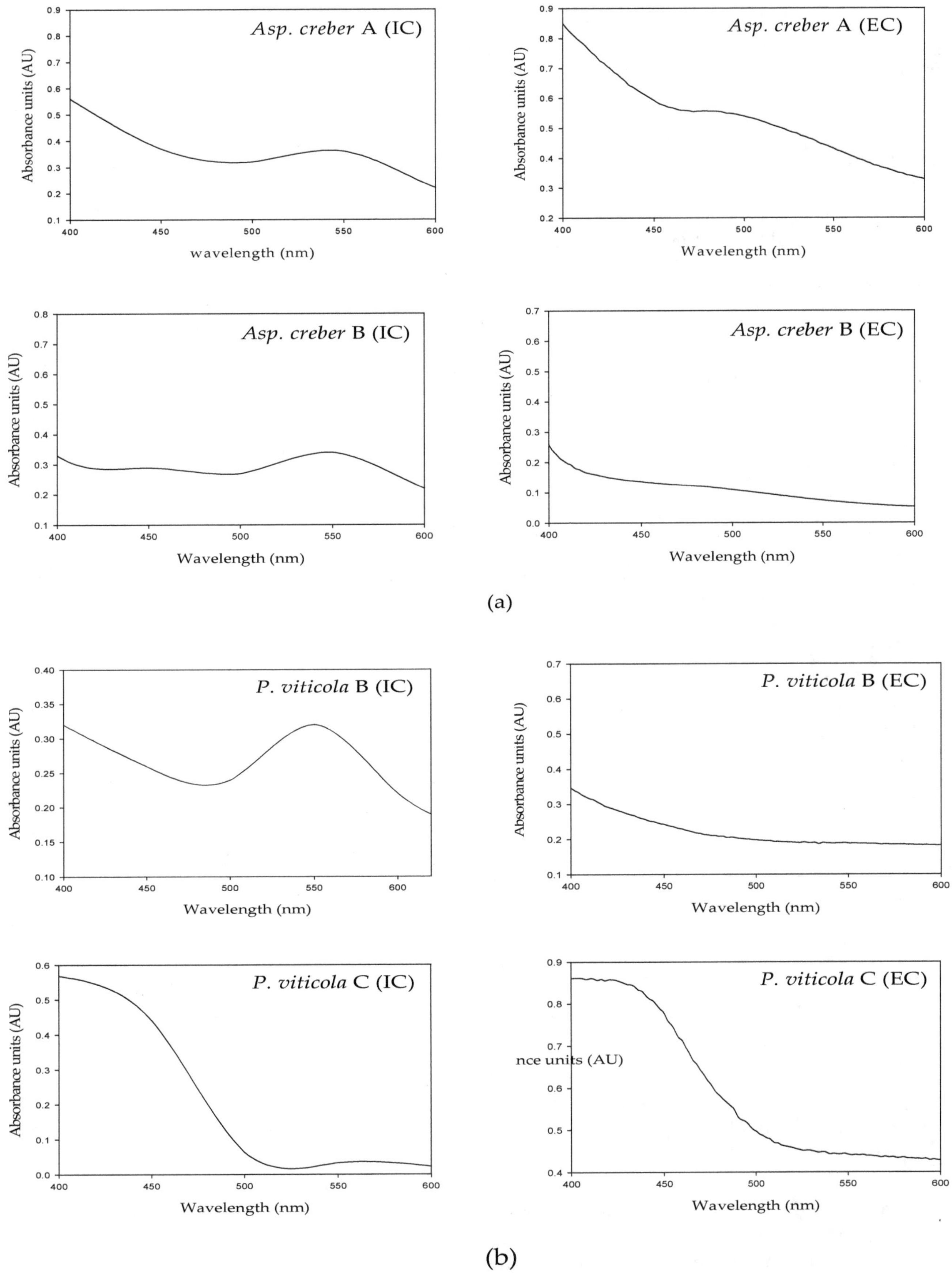

Figure 4. *Cont.*

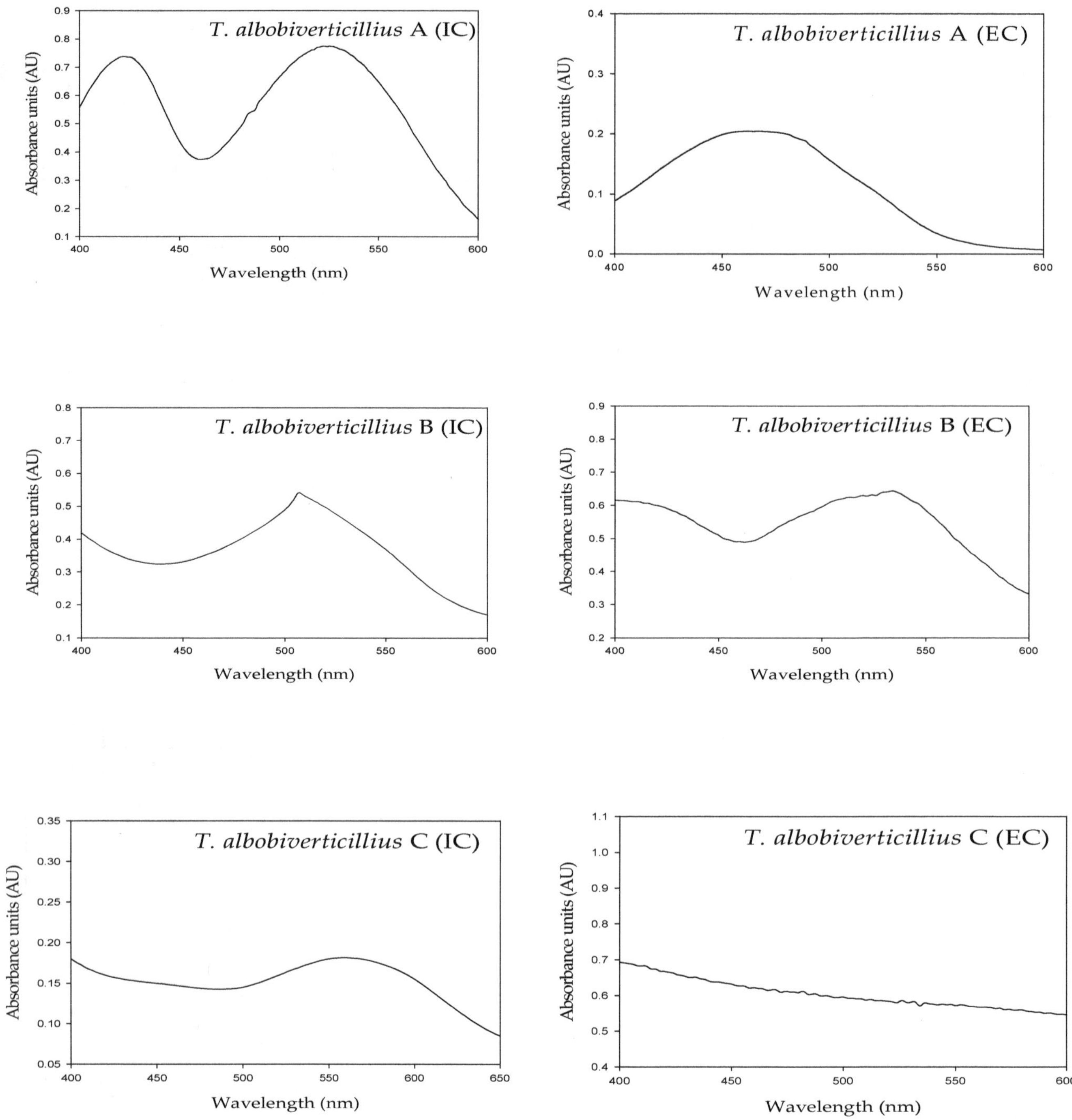

(c)

Figure 4. Intracellular (IC) and extracellular (EC) UV-visible spectra of: (**a**) *Aspergillus creber* A and B; (**b**) *Penicillium viticola* B and C; and (**c**) *Talaromyces albobiverticillius* A–C cultures in potato dextrose broth (7 days).

However, slight variations were noticed between extra- and intracellular liquids: in *A. creber* A as an example, extracellular maximum absorbance was around 470 nm (yellow-orange hue) instead of 550 nm (red shade) for intracellular liquid (Figure 4a). These slight variations however indicate that intra- and extracellular solutions may contain different assortments of colored compounds, in different proportions, resulting in different hues (Figure 5).

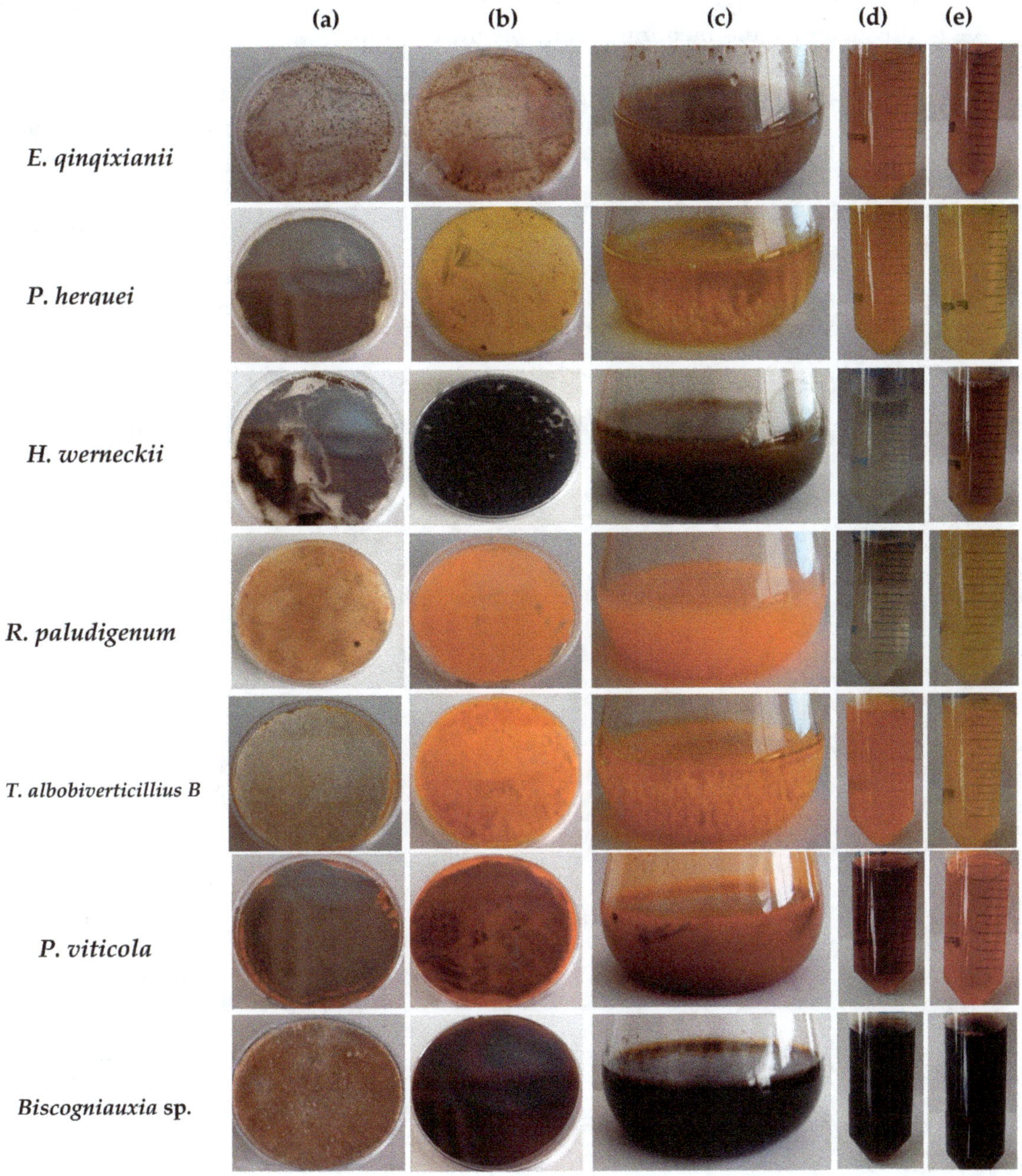

Figure 5. Colors observed in different fungal species: (**a**) obverse face on PDA; (**b**) reverse face on PDA; (**c**) culture in PDB (seven days); (**d**) extract of intracellular pigments (Ethanol/water 50/50) (IC); and (**e**) filtrate from liquid culture (EC).

Differences were also observed among the spectral profiles of different isolates belonging to the same species. As shown from the intracellular profiles of *A. creber* A and B, *P. viticola* B and C, and *T. albobiverticillius* A–C (Figure 4a–c, respectively, and Table 4), maximal absorbance areas differed in the visible region (510–560 nm for *P. viticola* B and 420–450 nm for *P. viticola* C; and 422–525 nm for *T. albobiverticillius* A, 500 nm for *T. albobiverticillius* B and 520–580 nm for *T. albobiverticillius* C), but, for *A. creber* A and B, the spectra looked similar (Figure 4a,b). Similar variation was stated between the extracellular profiles.

These results clearly imply that isolates from a same species produce and secrete different pigments and therefore have different behavior towards colored compound production.

Table 4. Summary of main peaks (λ_{max}) noticed in 10-days old culture of *Talaromyces albobiverticillius* isolates A–C cultivated in liquid medium (potato dextrose broth).

T. albobiverticillius	Sample	Peaks in the UV Region (nm)		Peaks in the Visible Region (nm)	
		200–250	250–300	300–400	>400
A	IC	235	286	362	422, 425, 511, 525
	EC		265	365	458, 469.8, 480
B	IC	232	268, 292		410, 440, 460, 500
	EC		288		412, 524, 532
C	IC	222	283	385	520–580
	EC		283	370, 385	436

3.4. Evaluation of Intracellular and Extracellular Contents in Pigments

The amount of pigments in IC and EC solutions, expressed in mg eq. Purpurin L^{-1}, are presented in Figure 6.

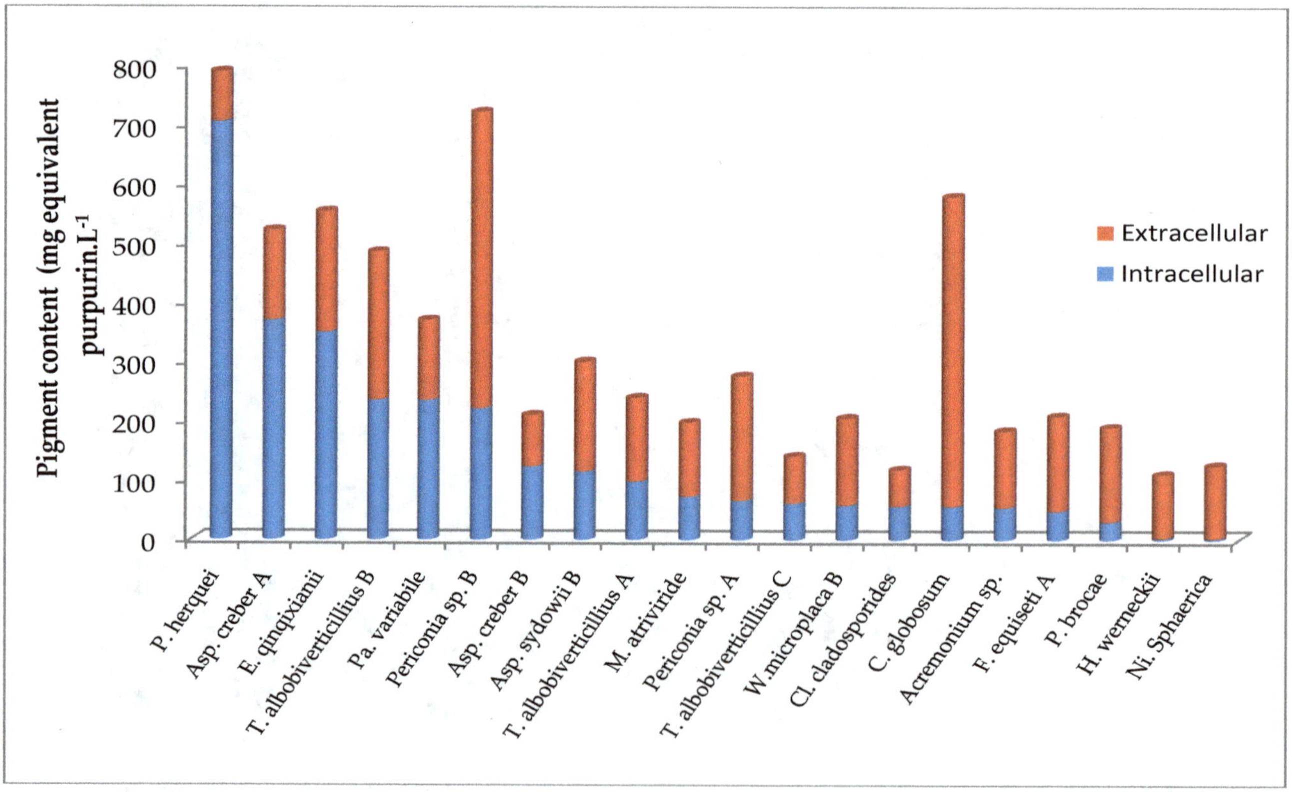

Figure 6. Colored compounds in biomass (intracellular, IC) and culture filtrate (extracellular, EC) for 20 isolates of marine derived fungi isolated around La Réunion Island, in mg eq. purpurin L^{-1} of culture medium (potato dextrose broth, absorbance at 254 nm).

In regard with the diversity of isolates and colored compounds involved in this study, instead of the intensity of the color, the values express the global amount of polyketides compounds produced by each isolate, hues ranging from pale green, light yellow to dark red or maroon.

In the extracellular samples, the maximum amount was produced by *Chaetomium globosum* (521.44 mg equivalent purpurin L^{-1}), followed by *Periconia* sp. B (498.39 mg equivalent purpurin L^{-1}). For intracellular samples, the maximum levels were measured for *P. herquei*, *A. creber* A, *E. qinqixianii* with 704.55, 371.18 and 350.93 mg equivalent purpurin L^{-1}, respectively.

The amount of intracellular content was significantly lower than the one of the extracellular content in this population (n = (20,20), $V = 10$, $P = 6 \times 10^{-4}$). However, looking at each isolate separately, the amount of intracellular pigments was significantly higher than the extracellular one for *P. herquei* (704.55 vs. 84 mg equivalent purpurin L^{-1}), and *A. creber* A (371.18 vs. 151.11) and B (125.25 vs. 86.35), and *E. qinqixianii* (350.93 vs. 202.74).

4. Discussion

4.1. Biodiversity of Marine-Derived Fungi around La Réunion Island

From the sampling in La Réunion Island marine biotopes, 31 different species distributed in 25 genera were identified as pigment producers. The identification of isolates collected in coral reefs and lava flows of La Réunion Island coincides with identifications conducted from various marine environments. Indeed, the majority of the studied fungi, such as those sampled from north of the Indian Ocean, belong to the phylum Ascomycetes. The fungi of the genus *Aspergillus*, particularly *A. sydowii*, are also found in the Caribbean corals (*Porites lobata*), Polynesia, and in sediments off the coast of India [46,47]. *Penicillium citrinum* was isolated from the red algae *Actinotrichia fragilis*, from sponges, and the species was also found on other substrates such as hard substrate or water [48–51]. The genera *Penicillium*, *Cladosporium*, *Chaetomium*, and *Fusarium*, and species *Nigrospora oryzae* and *Hortea werneckii*, have been identified in marine sediments collected at different depths in the central basin of the Indian Ocean and considered to be coral pathogens [8,52]. Fungi, namely *Alternaria* sp., *Acremonium* sp. and *Rhodotorula mucilaginosa*, were isolated from salt lakes in Antarctica, as were *P. chrysogenum* and *P. crustosum* [53]. *Rhodotorula mucilaginosa* was also found in the sediments of central Indian basin [52].

The diversity within the isolated fungal population was crucial while comparing the ability of pigment production [54]. However, in our samples, the highest diversity of pigmented fungi was revealed from the water column (13 species) and from hard substrates (limestone or lava flow) (11 species). If the water column can be suspected of carrying a multitude of fungal propagules originating from terrestrial environments, hard substrates are probably more representative of marine and marine-derived biodiversity.

Our study demonstrates that the living coral *Pocillopora* sp. shelters fungi from the genera *Aspergillus* (*A. creber*, *A. sydowii* and *Eurotium amstellodami*) and *Penicillium* (*P. viticola*), as well as *Hypocrea koningii* and *Acremonium* sp. Widely disseminated on land, this mainly saprophytic genus *Acremonium* sp., has already been isolated from marine environments (sea fans, sea water, sea cucumbers, and intertidal sediment samples) [55–57]. These fungi, were extracted from the inner parts of the coral structure. They are then supposed to be at least endophytic species for this coral genus.

The coral rubbles (dead part of corals) from our samples contained colored *Penicillium* (*P. herquei*) or related species *T. albobiverticillius* as well as an isolate of *Chaetomium globosum*. *Chaetomium globosum* is a common fungal species from soil and environment.

Most of the fungi we identified can also be found on land, in soil, on plants or insects, but some of them have rarely been isolated from marine environments such as *Whalleya microplaca*, *Biscogniauxia* sp., *Paraconiothyrium variabile*, *Myrothecium atroviride*, *Nectria haematococca*, *Peyronellaea glomerata*, *Epicoccum sorghi*, *Sporisorium exsertum* and *Periconia* sp.. From our study, the genera *Aspergillus* and *Penicillium* or the close ones such as *Talaromyces*, *Emericella* or *Eurotium* (from the Trichocomaceae family) are much more diverse than others in these tropical marine biotopes (12 different species), and are represented in several types of samples and locations. These aerobic and xerophilic species are well-known for populating dry and/or salty biotopes. However, their ability to subsist or develop underwater, with widely varying oxygenation conditions is less known. These cosmopolitan fungi are well-known to produce a wide range of secondary metabolites such as polyketide-based pigments in solid and liquid cultures. Overall, in our study, some fungal species (*T. albobiverticillius* or *N. haematococca*) were identified from different types of samples in the same area. Some others (*A. sydowii*, *P. viticola*) appeared in separate locations. These fungi found in several sample types and/or in different locations may be considered as frequent in marine environment around La Réunion Island.

4.2. Qualitative Aspect of the Pigment Production

For marine-derived isolates, two statuses lead to particular behaviors and products: the challenge of facing unusual living conditions (exogenous fungi) and the use of specific procedures naturally adapted to the marine niches (for instance fungal endophytes of marine microorganisms, i.e., indigenous micromycetes, naturally selected by aquatic environments).

Overall, in unusual biotopes (sometimes extreme), the fungal species with pigmented cell walls (in the spores and/or mycelium), are clearly able to tolerate dehydration-hydration cycles or high solar radiations, better than the moniliaceous fungi, whose cells are devoid of pigments. These aromatic compounds, as melanin, sporopollenin (brown product of oxydative polymerization of β-carotene) or cycloleucomelone (terphenylquinone), often show significant antioxidant activities, and are bound to protect the biological structures, giving them an excellent durability and a high potential for survival in hostile environments [58,59].

From the available literature, the microorganisms of the genus *Trichoderma* are frequent in marine environments and some terrestrial strains are able to produce anthraquinone-like compounds [60]. Indeed, isolates of the family Hypocreaceae (*Trichoderma, Hypocrea* and *Acremonium*) are also represented in our study and exhibit orange to purple hues. Some strains of the common soil fungus *Cladosporium cladosporioides*, also isolated from our samples with green shades, have already been studied for their production of intracellular melanin [61].

The most important colored compounds produced by *Aspergillus* and *Emericella* species are respectively, hydroxyanthraquinones and azaphilone pigments, exhibiting a very wide range of hues. Furthermore, *A. sydowii* and *Eurotium amstelodami* isolated from La Réunion Island showed red and yellow colors respectively, as produced by their terrestrial counterparts [62].

Penicillium species and related ones seem to adjust easily to multiple conditions and to be a source of original compounds as they appear among the most chemically inventive fungi. In *Penicillium* and *Talaromyces* species, polyketide-based pigments are also very common, and, particularly, the azaphilones, such as the derivatives of monascorubrin and rubropunctatin [63]. *Monascus*-like azaphilone pigments such as *N*-glutarylmonascorubramine, *N*-glutarylrubropunctamine, monascorubramine homologues PP-V [(10Z)-12-carboxyl-monascorubramine] and PP-R [(10Z)-7-(2-hydroxyethyl)-monascorubramine] are frequently identified in their cultures [64,65]. However the commercial production of red anthraquinoid pigments (Arpink Red™, Natural Red™) has already been carried out with *P. oxalicum var. armeniaca* [1]. The most common hues produced by both genera include yellow, red, orange and reddish-brown. Nevertheless, it was found that the yellow pigments seem predominant in most of the *Penicillium* species, while *Talaromyces* species mainly produce red pigments with few synthesizing yellow compounds of azaphilone series [66]. The colored molecules sometimes demonstrate mycotoxic activities such as rubratoxins A and B, rugulovasins and luteoskyrin [67].

Some strains of the widespread *Acremonium* sp. produce the yellow oosporein (chaetomidin) (biquinone, benzoquinone) and also some toxic compounds as diterpene glycosides [68].

Chaetomium globosum, isolated from the coral rubbles biosynthesizes maroon pigments in the culture conditions of our experiment. Many members of the family produce metabolites with antifungal properties. *C. globosum* is already known to biosynthesize yellow azaphilones named chaetoviridins (A–D), antifungal compounds involved in the induction of chlamydospores-like cells [69]. It also produces nitrogenous azaphilones (4′-epi-*N*-2-hydroxyethyl-azachaetoviridin A, and *N*-2-butyric-azochaetoviridin E) and isochromophilone XIII, with orange to red hues. Some strains generate pigmented chaetoglobins, chaetoglobosins, chaetomugilins, and seco-chaetomugilins, while others can secrete a purple pigment called cochliodinol [70–73].

Associated with lava flows, *Fusarium equiseti* belongs to a group of widespread plant pathogens, but marine-derived *Fusarium* strains are also frequent in mangroves or associated with marine organisms. These are already known to produce original colored anthraquinoid compounds (5-acetyl-2-methoxy-1,4,6-trihydroxy-anthraquinone;6,8-dimethoxy-1-methyl-2-(3-oxobutyl)-anthraquinone and fusaquinones) [19]. Among the *Fusarium* secondary metabolites, numerous polyketide pigments have already been identified, such as naphthoquinone pigments which are the most abundant (bikaverin, nor-bikaverin, javanicin, anhydrojavanicin, fusarubin, anhydrofusarubin, bostrycoidin, and novarubin) and the hydroxyanthraquinones emodin, physcion, dermolutein, chrysophanol, erythroglaucin, dermocybin, dermorubin, tritisporin, cynodontin, helminthosporin or aurofusarin (review in [19,21]). All these molecules develop a palette of colors, ranging from yellow to purple or brown. Some species are also able to produce orange carotenoids (neurosporaxanthin by *F. fujikuroi*) [74]. The putative carcinogen, fusarin C, apicidin F, fujikurins, the perithecal pigments fusarubins as well as the mycelial pigment bikaverin are also produced in the family.

From our work, *Periconia* sp. A isolate produced an impressive violet hue in PDB culture. *Periconia* is a cosmopolitan genus, often found in soil, and decaying herbs and forages. Some *Periconia* strains were nevertheless identified from marine environments (*P. abyssa* (deep sea), *P. byssoides* (sea slug *Aplysia kurodai*)) [75–77]. They attract interest because of the production of promising anti-cancer drugs, such as the carbosugar pericosine A. Some strains may produce an unidentified hepatoxin.

4.3. Quantitative Aspect of Pigment Production

As a promising factor, several of the marine-derived fungi isolated in this study had the ability to grow and biosynthesize pigments in unsalted synthetic conditions (e.g., Czapek Dox medium, PDB). During the period of fermentation, the pigment production started between Day 1 and Day 4 for the majority of isolates such as *Aspergillus, Eurotium, Fusarium, Nigrospora, Pencillium* and *Talaromyces*. For some fungi, the detection of the pigment production was notably delayed (e.g., *Acremonium, Epicoccum*, or *Myrothecium*). This might be due to the low level of pigment producing ability of the fungi or due to unfavorable environmental conditions for pigment production such as pH, temperature, nutrient sources, osmolarity and illumination conditions [78].

Considering the visual observation of pigment color in flasks and the respective UV-visible spectra, fungi belonging to the same species may produce different colored mixtures (e.g., *Aspergillus creber* A and B or *Talaromyces alboverticillius* A–C). They may then belong to different varieties and thus produce pigments of distinct natures. The slight variations observed between intra- and extra-cellular solutions also indicate that the solutions may contain different assortments of colored compounds, in different proportions, resulting in different hues.

From these findings, it is understood that a higher quantity of pigments has been mainly purified from extracellular filtrates in a significant manner (11/21 isolates). In our experimental conditions, the maximum pigment production was obtained in the extracellular samples for *C. globosum* and *Periconia* sp. *B*. On average, the values measured in the cells were significantly lower; indicating that pigments secretion in the liquid medium seems a widespread behavior in the conditions of the experiment. Only the isolate *P. herquei* had a very high level of intracellular pigment biosynthesis (704.55 mg equivalent purpurin L^{-1}). Nevertheless, for high intracellular pigment production from biomass, *A. creber* A and *E. qinqixianii* present a true production potential. On the other hand, the extraction of intracellular colored compounds appeared sometimes not completely effective. The fungal biomass was still colored even after extraction. The efficiency of the extraction process could probably be improved to recover higher pigment quantities from intracellular samples [79].

This work highlights different behaviors of fungal isolates towards the secretion of colored molecules compared to internal storage. Anyway, the production of secondary metabolites often occurs after fungal growth has ceased, as a result of nutrient limitation coupled with excess carbon availability. This makes it possible to manipulate their formation [80,81].

5. Conclusions

Marine and marine-derived fungi are promising resources for the production of new metabolites of interest, and, among them, pigments are attractive [82–84]. The potential of marine-derived microorganisms to produce unique and original molecules may come from specific metabolic or genetic adaptation appearing to meet very specific combinations of physical and chemical parameters (high salinity, low O_2 penetration, low temperature, limited light access and high pressure). Based on this statement, our study explores, for the first time, the biodiversity of fungi from marine environments around La Réunion Island, Indian Ocean, along with the ability of the isolates to produce pigments. The potentiality of these marine derived isolates to secrete pigments or to concentrate colored compounds inside the cells was highlighted. Several isolates collected from lava flows, hard substrates sediments and corals (living or dead) turned out to be the interesting producers of intense colors on PDA culture medium. The main types identified, *Aspergillus*, *Penicillium* and related genera, are also found in other marine regions (such as Polynesia or along the coast of India). However, a great biodiversity (31 species) emphasizes the range of possible hues and molecules susceptible to be isolated. The majority of the isolates, probably marine optional, may also be able to grow in synthetic media, devoid of sea salts and may show the competence of producing pigments in an industrial scale. The most promising pigmented products, probably of intense red or purple hues, which seem to consist in mixtures, will be subjected to purification and further analyses by analytical techniques such as liquid chromatography–mass spectrometry/time-of-flight (LC-MS/TOF) and Nuclear Magnetic Resonance (NMR). The interesting isolates will also be subjected to further analyses to determine their ability as antibiotics or for enzyme production.

Acknowledgments: The authors are grateful to Regional Council of La Réunion Island for financial support. Thanks are also given to BIOLAVE program and Quod Jean Pascal.

Author Contributions: Mireille Fouillaud conceived, designed and performed the experiments. Mireille Fouilaud and Pascale Cuet collected the samples. Melissa Llorente, Hélène Magalon and Mekala Venkatachalam performed the molecular analysis and analyzed the genetic data. Mireille Fouillaud, Mekala Venkatachalam and Laurent Dufossé contributed to write the paper. The authors would also like to thank Cathie Milhau from ESIROI and Patricia Clerc from LCSNSA, of Université de La Réunion, for their logistic and technical help; and Gary Mares for his timely help on data analysis.

References

1. Dufossé, L.; Galaup, P.; Yaron, A.; Arad, S.M.; Blanc, P.; Chidambara Murthy, K.N.; Ravishankar, G.A. Microorganisms and microalgae as sources of pigments for food use: A scientific oddity or an industrial reality? *Trends Food Sci. Technol.* **2005**, *16*, 389–406. [CrossRef]
2. Mayer, A.M.; Rodriguez, A.D.; Taglialatela-Scafati, O.; Fusetani, N. Marine pharmacology in 2009–2011: Marine compounds with antibacterial, antidiabetic, antifungal, anti-inflammatory, antiprotozoal, antituberculosis, and antiviral activities; affecting the immune and nervous systems, and other miscellaneous mechanisms of action. *Mar. Drugs* **2013**, *11*, 2510–2573. [PubMed]
3. Mapari, S.A.S.; Nielsen, K.F.; Larsen, T.O.; Frisvad, J.C.; Meyer, A.S.; Thrane, U. Exploring fungal biodiversity for the production of water-soluble pigments as potential natural food colorants. *Curr. Opin. Biotechnol.* **2005**, *16*, 231–238. [CrossRef] [PubMed]
4. Hohmann, C.; Schneider, K.; Bruntner, C.; Irran, E.; Nicholson, G.; Bull, A.T.; Jones, A.L.; Brown, R.; Stach, J.E.; Goodfellow, M.; et al. Caboxamycin, a new antibiotic of the benzoxazole family produced by the deep-sea strain *Streptomyces* sp. Ntk 937. *J. Antibiot.* **2009**, *62*, 99–104. [CrossRef] [PubMed]
5. Costantino, V.; Fattorusso, E.; Mangoni, A.; Perinu, C.; Cirino, G.; De Gruttola, L.; Roviezzo, F. Tedanol: A potent anti-inflammatory ent-pimarane diterpene from the caribbean sponge *Tedania ignis*. *Bioorg. Med. Chem.* **2009**, *17*, 7542–7547. [CrossRef] [PubMed]

6. Bonugli-Santos, R.C.; dos Santos Vasconcelos, M.R.; Passarini, M.R.Z.; Vieira, G.A.L.; Lopes, V.C.P.; Mainardi, P.H.; dos Santos, J.A.; de Azevedo Duarte, L.; Otero, I.V.R.; da Silva Yoshida, A.M.; et al. Marine-derived fungi: Diversity of enzymes and biotechnological applications. *Front. Microbiol.* **2015**, *6*, 269. [CrossRef] [PubMed]
7. Panno, L.; Bruno, M.; Voyron, S.; Anastasi, A.; Gnavi, G.; Miserere, L.; Varese, G.C. Diversity, ecological role and potential biotechnological applications of marine fungi associated to the seagrass *Posidonia oceanica*. *New Biotechnol.* **2013**, *30*, 685–694. [CrossRef] [PubMed]
8. Cathrine, S.J.; Raghukumar, C. Anaerobic denitrification in fungi from the coastal marine sediments off Goa, India. *Mycol. Res.* **2009**, *113*, 100–109. [CrossRef] [PubMed]
9. Holguin, G.; Vazquez, P.; Bashan, Y. The role of sediment microorganisms in the productivity, conservation, and rehabilitation of mangrove ecosystems: An overview. *Biol. Fertil. Soils* **2001**, *33*, 265–278. [CrossRef]
10. Bugni, T.S.; Ireland, C.M. Marine-derived fungi: A chemically and biologically diverse group of microorganisms. *Nat. Prod. Rep.* **2004**, *21*, 143–163. [CrossRef] [PubMed]
11. Jones, E.B.G.; Pang, K.L. *Marine Fungi: And Fungal-Like Organisms*; De Gruyter: Berlin, Germany, 2012.
12. Kathiresan, K.; Bingham, B.L. Biology of mangroves and mangrove ecosystems. In *Advances in Marine Biology*; Southward, A., Young, C., Fuiman, L., Tyler, P., Eds.; Academic Press: San Diego, CA, USA, 2001; Volume 40, pp. 81–251.
13. Kohlmeyer, J.; Kohlmeyer, E. *Marine Mycology*; Elsvier Inc.: London, UK, 1979; p. 704.
14. Jones, E.B.G. Marine fungi: Some factors affecting biodiversity. *Fungal Divers.* **2000**, *4*, 53–73. [CrossRef]
15. Saleem, M.; Nazir, M. Bioactive natural products from marine-derived fungi: An update. In *Studies in Natural Products Chemistry*; Atta-ur-Rahman, Ed.; Elsevier: Amsterdam, The Netherlands, 2015; Volume 45, pp. 297–361.
16. Imhoff, J.F. Natural products from marine fungi—Still an underrepresented resource. *Mar. Drugs* **2016**, *14*, 19. [CrossRef] [PubMed]
17. Debashish, G.; Malay, S.; Barindra, S.; Joydeep, M. Marine enzymes. *Adv. Biochem. Eng./Biotechnol.* **2005**, *96*, 189–218.
18. Zhang, C.; Kim, S.K. Application of marine microbial enzymes in the food and pharmaceutical industries. *Adv. Food Nutr. Res.* **2012**, *65*, 423–435. [PubMed]
19. Fouillaud, M.; Venkatachalam, M.; Girard-Valenciennes, E.; Caro, Y.; Dufossé, L. Anthraquinones and derivatives from marine-derived fungi: Structural diversity and selected biological activities. *Mar. Drugs* **2016**, *14*, 64. [CrossRef] [PubMed]
20. Fouillaud, M.; Venkatachalam, M.; Girard-Valenciennes, E.; Caro, Y.; Dufossé, L. Marine-derived fungi producing red anthraquinones: New resources for natural colors? In Proceedings of the 8th International Conference of Pigments in Food, "Coloured Foods for Health Benefits", Cluj-Napoca, Romania, 28 June—1 July 2016.
21. Caro, Y.; Venkatachalam, M.; Lebeau, J.; Fouillaud, M.; Dufossé, L. Pigments and colorants from filamentous fungi. In *Fungal Metabolites*; Merillon, J.-M., Ramawat, G.K., Eds.; Springer: Cham, Switzerland, 2016; pp. 1–70.
22. Ebel, R. Natural product diversity from marine fungi. In *Comprehensive Natural Products II: Chemistry and Biology*; Mander, L., Liu, H.-W., Eds.; Elsevier: Oxford, UK, 2010; Volume 2, pp. 223–262.
23. Calvo, A.M.; Wilson, R.A.; Bok, J.W.; Keller, N.P. Relationship between secondary metabolism and fungal development. *Microbiol. Mol. Biol. Rev.* **2002**, *66*, 447–459. [CrossRef] [PubMed]
24. Margalith, P. *Pigment Microbiology*; Springer: London, UK; New York, NY, USA, 1992; p. 156.
25. Demain, A.L.; Fang, A. The natural functions of secondary metabolites. In *History of Modern Biotechnology I*; Fiechter, A., Ed.; Springer: Berlin/Heidelberg, Germany, 2000; pp. 1–39.
26. Réunion's Coral Reef. Available online: https://en.wikipedia.org/wiki/R%C3%A9union%27s_coral_reef (accessed on 1 May 2017).
27. Peyrot-Clausade, M.; Chazottes, V.; Pari, N.; Peyrot-Clausade, M.; Chazottes, V.; Pari, N. Bioerosion in the carbonate budget of two indo-pacific reefs: La Réunion (Indian Ocean) and moorea (Pacific Ocean). *Bull. Geol. Soc. Denmark* **1999**, *1999*, 1–30.

28. Conand, C.; Chabanet, P.; Cuet, P.; Letourneur, Y. The carbonate budget of a fringing reef in La Reunion Island (Indian Ocean): Sea urchin and fish bioerosion and net calcification. In Proceedings of the 8th International Coral Reef Symposium, Panama City, Panama, 24–29 June 1997; Lessios, H.A., Macintyre, I.G., Eds.; pp. 953–958.
29. Naim, O.; Cuet, P.; Mangar, V. Coral reefs of the mascarene archipelago. In *Coral Reefs of the Indian Ocean: Their Ecology and Conservation*; McClanahan, T.R., Sheppard, C., Obura, D.O., Eds.; Oxford University Press: New York, NY, USA, 2000; pp. 353–381.
30. Turner, J.; Klaus, R. Coral reefs of the mascarenes, western indian ocean. *Philos. Trans. R. Soc. Lond. A Math. Phys. Eng. Sci.* **2005**, *363*, 229–250. [CrossRef] [PubMed]
31. Montaggioni, L.; Faure, G. *Les Récifs Coralliens des Mascareignes (Océan Indien)*; Université Francaise de l'Océan Indien, Centre Universitaire de La Réunion: Réunion, France, 1980; p. 151.
32. Sanders, E.R. Aseptic laboratory techniques: Plating methods. *J. Vis. Exp. JoVE* **2011**, *63*, e3064. [CrossRef] [PubMed]
33. Jong, S.; Dugan, F.; Edwards, M. *ATCC Filamentous Fungi*, 19th ed.; Rockville, MD American Type Culture Collection: Manassas, VA, USA, 1996.
34. Dahmen, H.; Staub, T.; Schwinn, F. Technique for long-term preservation of phytopathogenic fungi in liquid nitrogen. *Phytopathology* **1983**, *73*, 241–246. [CrossRef]
35. Knebelsberger, T.; Stoger, I. DNA extraction, preservation, and amplification. *Methods Mol. Biol.* **2012**, *858*, 311–338. [PubMed]
36. Toju, H.; Tanabe, A.S.; Yamamoto, S.; Sato, H. High-coverage its primers for the DNA-based identification of ascomycetes and basidiomycetes in environmental samples. *PLoS ONE* **2012**, *7*, e40863. [CrossRef] [PubMed]
37. Samson, R.A.; Visagie, C.M.; Houbraken, J.; Hong, S.B.; Hubka, V.; Klaassen, C.H.W.; Perrone, G.; Seifert, K.A.; Susca, A.; Tanney, J.B.; et al. Phylogeny, identification and nomenclature of the genus aspergillus. *Stud. Mycol.* **2014**, *78*, 141–173. [CrossRef] [PubMed]
38. Samson, R.A.; Yilmaz, N.; Houbraken, J.; Spierenburg, H.; Seifert, K.A.; Peterson, S.W.; Varga, J.; Frisvad, J.C. Phylogeny and nomenclature of the genus talaromyces and taxa accommodated in penicillium subgenus biverticillium. *Stud. Mycol.* **2011**, *70*, 159–183. [CrossRef] [PubMed]
39. White, T.J.; Bruns, T.; Lee, S.; Taylor, J. Amplification and direct sequencing of fungal ribosomal RNA genes for phylogenetics. *PCR Protoc. Guide Methods Appl.* **1990**, *18*, 315–322.
40. Romanelli, A.M.; Sutton, D.A.; Thompson, E.H.; Rinaldi, M.G.; Wickes, B.L. Sequence-based identification of filamentous basidiomycetous fungi from clinical specimens: A cautionary note. *J. Clin. Microbiol.* **2010**, *48*, 741–752. [CrossRef] [PubMed]
41. Sutton, D.A.; Marín, Y.; Thompson, E.H.; Wickes, B.L.; Fu, J.; García, D.; Swinford, A.; de Maar, T.; Guarro, J. Isolation and characterization of a new fungal genus and species, aphanoascella galapagosensis, from carapace keratitis of a galapagos tortoise (chelonoidis nigra microphyes). *Med. Mycol.* **2013**, *51*, 113–120. [CrossRef] [PubMed]
42. Zhou, H.; Li, Y.; Tang, Y. Cyclization of aromatic polyketides from bacteria and fungi. *Nat. Prod. Rep.* **2010**, *27*, 839–868. [CrossRef] [PubMed]
43. Caro, Y.; Anamale, L.; Fouillaud, M.; Laurent, P.; Petit, T.; Dufosse, L. Natural hydroxyanthraquinoid pigments as potent food grade colorants: An overview. *Nat. Prod. Bioprospect.* **2012**, *2*, 174–193. [CrossRef]
44. Machatová, Z.; Barbieriková, Z.; Poliak, P.; Jančovičová, V.; Lukeš, V.; Brezová, V. Study of natural anthraquinone colorants by epr and uv/vis spectroscopy. *Dyes Pigments* **2016**, *132*, 79–93. [CrossRef]
45. Geyer, C.J. Fuzzy *p*-Values and Ties in Nonparametric Tests. Avaliable online: http://www.stat.umn.edu/geyer/fuzz (accessed on 30 June 2017).
46. Golubic, S.; Radtke, G.; Le Campion-Alsumard, T. Endolithic fungi in marine ecosystems. *Trends Microbiol.* **2005**, *13*, 229–235. [CrossRef] [PubMed]
47. Priess, K.; Le Campion-Alsumard, T.; Golubic, S.; Gadel, F.; Thomassin, B. Fungi in corals: Black bands and density-banding of *Porites lutea* and *P. lobata* skeleton. *Mar. Biol.* **2000**, *136*, 19–27. [CrossRef]
48. Nicoletti, R.; Trincone, A. Bioactive compounds produced by strains of penicillium and *Talaromyces* of marine origin. *Mar. Drugs* **2016**, *14*, 37. [CrossRef] [PubMed]
49. Tsuda, M.; Kasai, Y.; Komatsu, K.; Sone, T.; Tanaka, M.; Mikami, Y.; Kobayashi, J.i. Citrinadin a, a novel pentacyclic alkaloid from marine-derived fungus *Penicillium citrinum*. *Org. Lett.* **2004**, *6*, 3087–3089. [CrossRef] [PubMed]

50. Malmstrøm, J.; Christophersen, C.; Frisvad, J.C. Secondary metabolites characteristic of *Penicillium citrinum*, *Penicillium steckii* and related species. *Phytochemistry* **2000**, *54*, 301–309. [CrossRef]
51. Endo, A.; Kuroda, M.; Tsujita, Y. Ml-236a, ml-236b, and ml-236c, new inhibitors of cholesterogensis produced by *Penicillium citrinum*. *J. Antibiot.* **1976**, *29*, 1346–1348. [CrossRef] [PubMed]
52. Singh, P.; Raghukumar, C.; Verma, P.; Shouche, Y. Assessment of fungal diversity in deep-sea sediments by multiple primer approach. *World J. Microbiol. Biotechnol.* **2012**, *28*, 659–667. [CrossRef] [PubMed]
53. Brunati, M.; Rojas, J.L.; Sponga, F.; Ciciliato, I.; Losi, D.; Gottlich, E.; de Hoog, S.; Genilloud, O.; Marinelli, F. Diversity and pharmaceutical screening of fungi from benthic mats of antarctic lakes. *Mar. Genom.* **2009**, *2*, 43–50. [CrossRef] [PubMed]
54. Yahr, R.; Schoch, C.L.; Dentinger, B.T. Scaling up discovery of hidden diversity in fungi: Impacts of barcoding approaches. *Philos. Trans. R. Soc. B* **2016**, *371*, 20150336. [CrossRef] [PubMed]
55. An, X.; Feng, B.-M.; Chen, G.; Chen, S.-F.; Wang, H.-F.; Pei, Y.-H. Isolation and identification of two new compounds from marine-derived fungus *Acremonium fusidioides* rz01. *Chin. J. Nat. Med.* **2016**, *14*, 934–938. [CrossRef]
56. Afiyatullov, S.S.; Kalinovsky, A.I.; Antonov, A.S.; Zhuravleva, O.I.; Khudyakova, Y.V.; Aminin, D.L.; Yurchenko, A.N.; Pivkin, M.V. Isolation and structures of virescenosides from the marine-derived fungus *Acremonium striatisporum*. *Phytochem. Lett.* **2016**, *15*, 66–71. [CrossRef]
57. Gallardo, G.L.; Butler, M.; Gallo, M.L.; Rodríguez, M.A.; Eberlin, M.N.; Cabrera, G.M. Antimicrobial metabolites produced by an intertidal acremonium furcatum. *Phytochemistry* **2006**, *67*, 2403–2410. [CrossRef] [PubMed]
58. Hiort, J.; Maksimenka, K.; Reichert, M.; Perovic-Ottstadt, S.; Lin, W.H.; Wray, V.; Steube, K.; Schaumann, K.; Weber, H.; Proksch, P.; et al. New natural products from the sponge-derived fungus *Aspergillus niger*. *J. Nat. Prod.* **2004**, *67*, 1532–1543. [CrossRef] [PubMed]
59. Pagano, M.C.; Rosa, L.H. Fungal molecular taxonomy. In *Fungal Biomolecules*; John Wiley & Sons, Ltd.: Chichester, UK, 2015; pp. 311–321.
60. Slater, G.; Haskins, R.; Hogge, L.; Nesbitt, L. Metabolic products from a *Trichoderma viride* pers. Ex fries. *Can. J. Chem.* **1967**, *45*, 92–96. [CrossRef]
61. Duran, N.; Teixeira, M.F.; De Conti, R.; Esposito, E. Ecological-friendly pigments from fungi. *Crit. Rev. Food Sci. Nutr.* **2002**, *42*, 53–66. [CrossRef] [PubMed]
62. Butinar, L.; Frisvad, J.C.; Gunde-Cimerman, N. Hypersaline waters–a potential source of foodborne toxigenic *Aspergilli* and *Penicillia*. *FEMS Microbiol. Ecol.* **2011**, *77*, 186–199. [CrossRef] [PubMed]
63. Woo, P.C.; Lam, C.W.; Tam, E.W.; Lee, K.C.; Yung, K.K.; Leung, C.K.; Sze, K.H.; Lau, S.K.; Yuen, K.Y. The biosynthetic pathway for a thousand-year-old natural food colorant and citrinin in *Penicillium marneffei*. *Sci. Rep.* **2014**, *4*, 6728. [CrossRef] [PubMed]
64. Arai, T.; Koganei, K.; Umemura, S.; Kojima, R.; Kato, J.; Kasumi, T.; Ogihara, J. Importance of the ammonia assimilation by *Penicillium purpurogenum* in amino derivative monascus pigment, PP-V, production. *AMB Express* **2013**, *3*, 19. [CrossRef] [PubMed]
65. Ogihara, J.; Kato, J.; Oishi, K.; Fujimoto, Y. Pp-r, 7-(2-hydroxyethyl)-monascorubramine, a red pigment produced in the mycelia of *Penicillium* sp. AZ. *J. Biosci. Bioeng.* **2001**, *91*, 44–47. [CrossRef]
66. Frisvad, J.C.; Yilmaz, N.; Thrane, U.; Rasmussen, K.B.; Houbraken, J.; Samson, R.A. *Talaromyces atroroseus*, a new species efficiently producing industrially relevant red pigments. *PLoS ONE* **2013**, *8*, e84102. [CrossRef] [PubMed]
67. Yilmaz, N.; Houbraken, J.; Hoekstra, E.S.; Frisvad, J.C.; Visagie, C.M.; Samson, R.A. Delimitation and characterisation of *Talaromyces purpurogenus* and related species. *Persoonia* **2012**, *29*, 39–54. [CrossRef] [PubMed]
68. Thomson, R.H. *Naturally Occurring Quinones IV: Recent Advances*; Blackie Academic & Professional: London, UK; New York, NY, USA, 1997.
69. Takahashi, M.; Koyama, K.; Natori, S. Four new azaphilones from *Chaetomium globosum* var. Flavo-viridae. *Chem. Pharm. Bull.* **1990**, *38*, 625–628. [CrossRef]
70. McMullin, D.R. Structural Characterization of Secondary Metabolites Produced by Fungi Obtained from Damp Canadian Building. Ph.D. Thesis, Ottawa-Carleton University, Ottawa, ON, Canada, 2008.

71. McMullin, D.R.; Sumarah, M.W.; Miller, J.D. Chaetoglobosins and azaphilones produced by Canadian strains of *Chaetomium globosum* isolated from the indoor environment. *Mycotoxin Res.* **2013**, *29*, 47–54. [CrossRef] [PubMed]
72. Ming Ge, H.; Yun Zhang, W.; Ding, G.; Saparpakorn, P.; Chun Song, Y.; Hannongbua, S.; Xiang Tan, R. Chaetoglobins A and B, two unusual alkaloids from endophytic *Chaetomium globosum* culture. *Chem. Commun.* **2008**, 5978–5980. [CrossRef] [PubMed]
73. Brewer, D.; Jerram, W.A.; Taylor, A. The production of cochliodinol and a related metabolite by chaetomium species. *Can. J. Microbiol.* **1968**, *14*, 861–866. [CrossRef] [PubMed]
74. Prado-Cabrero, A.; Schaub, P.; Diaz-Sanchez, V.; Estrada, A.F.; Al-Babili, S.; Avalos, J. Deviation of the neurosporaxanthin pathway towards beta-carotene biosynthesis in *Fusarium fujikuroi* by a point mutation in the phytoene desaturase gene. *FEBS J.* **2009**, *276*, 4582–4597. [CrossRef] [PubMed]
75. Kohlmeyer, J. New genera and species of higher fungi from the deep sea (1615–5315 m). *Revue de Mycologie* **1977**, *41*, 189–206.
76. Usami, Y.; Ichikawa, H.; Arimoto, M. Synthetic efforts for stereo structure determination of cytotoxic marine natural product pericosines as metabolites of *Periconia* sp. From sea hare. *Int. J. Mol. Sci.* **2008**, *9*, 401–421. [CrossRef] [PubMed]
77. Dighton, J.; White, J.F. *The Fungal Community: Its Organization and Role in the Ecosystem*, 3rd ed.; CRC Press: Boca Raton, FL, USA, 2005.
78. Ogbonna, C.N. Production of food colourants by filamentous fungi. *Afr. J. Microbiol. Res.* **2016**, *10*, 960–971.
79. Kaufmann, B.; Christen, P. Recent extraction techniques for natural products: Microwave-assisted extraction and pressurised solvent extraction. *Phytochem. Anal. PCA* **2002**, *13*, 105–113. [CrossRef] [PubMed]
80. Debbab, A.; Aly, A.H.; Proksch, P. Bioactive secondary metabolites from endophytes and associated marine derived fungi. *Fungal Divers.* **2011**, *49*, 1–12. [CrossRef]
81. Gunatilaka, A.L.; Wijeratne, E.K. *Natural Products from Bacteria and Fungi*; Elsevier: Amsterdam, The Netherlands, 2000.
82. Kim, S.-K. *Marine Microbiology: Bioactive Compounds and Biotechnological Applications*; Wiley & Sons: Weinheim, Germany, 2013; p. 550.
83. Kjer, J.; Debbab, A.; Aly, A.H.; Proksch, P. Methods for isolation of marine-derived endophytic fungi and their bioactive secondary products. *Nat. Protoc.* **2010**, *5*, 479–490. [CrossRef] [PubMed]
84. Rateb, M.E.; Ebel, R. Secondary metabolites of fungi from marine habitats. *Nat. Prod. Rep.* **2011**, *28*, 290–344. [CrossRef] [PubMed]

8

Does Arbuscular Mycorrhiza Determine Soil Microbial Functionality in Nutrient-Limited Mediterranean Arid Ecosystems?

Neji Mahmoudi [1], Teresa Dias [2], Mosbah Mahdhi [3], Cristina Cruz [2], Mohamed Mars [1] and Maria F. Caeiro [4,*]

1 Unité de Recherche Biodiversité et Valorisation des Bio-ressources en Zones Arides (BVBZA), Faculté des Sciences de Gabès, Cité Erriadh Zrig 6072, Tunisie; neji.2012@yahoo.com (N.M.); Mohamed.Mars@fsg.rnu.tn (M.M.)

2 Centre for Ecology, Evolution and Environmental Changes (cE3c), Faculdade de Ciências da Universidade de Lisboa, Edifício C2, Piso 5, Sala 2.5.03 Campo Grande, 1749-016 Lisboa, Portugal; mtdias@fc.ul.pt (T.D.); cmhoughton@fc.ul.pt (C.C.)

3 Center for Environmental Research and Studies, Jazan University, Jazan-P.O. Box 114, Jazan 82817, Saudi Arabia; mmahdhi@jazanu.edu.sa

4 Centro de Estudos do Ambiente e do Mar (CESAM), Faculdade de Ciências da Universidade de Lisboa, Edifício C2, Piso 2, Campo Grande, 1749-016 Lisboa, Portugal

* Correspondence: mfcaeiro@fc.ul.pt

Abstract: Arbuscular mycorrhizal fungi (AMF) are determinant for the performance of plant communities and for the functionality of terrestrial ecosystems. In natural ecosystems, grazing can have a major impact on mycorrhizal fungi and consequently on plant growth. The objective of this study was to evaluate the statements referred above in Mediterranean arid areas in Tunisia. Root samples and rhizosphere soils of five dominant herbaceous plants were studied at six distinct arid sites differing on soil proprieties and grazing intensity. At each site, chemical and dynamic properties of the soil were characterized as well as the AMF colonization intensity and the soil functionality. Results showed that the mycorrhizal frequency and intensity and spore density, varied between plants in the same site and, for each plant, between sites and evidenced a positive effect of mycorrhized plants on soil microbial activity. Grazing and soil properties strongly affected AMF composition and the soil microbial and biochemical dynamics, which presented the lowest values at the sites with the highest grazing intensities. In conclusion, these results demonstrate that AMF improve soil biological properties, supporting the hypothesis that mycorrhiza and grazing compete for plant photosynthates, and highlight the importance of mycorrhizal symbiosis towards soil functionality under arid conditions.

Keywords: arbuscular mycorrhizal fungi; arid areas; biological properties; conserved areas; grazing; mycorrhiza

---

1. Introduction

Arid and semi-arid regions of the world are considered as being particularly vulnerable to climate change [1]. They are already climatically stressed with high temperatures, low rainfall and long dry seasons. These ecosystems are highly dynamic, with bursts of productivity in the wet season of some years, and very low productivity in dry years. The arid and semi-arid regions of the world have been subjected to accelerated desertification due to increasing grazing intensity, decreased rainfall, higher temperatures and prolonged periods of drought [2,3]. These pressures, associated with increased anthropogenic impacts and climate changes, caused the decline of forests, regression and extinction of many pastoral and forage species, and accelerated soil degradation and change of soil microbial communities [1].

In most cases, the degradation process starts with the disruption of functional networks that protect and alleviate stress, conferring to the organisms involved protection to global changes [4], including land use changes. Plants are well known for their symbioses with nitrogen fixing bacteria and arbuscular mycorrhiza fungi (AMF). These are just two examples of a vast range of symbiotic relationships that consist the plant microbiome and modulate plant phenotype and, thus, plant fitness. In this context, the microbial community of the rhizosphere [5] is of great importance to plant performance, playing a crucial role in ecosystem functioning. The microbial activities of the rhizosphere also determine the bioavailability of nutrients and, therefore, soil fertility. Plants interact with guilds of these beneficial microorganisms living in their rhizosphere, promoting their growth and development [6].

Arbuscular mycorrhiza involves reciprocal complementary resource exchanges between plant roots and soil fungi and is widespread in natural ecosystems [7]. When colonizing the roots, AMF develop intra- and extra-radical mycelium. The hyphae of the intra-radical mycelium colonize the cells of the root cortex and penetrate the periplasmic space where they develop vesicles and arbuscules that are the structures responsible for most of the exchanges between the fungi and the plant [8,9]. The extra-radical mycelium spreads its hyphae beyond the root surface and colonizes the surrounding substratum, increasing the volume of soil explored, and creating a privileged space for microbial development. As part of the interaction, the host plant provides the fungus with carbon in exchange for nutrients and water is taken up by the fungus [8,10]. Thus, mycorrhiza plays a crucial role in terrestrial ecosystem functioning, especially in arid or semi-arid areas, where root exudates are the major carbon source supporting soil microbial activities [11]. Apart from the nutritional benefits, mycorrhization tend to increase plant tolerance to other stress (biotic and abiotic) conditions [12].

Despite the major ecological importance of AMF, there is a lack of knowledge on the effect of AMF on soil microbial properties and biological activities, or the effect of ecosystem management (including grazing) on mycorrhization. The complex herbivore–plant–AMF is dependent on the species involved in these interactions, the intensity and frequency of grazing, vegetation, and topography [13,14]. Some studies indicate that grazing influences soil biochemical activity, usually through the degradation of soil structure by trampling [15] and the imbalance of soil chemical, microbial and biochemical properties [16].

There is limited information about the effects of grazing on AMF communities and on soil microbial composition and function, especially for low fertility soils, as found in most Mediterranean arid and semi-arid regions. This work addressed both questions in three arid and semi-arid regions of Tunisia, by evaluation of grazing intensity on AMF–plant interactions, and the consequent impact on soil microbial communities.

2. Materials and Methods

2.1. Study Site

This study was developed at three conserved natural areas (Figure 1) in the arid and semi-arid ecosystem of Tunisia, under Mediterranean climate: Bou-Hedma National Park (Sidi Bouzid coordinates, 34°39′ N, 94°8′ E), Zarat-Gabes protected area (Gabes coordinates, 33°41′ N, 10°23′ E) and Oued Dkouk Natural Reserve (Tataouine coordinates, 32°37′ N, 10°18′ E). The annual temperature range is very high in these areas, with minimal and maximal monthly temperature means of 17 °C (January) and 36 °C (August), respectively. The mean annual rainfall in these ecosystems varies between 100 and 260 mm.

Figure 1. Location of the three conserved natural areas in the arid Mediterranean ecosystem of Tunisia.

Experiments were carried out in three natural protected areas of Tunisia (Bou-Hedma, Zarat, and Oued Dkouk) (Table 1). At each natural park samples were taken in two sites, one inside and the other outside the protected area. The studied sites differed in grazing intensity and soil type (Table 1). The three sites inside the conserved areas were subjected to a light grazing, while the other three sites outside were subjected to more intensive grazing by domestic herds of sheep, goats and camels.

Table 1. Details of the six sampling sites at the three conserved natural areas.

Conserved Areas		Site	Vegetation Type	Grazing Intensity
Bou-Hedma	inside	1	*Acacia* spp. and spontaneous *herbaceous plants*	Light
	outside	2	Cultivated and spontaneous herbaceous plants	Intensive
Zarat	inside	3	*Acacia* spp. and spontaneous herbaceous plants	Light
	outside	4	Cultivated and spontaneous herbaceous plants	Intensive
Oued Dkouk	inside	5	Spontaneous herbaceous plants	Light
	outside	6	Spontaneous herbaceous plants	Intensive

2.2. Roots and Soil Sampling

Five plant species (common to all study sites) and three plants per species per site were analyzed for AMF colonization. The herbaceous plants collected were: *Lotus creticus* (Fabaceae), *Medicago truncatula* (Fabaceae), *Astragalus corrugatus* (Fabaceae), *Malva aegyptiaca* (Malvaceae), *Diplotaxis simplex* (Cruciferacea).

Plant roots were carefully collected in order to access the fine active roots where mycorrhiza colonization occurs. Simultaneously, rhizosphere soil of each plant was also collected. For each site, a pooled soil sample (composed of 5 soil cores with 10 cm diameter and 20 cm length) was collected in an area without vegetation and used as a control (bulk soil) for the influence of the plant in the dynamic soil characteristics. Soils were sieved (2 mm) to remove the remains of plants, gravel and earthworms, and stored at 4 °C for further analysis. Soil samples were analyzed in triplicates.

2.3. AMF Colonization Status

According to the methods of Phillips and Hayman [17], AMF colonization was evaluated after observation of 30 root fragments per plant. Of each plant species, root segments (1–2 cm length) were submerged in 10% potassium hydroxide (KOH) at 90 °C for 45 min. After bleaching and acidification steps with hydrogen peroxide (H_2O_2) and hydrogen chloride (HCl), respectively, root pieces were colored with 1% Trypan Blue solution. The duration of each step varied among plant species, according to the respective root diameter and surface root characteristics. For each plant species from each site, 30 root fragments were placed on slides and preserved with lactoglycerol. In all the stained roots, the presence of hyphae, vesicles, and arbuscules inside the root were viewed through a microscope (Nikon, Tokyo, Japan) at 400× magnification. The frequency (F%): number of colonized roots/total number of observed roots) and intensity of mycorrhization (M%): proportion, in percentage, of the root colonized by AMF) were calculated using the Mycocalc program (http://www2.dijon.inra.fr/mychintec/Mycocalcprg/download.html)

2.4. *Quantification of AMF Spore Density*

Using the wet sieving method from Gerdemann and Nicolson [18], AMF spores occurring in soil samples were extracted and quantified. Three nested sieves of 1000, 100, and 32 μm were used. For each soil sample, quantities of 100 g were submerged in 1 L of tap water. After a stirring step, the supernatant was sieved through the nested sieves. The spores that hold on to the two sieves of 100 and 32 μm were recovered in 5 mL centrifuge tubes. After a step of centrifugation on a viscosity gradient (sucrose solution at 60%), the supernatant retained was rinsed with distilled water to remove the sucrose solution. Retrieved AMF spores of each soil sample were counted under a stereomicroscope (40× magnification) and average numbers were calculated per 100 g of dry soil.

2.5. *Soil Analysis: Physical and Chemical Properties*

Soil pH and electrical conductivity were determined in a 1:1 (*v/w*) water: bulk soil suspension using a pH meter (Matest, Treviolo, Italy) and a conductivity meter (Bibby Scientific, Bibby Scientific, Staffordshire, UK), respectively [19]. Soil texture, which represents the granulometric distribution of its constituents (the proportion between small particles: clay, silt and sand), was calculated using the Robinson's pipette method [20]. The other main physicochemical soil characteristics (organic matter, total nitrogen, total phosphorus and water content) were determined by conventional analyses performed by the Soil Analysis Laboratory in the Regional Commissariat for Agricultural Development, in Gabes. The determination of the organic matter was carried out indirectly, starting from the determination of the organic carbon content of soil. The determination of total phosphorus was evaluated by a degradation acid reaction step followed by a dosing step carried out in an automated spectrophotometer (Shimadzu, Kyoto, Japan). The total nitrogen was determined following the Kjeldahl method.

2.6. *Microbiological and Biochemical Properties*

For the impact of AMF on microbiological parameters, the carbon of the microbial biomass (Cmic) present in the plant rhizosphere and bulk soil were evaluated following the "fumigation–extraction" technique [21]. This method is based on three essential steps: a fumigation step with chloroform, incubation in a 10-days fumigation period, and an extraction step with ninhydrin-N reactive and potassium chloride (KCl). This technique has been used to provide rapid and accurate measurements of soil biomass -C and -N.

To evaluate the effect of AMF on biochemical properties, alkaline phosphatase and β-glucosidase activities were calculated and evaluated according to the method of Caravaca et al. [22] in a spectrophotometer (Shimadzu, Japan) at 398 nm. The dehydrogenase activity was determined as described by Garcia et al. [23] in a spectrophotometer (Shimadzu, Japan) at 490 nm.

2.7. *Statistical Analyses*

Analyses of variance (ANOVA) for repeated measures using the XLSTAT (v2010.5.04) software (Addinsoft, New York, NY, USA) were ascertained to test the effect and the significant difference between the studied parameters. Least significant difference values at the 5% levels of significance ($p \leq 0.05$) were calculated to assess differences between different values.

To evaluate the effects of the grazing parameter (explanatory variable) on mycorrhizal properties (F%, M% and number of spores) (response variables), ANOVA and Canonical Correlation Analysis (CCorA) were applied.

To evaluate the relationships between soil physical and chemical parameters and mycorrhizal properties (F%, M% and number of spores), a principal-component analysis (PCA) was applied.

Pearson's correlation was used to determine relationships between variables: mycorrhizal properties and microbiological and biochemical parameters.

To model the relationships between mycorrhizal properties and parameters of soil microbial and biochemical activities by linear regression, XLSTAT (v2010.5.04) software was used.

3. Results

3.1. *Physical and Chemical Properties of Soils*

The more resilient soil characteristics such as soil texture (Table 2) and pH did not show big discrepancies among sites inside and outside the protected areas. However, differences were obtained for the more responsive soil dynamic characteristics: electrical conductivity (E.c), total nitrogen (T.N), total phosphorus (T.P), organic matter (Org. Mat), and water content (Wat. Con). In general, the studied sites had an alkaline pH ranging from 8.0 to 8.4. The highest percentages of soil organic matter were observed in Site 1 (2.6%) as well as inside the other protected areas. Identical patterns were observed for the total nitrogen, the highest level (194 ppm) also being observed in Site 1, followed by Sites 3 and 5 (inside the protected areas). For the total phosphorus content, it was the contrary: the highest values were registered in the sites outside the protected areas (15.3 ppm was the highest value, registered in Site 6). Water content (varying between 1.3 to 3.3%) and electrical conductivity (ranging from 1.3 to 2.5 s·m^{-1}) always presented the highest values in the sites inside each protected area.

Table 2. Physical and chemical properties of the six sampling sites from the three conserved ecosystems.

Parameters	Bou-Hedma		Zarat		Oued Dkouk	
	Inside	Outside	Inside	Outside	Inside	Outside
	Site 1	Site 2	Site 3	Site 4	Site 5	Site 6
Clay (%)	11.0 ± 0.2 ^{b}	12.8 ± 0.1 ^{a}	9.2 ± 0.2 ^{d}	10.2 ± 0.1 ^{c}	6.5 ± 0.2 ^{f}	7.0 ± 0.1 ^{e}
Silt (%)	23.5 ± 1.1 ^{d}	26.1 ± 0.1 ^{b}	25.1 ± 0.2 ^{c}	34.7 ± 0.3 ^{a}	16.4 ± 0.7 ^{e}	15.3 ± 0.5 ^{f}
Sand (%)	65.8 ± 0.2 ^{c}	60.0 ± 1.2 ^{e}	64.7 ± 1.1 ^{d}	55.0 ± 1.4 ^{f}	77.0 ± 1.1 ^{b}	78.6 ± 1.1 ^{a}
pH	8.1 ± 0.1 ^{b}	8.1 ± 0.1 ^{b}	8.0 ± 0.1 ^{b}	8.1 ± 0.1 ^{b}	8.4 ± 0.1 ^{a}	8.4 ± 0.1 ^{a}
E.c (s·m^{-1})	2.5 ± 0.3 ^{a}	1.8 ± 0.2 ^{c}	2.4 ± 0.3 ^{a}	1.6 ± 0.2 ^{d}	2.0 ± 0.1 ^{b}	1.3 ± 0.1 ^{e}
T.N (ppm)	191.0 ± 23 ^{a}	150.0 ± 10 ^{d}	174.0 ± 15 ^{b}	130.0 ± 10 ^{e}	160.0 ± 10 ^{c}	90.0 ± 5 ^{f}
T.P (ppm)	7.0 ± 0.1 ^{f}	10.6 ± 0.2 ^{c}	8.2 ± 0.2 ^{e}	12.4 ± 0.2 ^{b}	10.0 ± 0.2 ^{d}	15.3 ± 0.6 ^{a}
Org. Mat (%)	2.6 ± 0.2 ^{a}	1.4 ± 0.1 ^{d}	2.0 ± 0.1 ^{b}	1.2 ± 0.1 ^{e}	1.6 ± 0.3 ^{c}	0.8 ± 0.1 ^{f}
Wa. Cont (%)	3.3 ± 0.1 ^{a}	2.3 ± 0.1 ^{d}	2.9 ± 0.2 ^{b}	2.0 ± 0.1 ^{e}	2.5 ± 0.1 ^{c}	1.4 ± 0.3 ^{f}

E.c: electrical conductivity; T.N: total nitrogen; T.P: total phosphorus; Org. Mat: organic matter; Wat. Con: water content. Letters a–f: significant differences ($p < 0.05$); mean and standard error values ($n = 3$).

3.2. *AMF Colonization of Plant Roots*

Three plants and 30 root fragments of each one of the five plant species were analyzed per site. Direct observation of the root samples under the microscope showed that the roots of all studied plants, except those of *Diplotaxis simplex*, were colonized by AMF. All the structures characteristic of root colonization by AMF (intracellular aseptate hyphae, vesicles and arbuscules) were observed. The highest mycorrhizal frequency (F%)) was observed for *M. truncatula* in all the studied sites. At each site, the mycorrhizal frequency (F%)) varied among the plant species (Figure 2). The plants with the highest AMF root colonization belong to the Fabaceae family (*M. truncatula*, *A. corrugatus* and *L. creticus*) in all the studied sites, always with higher values in the site inside each conserved area.

There were significant differences ($p < 0.001$) in the mycorrhizal intensity (M%) among sites, again with highest M% for *M. truncatula* as well as, for each plant species, for the plants in the sites inside the conserved areas (Figure 2).

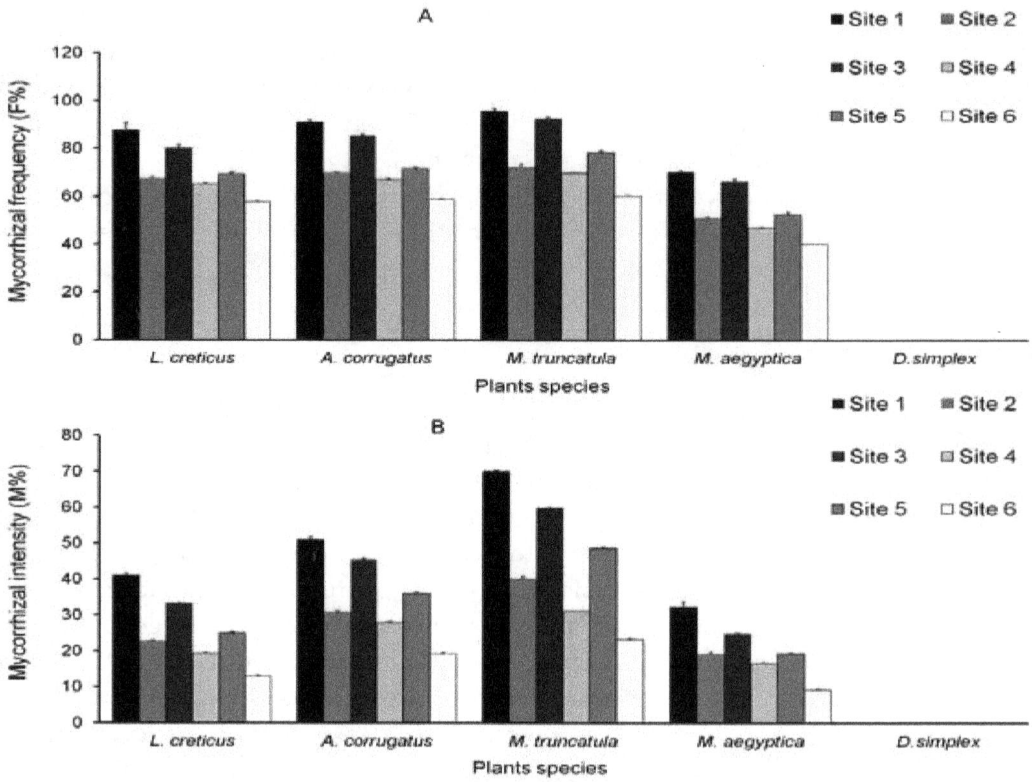

Figure 2. Mycorrhizal frequency (**A**) and Mycorrhizal intensity (**B**) of the five herbaceous plants in the different studied sites. Data are reported as mean (±SE) of three replicates per sample.

3.3. Densities of Spore Populations in the Studied Soils

The density of AMF spores isolated from the rhizosphere of the sampled plants varied between 856 (*M. truncatula*) in Site 1 and 81 spores/100 g of soil (*D. simplex*) in Site 6 (Figure 3). Therefore, spore density varied significantly ($p < 0.001$) among the studied sites; the maximum values were recorded in the rhizosphere of plants from sites 1, 3 and 5 (inside the conserved areas and lightly grazed) and the minimum values in sites 2, 4 and 6 (outside the conserved areas and intensively grazed). Bare areas (bulk soil), followed by *D. simplex* rhizosphere soil, always presented the lowest values for the six sites (Figure 3).

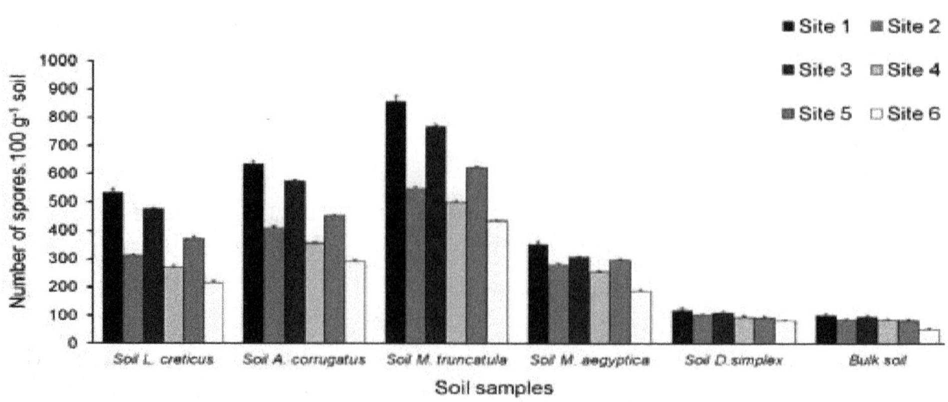

Figure 3. Distribution of AMF spores in the different rhizosphere soils. Data are reported as mean (±SE) of three replicates per sample.

There was a clear relationship between the intensity of root cortex AMF colonization (M%), mycorrhizal frequency (F%), and the density of AMF spores in the rhizospheres (Figures 2 and 3). Furthermore, the highest spore density was registered in the rhizosphere of the plant species with higher mycorrhiza frequency and intensity (*M. truncatula*).

3.4. *Effect of the Grazing Intensity on the Different AMF Parameters*

The grazing intensity strongly affected the mycorrhizal colonization and density of spores (Table 3), in accordance with the lowest values always registered, either for the plant mycorrhizal status or for AMF spore populations, in samples from sites outside the protected areas (Figures 2 and 3). This explains the negative effect of the grazing intensity on the different mycorrhizal parameters (Figure 4).

Table 3. Two-factor ANOVA analysis of the impact of grazing intensity on mycorrhizal parameters.

Factor	Grazing Intensity	
	F-Ratio	***p*-Value**
Mycorrhizal frequency (F%)	3.52	0.07 ns
Mycorrhizal intensity (M%)	11.26	0.001 ***
AMF spore density	10.55	0.002 **

*** Significant at $p < 0.001$, ** Significant at $p < 0.01$, ns: non-significant.

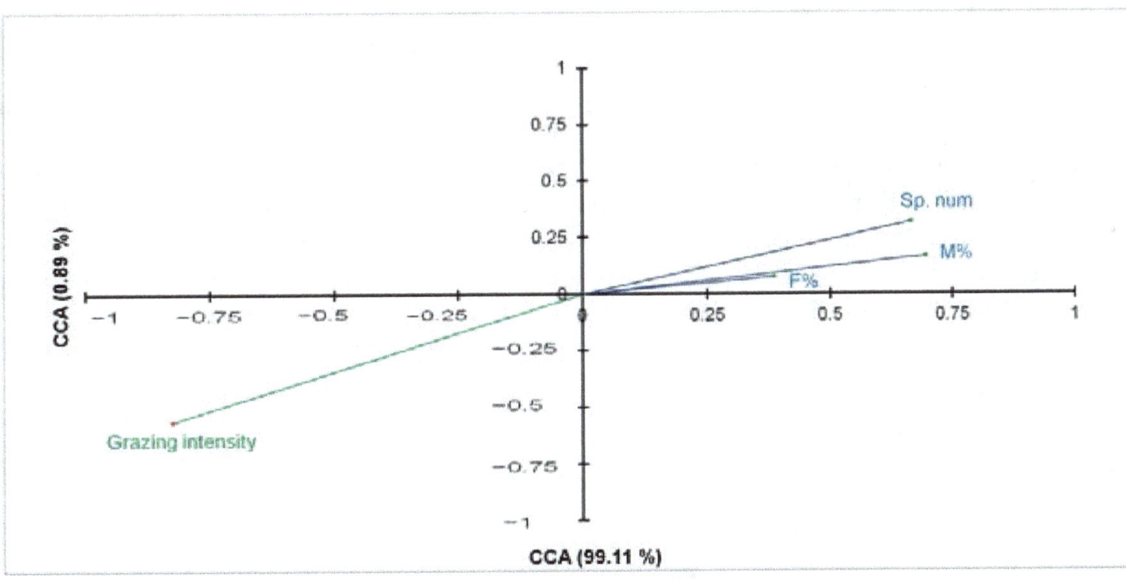

Figure 4. Results of a Canonical Correlation Analysis (CCorA) for the relationships between grazing intensity and AMF properties. Sp. num: number of spores; F%: mycorrhizal frequency; M%: mycorrhizal intensity.

3.5. *Effect of Soil Properties on the Different AMF Parameters*

According to the results of the Principal-component analysis (PCA) (Figure 5), the highest values of AMF colonization and number of spores were found in Site 1 at Bou-Hedma followed by Site 3 at Zarat. The major physical and chemical parameters of these sites (Org. Mat, E.c, Wat. Con, T.N) were in a strongly positive correlation with the different mycorrhizal properties. In contrast, the total phosphorus (T.P) available negatively affected all the mycorrhizal parameters and was considered a limited factor to these parameters (Figure 5).

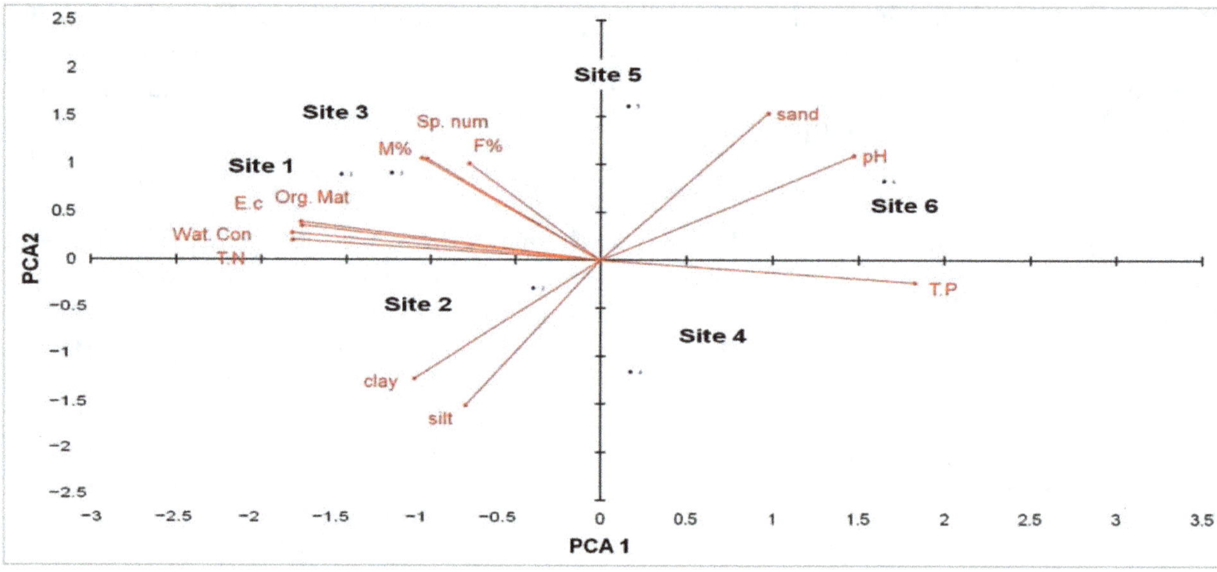

Figure 5. Results of a principal-component analysis (PCA) showing the relationships between soil properties and AMF parameters. E.c: electrical conductivity; T.N: total nitrogen; T.P: total phosphorus; Org. Mat: organic matter; Wat. Con: water content; Sp. num: number of spores; F%: mycorrhizal frequency; M%: mycorrhizal intensity.

3.6. Microbiological and Biochemical Properties of the Soils and Impact of AMF on Soil Microbial Communities

We found significant effects ($p < 0.001$) of the mycorrhizal plants on Cmic (Figure 6), this meaning that plants are affecting soil microbial biomass and consequently potential soil dynamics. The Cmic values were lower in bare areas (control soil) than in rhizosphere soils, the maximum values being recorded in the rhizospheres of the three Fabaceae plants *M. truncatula, A. corrugatus* and *L. creticus*. Cmic also markedly decreased in the intensively grazed sites (outside the protected areas), following the same trend of mycorrhizal frequency and intensity in root samples and of number of spores in rhizosphere soils (Figures 2, 3, 6 and 7).

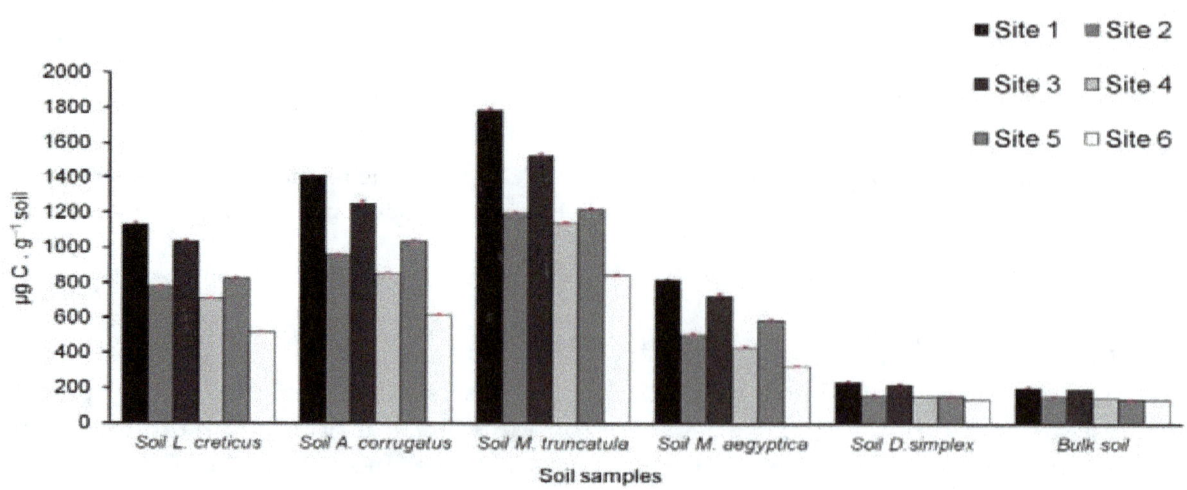

Figure 6. Microbial biomass carbon (Cmic) in the different sampled soils. Data are reported as mean (±SE) of three replicates per sample.

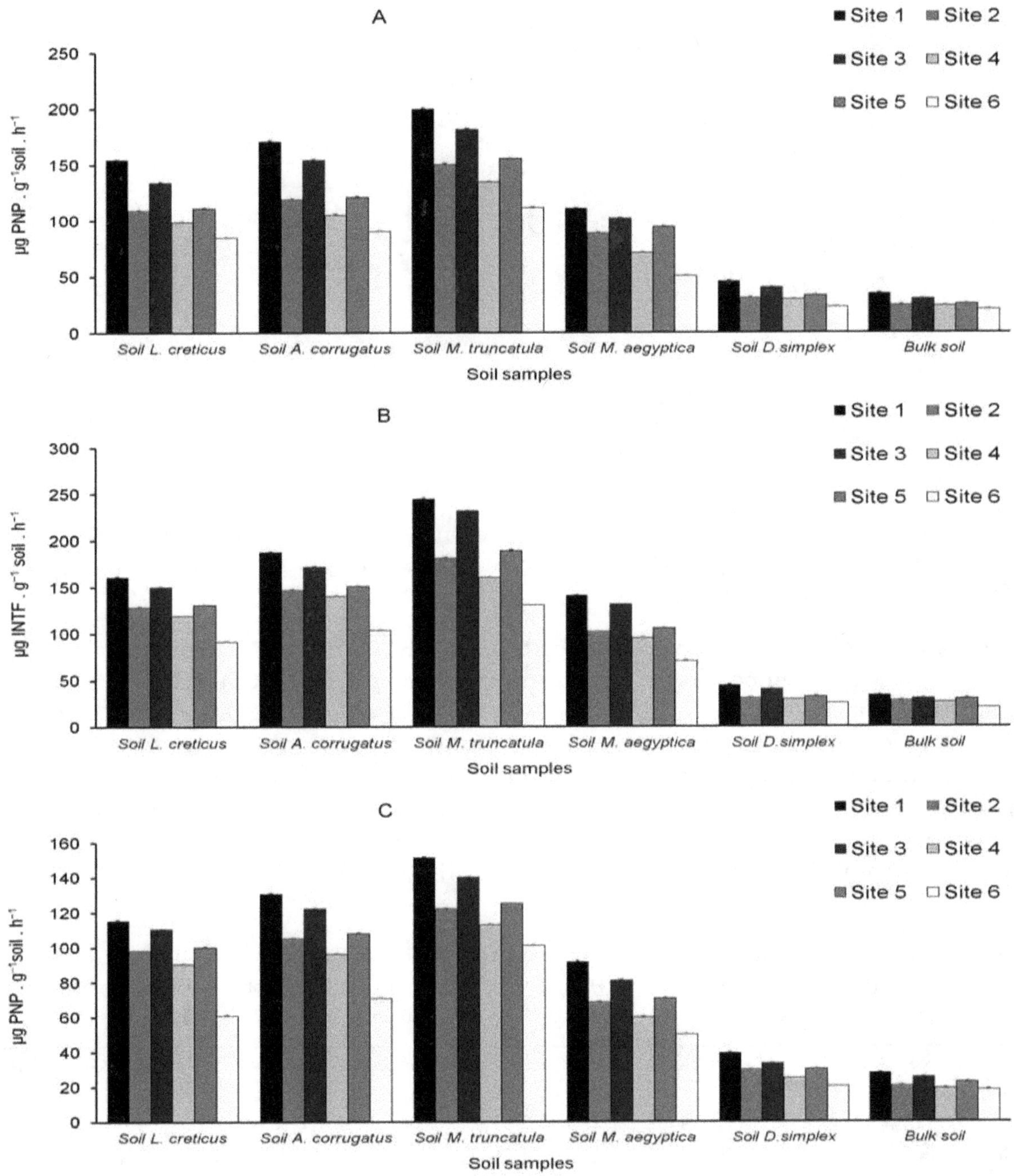

Figure 7. Phosphatase (**A**), dehydrogenase (**B**), β-glucosidase (**C**) activities in the studied soils. PNP: p nitrophenol, INTF: iodonitrotetrazolium formazan. Data are reported as mean (±SE) of three replicates per sample.

The highest activities of the three enzymes evaluated (phosphatase, dehydrogenase and β-glucosidase) (Figure 7) were observed in the rhizospheres of the three Fabaceae plants *M. truncatula*, *A. corrugatus* and *L. halophilus*, following the same trend of Cmic for each plant species and each site (Figure 6) as well as the values of mycorrhizal frequency and intensity and number of spores (Figures 2, 3 and 7), evidencing activities consistently greater in lightly grazed sites in comparison with those that were intensively grazed (Figures 2, 3, 7 and 8). The ANOVA analysis evidenced significant differences in these values.

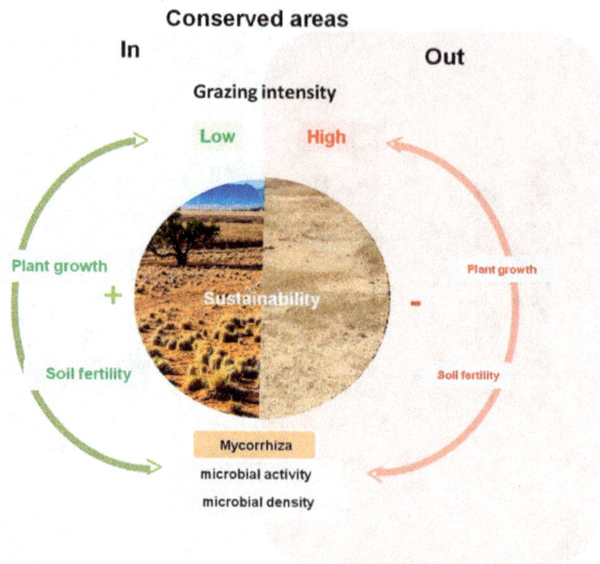

Figure 8. Schematic representation of the AMF effects on soil fertility and ecosystem stability under lightly grazed sites (inside the protected areas) versus intensively grazed sites (outside the protected areas), for arid Mediterranean ecosystems.

A clear relationship was observed between the AMF parameters (frequency and intensity of colonization of roots and number of AMF spores in the rhizosphere) and microbial activity (microbial biomass and enzyme activities) (Table 4, Figures S1 and S2). Based on the Pearson correlation coefficient, positive correlations with a highly significant value ($p < 0.001$) were observed between all AMF parameters (F%, M% and number of spores) and microbial activity (microbial biomass and enzyme activities: dehydrogenase, phosphatase and the β-glucosidase). A lower significant value ($p < 0.05$) was only observed in Site 6 (outside Oued Dkouk Natural Reserve) between the frequency of AMF colonization and two microbial parameters: microbial biomass and activity of β-glucosidase (Table 4).

Table 4. Pearson correlation coefficient between frequency of mycorrhization (F%), intensity of mycorrhization (M%) and density of AMF spores (Sp. num) and microbial activity expressed by microbial biomass (Cmic), phosphatase activity, dehydrogenase activity, and β-glucosidase activity.

Conserved Areas		Title	Cmic	Phosphatase	Deydrogenase	β-Glucosidade
Bou Hedma	**Site 1**	**F%**	0.93 ***	0.97 ***	0.95 ***	0.96 ***
		M%	0.99 ***	0.99 ***	0.99 ***	0.98 ***
		Sp. num	0.93 ***	0.97 ***	0.95 ***	0.96 ***
	Site 2	**F%**	0.93 ***	0.97 ***	0.96 ***	0.96 ***
		M%	0.98 ***	0.98 ***	0.99 ***	0.99 ***
		Sp. num	0.98 ***	0.96 ***	0.98 ***	0.98 ***
Zarat	**Site 3**	**F%**	0.95 ***	0.97 ***	0.96 ***	0.97 ***
		M%	0.99 ***	0.98 ***	0.99 ***	0.98 ***
		Sp. num	0.95 ***	0.97 ***	0.96 ***	0.97 ***
	Site 4	**F%**	0.92 ***	0.97 ***	0.97 ***	0.96 ***
		M%	0.96 ***	0.97 ***	0.99 ***	0.98 ***
		Sp. num	0.97 ***	0.96 ***	0.97 ***	0.95 ***
Oued Dkouk	**Site 5**	**F%**	0.96 ***	0.97 ***	0.97 ***	0.97 ***
		M%	0.99 ***	0.99 ***	0.98 ***	0.97 ***
		Sp. num	0.98 ***	0.97 ***	0.99 ***	0.98 ***
	Site 6	**F%**	0.90 *	0.94 ***	0.95 ***	0.91 *
		M%	0.98 ***	0.98 ***	0.99 ***	0.98 ***
		Sp. num	0.98 ***	0.99 ***	0.97 ***	0.95 ***

*** Significant at $p < 0.001$, * Significant at $p < 0.05$.

The results shown above, concerning the effect of grazing on AMF colonization (Table 3, Figure 4) and the impact of AMF on microbial communities and activities (Table 4, Figures S1 and S2), are summarized in Figure 8, which evidences the importance of AMF on soil functionality under arid Mediterranean ecosystems. Significant mycorrhizal colonization of roots was registered in lightly grazed sites, leading to high levels of microbial communities expressed by high values of biochemical activities. All these necessarily enhances soil fertility, better-adaptation and growth of plants and finally, in the stability of the ecosystem (Figure 8).

4. Discussion

Mediterranean arid and semi-arid ecosystems are characterized by high temperatures and drought for most of the year. These conditions limit plant establishment and growth and accelerate soil degradation and change microbial communities [1]. Mycorrhiza form communication pathways between plants and soil, influencing plant nutrient cycling, and restoring and maintaining soil fertility, thus influencing the microbial communities of the rhizosphere and extending the influence of plants to the soil. Several works emphasized the role of AMF in sustaining plant cover in semi-arid and arid ecosystems [4,8,10] as is the case of the following Mediterranean conserved areas in Tunisia: Bou-Hedma National Park, Zarat protected area and Oued Dkouk Natural Reserve.

Under natural conditions, about 90% of the terrestrial plants are mycorrhized, and AMF are found in all climates and ecosystems [10]. Due to their role on plant nutrition and defense, and to the importance of the extra-radical mycelium in the establishment of biological networks, AMF are determinant for the establishment and sustainability of plant communities and environment functioning [7]. However, the diversity of the AMF community and the intensity of arbuscular mycorrhizal colonization of natural vegetation is dependent on the availability of AMF spores and the mycorrhizal dependency of the plant species [24,25], as well as on the soil structure and management [26]. All these statements agree with the results of the present study, which evidenced colonization by AMF for all the herbaceous species, except for the Cruciferaceae *Diplotaxis simplex* (Figure 2). This was not surprising since Cruciferaceae are usually recognized as non-mycorrhizal plants [27]. However, distinct AMF colonization rates and intensities were observed for each plant species, which may be related to different levels of mycorrhizal dependence [28], and/or the availability of AMF spores. Higher mycorrhization rates were observed in the legume species *M. truncatula*, *A. corrugatus* and *L. creticus*, which presented high levels of mycorrhizal intensity, independent of the sampled site (Figure 2). This may highlight their mycorrhizal dependency and high demand for phosphorus (P) in comparison with plants from other families such as Poaceae [29].

In general, and particularly in semi-arid and arid ecosystem, AMF vary greatly with soil characteristics. Several biotic and abiotic factors may contribute to the distinct mycorrhizal intensities observed for the same plant species [30,31]. One abiotic factor that severely interferes with mycorrhization is the concentration of available phosphorus [10,32,33]. In the present study, soil phosphorus concentrations varied between sites, ranging from 7 to 15.3 ppm, and could explain the distinct mycorrhizal intensities observed for the same species in distinct sites (Figures 2 and 5), supporting that its availability is a crucial driver of mycorrhizal communities and activities [34].

Another factor frequently cited as responsible for lower levels of mycorrhiza formation is the availability of AMF spores, which is known to greatly vary in the ecosystems [35]. Sporulation in AMF occurs when the development of the mycelium begins to be limited by nutrients and is a highly carbon demanding process. This may explain why the number of AMF spores in the rhizosphere varies among plant species and, for the same plant species, among sites [36–38]. Apart from the rate of spore formation, the number of AMF spores in a soil also depends on the rates of spore germination and degradation [39]. As AMF are obligate biotrophs, the number of spores and propagules tends to be higher in the rhizosphere than in the bulk soil [12,40], and higher in the rhizosphere of plants with a higher intensity of AMF colonization (Figure 3). These differences are particularly evident in arid and semi-arid soils with high organic matter turnover rates and low organic matter content [41].

In low fertility soils, mycorrhiza are highways for nutrient and water transport, expanding the plant root system and the volume of soil exploited by the plant. However, similarly to roots, hyphae are leaky and lose nutrients into the hyphosphere, which will promote the development of selected microorganisms. In this context, it is expected that AMF play a crucial role in the biological characteristics of the rhizosphere [22,42] (Figure 6). The high values of Cmic observed in the rhizosphere soil of mycorrhizal plants imply that mycorrhiza contribute to improve the availability of carbon substrates to the microbial community of the rhizosphere [43,44]. These results support the hypothesis that AMF establish unique interactions with plant roots, conferring special characteristics to the rhizosphere [45], where several by product-based symbiosis and microbial loops may be assembled, contributing to improved carbon use efficiency. In this particular respect, the importance of AMF in promoting the development of the microbial community is confirmed by the low and comparable levels of microbial carbon observed in the soils without plant cover (control) and in the rhizosphere soil of the non-mycorrhizal plant *D. simplex* (Figure 6).

AMF increase the diversity of the carbon sources available to the microorganisms in the rhizosphere [46]—which is partly due to their nutritional mode; the excretion of catabolic enzymes to the surrounding medium, and to the direct access by the AMF to the plant carbon. Phosphatase and β-glucosidase are two of those catabolic enzymes, and their activities were in fact higher in the rhizosphere of the mycorrhizal plants than in that of the non-mycorrhizal plant studied or in the bulk soil (Figure 5). The importance of the soil microbial activity in association with several enzymes' activities was highlighted by the similarity in the activity patterns of the two hydrolytic enzymes (phosphatase and β-glucosidase) as well as of dehydrogenase, indicators of microbial activity [47], which can be inferred as being decreased under high grazing intensity (Figures S1 and S2).

At this point, it is clear that plant and AMF species are important modulators of rhizosphere characteristics (Figures 6–8). What is not clear is why biological indicators of soil characteristics are so distinct in the three sites outside the conserved areas. One key factor common to these sites may be limiting and related to mycorrhizal colonization. Grazing may affect mycorrhization (Table 3, Figure 4) directly in nutrient-limited ecosystems through a direct competition for carbon (Figure 8). Grazing was consistently associated with lower soil organic matter and increased P concentrations showing a strong impact on soil chemical properties, able to negatively influence mycorrhization in a direct way through the increase of P bioavailability. But in nutrient-limited ecosystems grazing per se may also affect mycorrhization (Table 3, Figure 4).

Plants are the primary producers of ecosystems, obtaining their biomass and energy from the carbon fixed by photosynthesis (Figure 8). Depending on the plant species and on the growth conditions, 20–50% of the newly photosynthetically fixed carbon may be transported to the roots and lost as rhizodeposition. In arid soils, this carbon is the main source of energy and biomass building material for the microbial community. Therefore, increasing grazing intensity decreases the carbon availability from plants and, consequently, also decreases the carbon available for mycorrhization [13,14]. An increase in grazing intensity has impacts on soil biological characteristics, with a significant decrease in the mycorrhizal intensity, as evidenced in the present study, for the three mycorrhizal plant species sampled in the sites outside the conserved areas (Table 3, Figure 5; Figure 8).

5. Conclusions

AMF is of vital importance in the arid Tunisian ecosystem, where soils are generally poor in organic matter and nutrients. High mycorrhization levels and high spore densities were revealed across the different sites, with particular incidence in the sites inside the conserved areas. Therefore, grazing intensity affected the potential beneficial influence of AMF on soil microbial processes. At this point, the importance of AMF to ecosystem dynamics is clear. What is not clear is how AMF are adapted to grazing intensity to allow ecosystem stability, particularly for situations that may be represented by the three sites outside the conserved areas. Therefore, future research should be focused on determining the ways through which AMF adapt to grazing intensity. This may be a key feature of ecosystem management, considering that AMF could be an important alternative for sustaining soil quality, and could be exploited as potential inoculants for rehabilitation and restoration programs in Mediterranean ecosystems.

Supplementary Materials:
Figure S1: Linear regression to model the correlations between mycorrhizal proprieties (mycorrhizal frequency, mycorrhizal intensity, and number of spores) and the microbiological parameter microbial biomass carbon. The analyses included all available data: from the rhizospheres of the four plants and from bulk soil, Figure S2: Linear regression to model the correlations between mycorrhizal proprieties (mycorrhizal frequency, mycorrhizal intensity, and number of spores) and biochemical activities: phosphatase (in μg PNP. g^{-1} soil·h^{-1}), dehydrogenase (μg INTF. g^{-1} soil·h^{-1}) and β-glucosidase (μg PNP. g^{-1} soil·h^{-1}). The analyses included all available data: from the rhizospheres of the four plants and from bulk soil.

Author Contributions: Conceptualization, M.M. (Mohamed Mars); methodology, N.M.; validation, M.M. (Mosbah Mahdhi) and T.D.; formal analysis, N.M.; investigation, N.M.; data curation, N.M.; writing—original draft preparation, N.M.; writing—review and editing, C.C. and M.F.C.; visualization, C.C. and M.F.C.; project administration, M.M. (Mohamed Mars); funding acquisition, C.C. and M.F.C. All authors have read and agreed to the published version of the manuscript.

Acknowledgments: We thank the personal staff of Bou-Hedma National Park, of Zarat-Gabes Protected area and of the Natural reserve of Oued Dkouk, for facilities and support for root and soil sampling. Thanks, are also due to Francisco Caeiro for providing helpful comments and English revision.

References

1. Martínez-García, L.; Miranda, J.; Pugnaire, F.I. Impacts of changing rainfall patterns on mycorrhizal status of a shrub from arid environments. *Eur. J. Soil Biol.* **2011**, *50*, 64–67. [CrossRef]
2. Mahmoudi, N.; Cruz, C.; Mahdhi, M.; Mars, M.; Caeiro, M.F. Arbuscular mycorrhizal fungi in soil, roots and rhizosphere of Medicago truncatula: Diversity and heterogeneity under semi-arid conditions. *PeerJ* **2019**. [CrossRef] [PubMed]
3. Mosbah, M.; Philippe, D.L.; Mohamed, M. Molecular identification of arbuscular mycorrhizal fungal spores associated to the rhizosphere of Retama raetam in Tunisia. *Soil Sci. Plant. Nutr.* **2018**, *64*, 335–341. [CrossRef]
4. Brundrett, M. Mycorrhizal associations and other means of nutrition of vascular plants: Understanding the global diversity of host plants by resolving conflicting information and developing reliable means of diagnosis. *Plant Soil* **2009**, *320*, 37–77. [CrossRef]
5. Van der Heijden, M.G.A.; Martin, F.M.; Selosse, M.A.; Sanders, J.R. Mycorrhizal ecology and evolution: The past, the present, and the future. *New Phytologist.* **2015**, *204*, 1406–1423. [CrossRef] [PubMed]
6. Barea, J.M.; Azcón, R.; Azcón-Aguilar, C. Mycorrhizosphere interactions to improve plant fitness and soil quality. Antonie van Leeuwenhoek. *Inter. J. Gen. Mol. Microb.* **2002**, *81*, 343–351.
7. Entry, J.A.; Rygiewicz, P.T.; Watrud, L.S.; Donnelly, P.K. Influence of adverse soil conditions on the formation and function of arbuscular mycorrhizas. *Adv. Environ. Res.* **2002**, *7*, 123–138. [CrossRef]
8. Smith, S.E.; Read, D.J. *Mycorrhizal Symbiosis*; Academic Press: San Diego, CA, USA, 1997.
9. Van Aarle, I.M.; Cavagnaro, T.R.; Smith, S.E.; Smith, F.A.; Dickson, S. Metabolic activity of Glomus intraradices in Arumand Paris-type arbuscular mycorrhizal colonization. *New Phytologist.* **2005**, *166*, 611–618. [CrossRef] [PubMed]
10. Smith, S.E.; Read, D.J. *Mycorrhizal Symbiosis*; Academic Press: San Diego, CA, USA, 2008.
11. Bago, B.; Vierheilig, H.; Piché, Y.; Ázcon-Aguilar, C. Nitrate depletion and pH changes induced by the extraradical mycelium of the arbuscular mycorrhizal fungus Glomus intraradices grown in monoxenic culture. *New Phytol.* **1996**, *133*, 273–280. [CrossRef] [PubMed]
12. Azcón-Aguilar, C.; Palenzuela, E.J.; Roldan, A.; Bautista, S.; Vallejo, R.; Barea, J.M. Analysis of the mycorrhizal potential in the rhizosphere of representative plant species from desertification-threatened Mediterranean shrub lands. *Appl. Soil Ecol.* **2003**, *22*, 29–37. [CrossRef]

13. Barto, E.K.; Rilling, M.C. Does herbivory really suppress mycorrhiza? A meta-analysis. *J. Ecol.* **2010**, *98*, 745–753. [CrossRef]
14. Mendoza, R.; Cabello, M.; Anchorena, J.; García, M.L. Soil parameters and host plants associated with arbuscular mycorrhizae in the grazed Magellanic steppe of Tierra del Fuego. *Agric. Ecosyst. Environ.* **2011**, *140*, 411–418. [CrossRef]
15. Raiesi, F.; Asadi, E. Soil microbial activity and litter turnover in native grazed and ungrazed rangelands in a semiarid ecosystem. *Biol. Fertil. Soil* **2006**, *43*, 76–82. [CrossRef]
16. Liao, J.D.; Boutton, T.W. Soil microbial biomass response to woody plant invasion of grassland. *Soil Biol. Biochem.* **2008**, *40*, 1207–1216. [CrossRef]
17. Phillips, J.M.; Hayman, D.S. Improved procedure for clearing roots and staining parasitic and vesicular arbuscular mycorrhizal fungi for rapid assessment of infection. *Trans. Br. Mycol. Soc.* **1970**, *55*, 158–161. [CrossRef]
18. Gerdemann, J.W.; Nicolson, T.H. Spores of mycorrhizal Endogone species extracted by wet sieving and decanting. *Trans. Br. Mycol. Soc.* **1963**, *46*, 235–244. [CrossRef]
19. AFNOR. *Recueil de normes françaises, qualité des sols, méthodes d'analyses*, 1st ed.; Association française de normalisation (Afnor): Paris, France, 1987.
20. Naanaa, W.; Susini, J. *Méthodes d'analyse physique et chimique des sols. ES 252, Direction des Sols*; Ministère de l'Agriculture: Tunis, Tunisie, 1988.
21. Amato, M.; Ladd, J.N. Assay for microbial biomass based on ninhydrin-reactive nitrogen in extracts of fumigated soils. *Soil Biol. Biochem.* **1988**, *20*, 107–114. [CrossRef]
22. Caravaca, F.; Alguacil, M.; Torres, P.; Roldán, A. Plant type mediates rhizospheric microbial activities and soil aggregation in a semiarid Mediterranean salt marsh. *Geoderma* **2005**, *12*, 338–375. [CrossRef]
23. Garcia, C.; Hernandez, T.; Costa, F. Potential use of dehydrogenase activity as an index of microbial activity in degraded soils. Commun. *Soil Sci. Plan.* **1997**, *28*, 123–134. [CrossRef]
24. Zhu, Y.G.; Laidlaw, A.S.; Christie, P.; Hammond, M.E. The specificity of arbuscular mycorrhizal fungi in perennial ryegrass, white clover pasture. *Agric. Ecosyst. Environ.* **2000**, *77*, 211–278. [CrossRef]
25. Duponnois, R.; Founoune, H.; Masse, D.; Pontanier, R. Inoculation of *Acacia holosericea* with ectomycorrhizal fungi in a semi-arid site in Senegal: Growth response and influences on the mycorrhizal soil infectivity after 2 years plantation. *For. Ecol. Manag.* **2005**, *207*, 351–362. [CrossRef]
26. Lekberg, Y.; Koide, R.T.; Rohr, J.R.; Aldrich-Wolfe, L.; Morton, J.B. Role of niche restrictions and dispersal in the composition of arbuscular mycorrhizal fungal communities. *J. Ecol.* **2007**, *95*, 95–105. [CrossRef]
27. Muchovej, R.M. *Importance of Mycorrhizae for Agricultural Crops, SS-AGR-170*; Agronomy Department, Florida Cooperative Extension Service; Institute of Food and Agricultural Sciences, University of Florida: Gainesville, FL, USA, 2004.
28. Strullu, D.G. *Les mycorhizes des arbres et des plantes cultivées*; Techniques et Documentation Lavoisier: Paris, France, 1991.
29. Bagayogo, M. *Site Specific Effects of Cereals/Legume Rotations in West Africa: Soil Mineral Nitrogen, Mycorrhizae and Nematodes*; Verlag Graner: Stuttgard, Germany, 1999.
30. Henriques, R.P.; Hay, J.D. The plant communities of a foredune in southeastern Brazil. *Can. J. Bot.* **1998**, *76*, 1323–1330.
31. Mahesh, V.; Selvaraj, T. Occurrence and distribution VA mycorrhizal fungi in the soils polluted with Tannery Effluent. *Adv. Biotech.* **2008**, *8*, 34–36.
32. Dickson, S.; Smith, S.E.; Smith, F.A. Characterization of two arbuscular mycorrhiza fungi in symbiosis with *Allium porum*: Colonization, plant growth and phosphate uptake. *New Phytol.* **1999**, *144*, 163–172. [CrossRef]
33. Bouamri, B.; Dalpé, Y.; Serrhini, M.N.; Bennani, A. Arbuscular Mycorrhizal fungi species associated with rhizosphere of Phoenix dactylifera L. in Morocco. *Afr. J. Boitech.* **2006**, *5*, 510–516.
34. Liu, W.; Zhang, Y.; Jiang, S.; Deng, Y.; Christic, P.; Murray, P.J.; Li, X.; Zhang, J. Arbuscular mycorrhizal fungi in soil and roots respond differently to phosphorus inputs in an intensively managed calcareous agricultural Soil. *Sci. Rep.* **2016**, *6*, 1–11. [CrossRef] [PubMed]
35. Li, L.; Yang, A.; Zhao, Z. Seasonality of arbuscular mycorrhizal symbiosis and dark septate endophytes in grassland site in Southwest China. *Microbial. Ecol.* **2005**, *54*, 367–373.
36. Houngnandan, P.; Yemadje, R.G.; Kane, A.; Boeckx, P.; Van Cleemput, O. Les glomales indigènes de la forêt claire à Isoberliniadoka (Craib et Stapf) à Wari-Maro au centre du Bénin. *Tropicultura* **2009**, *27*, 83–87.

37. Sghir, F.; Chliyeh, M.; Kachkouch, W.; Khouader, M.; OuazzaniTouhami, A.; Benkirane, R.; Douira, A. Mycorrhizal status of *Olea europaea* ssp. Oleaster in Morocco. *J. Appl. Biosci.* **2013**, *61*, 4478–4489. [CrossRef]
38. Al-Areqi, A.N.; Chliyeh, M.; Sghir, F.; Ouazzani, A.; Benkirane, R.T.; Douira, A. Diversity of Arbuscular mycorrhizal fungi in the rhizosphere of Coffea Arabica in the Republic of Yemen. *J. Appl. Biosci.* **2013**, *64*, 4888–4901. [CrossRef]
39. Gemma, J.N.; Koske, R.E.; Carreiro, M. Seasonal variation in spore abundance and dormancy of Gigasporagigantea in Mycorrhizal inoculum potential of a dune Soil. *Mycologia* **1989**, *80*, 211–216. [CrossRef]
40. Lovelock, C.E.; Andersen, K.; Morton, J.B. Arbuscular mycorrhizal communities in tropical forests are affected by host tree species and environment. *Oecologia* **2003**, *135*, 268–297. [CrossRef] [PubMed]
41. Mohammad, M.J.; Hamad, S.R.; Malkawi, H.I. Population of arbuscular mycorrhizal fungi in semi-arid environment of Jordan as influenced by biotic and abiotic factors. *J. Arid Environ.* **2003**, *53*, 409–417. [CrossRef]
42. Cruz, R.E.; Garcia, M.U. Nitrogen fixation and mycorrhizae in acacias on degraded grasslands. In *Tropical Acacias in East Asia and the Pacific*; Kamiset, A., Taylor, D.A., Eds.; Winrock Intenational Institute for Agriculture Research: Bangkok, Thailand, 1992; pp. 59–71.
43. Böhme, L.; Langer, U.; Böhme, F. Microbial biomass, enzyme activities and microbial community structure in two European long-term field experiments. *Agric. Ecosyst. Environ.* **2005**, *109*, 141–152. [CrossRef]
44. Flieβbach, A.; Oberholzer, H.R.; Gunst, L.; Mäder, P. Soil organic matter and biological soil quality indicators after 21 years of organic and conventional farming. *Agric. Ecosyst. Environ.* **2007**, *118*, 273–274. [CrossRef]
45. Dennis, P.G.; Miller, A.J.; Hirsch, P.R. Are root exudates more important than other sources of rhizodeposits in structuring rhizosphere bacterial communities? *Microbial. Ecol.* **2010**, *72*, 313–327. [CrossRef] [PubMed]
46. Rillig, M.C. Arbuscular mycorrhizae, glomalin and soil quality. *Can. J. Soil Sci.* **2004**, *84*, 355–363. [CrossRef]
47. Nannipieri, P.; Kandeler, E.; Ruggiero, P. Enzyme activities and microbiological and biochemical processes in Soil. In *Enzymes in the Environment*; Activity, Ecology and Applications; Burns, R.G., Dick, R.P., Eds.; Marcel Dekker: New York, NY, USA, 2002; pp. 1–33.

9

Carotenoid Biosynthesis in *Fusarium*

Javier Avalos [1,*], Javier Pardo-Medina [1], Obdulia Parra-Rivero [1], Macarena Ruger-Herreros [1], Roberto Rodríguez-Ortiz [1,2], Dámaso Hornero-Méndez [3] and María Carmen Limón [1]

1 Departamento de Genética, Facultad de Biología, Universidad de Sevilla, 41012 Sevilla, Spain; jpardo6@us.es (J.P.-M.); duly@us.es (O.P.-R.); macarenarugerherreros@gmail.com (M.R.-H.); lrrodriguezor@conacyt.mx (R.R.-O.); carmenlimon@us.es (M.C.L.)

2 Present Address: CONACYT-Instituto de Neurobiología-UNAM, Juriquilla, Querétaro 076230, Mexico

3 Departamento de Fitoquímica de los Alimentos, Instituto de la Grasa, CSIC, Campus Universidad Pablo de Olavide, 41013 Sevilla, Spain; hornero@ig.csic.es

* Correspondence: avalos@us.es.

Abstract: Many fungi of the genus *Fusarium* stand out for the complexity of their secondary metabolism. Individual species may differ in their metabolic capacities, but they usually share the ability to synthesize carotenoids, a family of hydrophobic terpenoid pigments widely distributed in nature. Early studies on carotenoid biosynthesis in *Fusarium aquaeductuum* have been recently extended in *Fusarium fujikuroi* and *Fusarium oxysporum*, well-known biotechnological and phytopathogenic models, respectively. The major *Fusarium* carotenoid is neurosporaxanthin, a carboxylic xanthophyll synthesized from geranylgeranyl pyrophosphate through the activity of four enzymes, encoded by the genes *carRA*, *carB*, *carT* and *carD*. These fungi produce also minor amounts of β-carotene, which may be cleaved by the CarX oxygenase to produce retinal, the rhodopsin's chromophore. The genes needed to produce retinal are organized in a gene cluster with a rhodopsin gene, while other carotenoid genes are not linked. In the investigated *Fusarium* species, the synthesis of carotenoids is induced by light through the transcriptional induction of the structural genes. In some species, deep-pigmented mutants with up-regulated expression of these genes are affected in the regulatory gene *carS*. The molecular mechanisms underlying the control by light and by the CarS protein are currently under investigation.

Keywords: neurosporaxanthin; xanthophyll; apocarotenoid; retinal; torulene; photoinduction; RING-Finger protein; carotenoid gene cluster

---

1. Introduction

The genus *Fusarium* is a complex group of phytopathogenic fungi, consisting of more than one thousand species [1], many of which have been the object of detailed attention [2]. Different *Fusarium* species were found to produce a large array of secondary metabolites. Some of them are pigments, and color variations have been used from long ago as distinctive traits of different strains (see, e.g., [3]). A most characteristic class of fungal pigments are the carotenoids, a family of lipophilic terpenoids ubiquitous in all major taxonomic groups [4,5]. The carotenoids are synthesized by all photosynthetic organisms, from cyanobacteria to higher plants, but they are also produced by diverse heterotrophic microorganisms, including fungi [6,7] and non-photosynthetic bacteria [8,9]. There are more than 750 natural carotenoids [5], providing typical yellowish, orange or reddish pigmentations to many plant organs, microorganisms and animals. With some notable exceptions in aphids [10], the animals are generally unable to synthesize carotenoids, and they get them through the diet. Carotenoids are especially relevant in photosynthetic species, where they play essential roles in light harvesting and photoprotection of the photosynthetic machinery [11] and in animals as a source for retinoids, i.e., retinal and retinoic acid [12]. However, the carotenoids do not seem to play essential roles in fungi, where their absence has no apparent phenotypic consequences apart of altered pigmentation [13].

As terpenoids, the synthesis of carotenoids derives from the condensation of C_5 isoprene units [4,5]. The universal terpenoid precursor is isopentenyl pyrophosphate (IPP), which can be synthesized either from mevalonate, produced from hydroxymethylglutaryl coenzyme A (HMG-CoA) (mevalonate pathway), or from derivatives of D-1-deoxyxylulose 1-phosphate, generated from the condensation of pyruvate and glyceraldehyde 3-phosphate (non-mevalonate pathway, [14]). In the cases investigated, fungal IPP is produced through the mevalonate pathway, while for carotenoids in bacteria and photosynthetic species IPP is produced through the non-mevalonate pathway, localized in plastids in the eukaryotic organisms. The early biosynthetic steps involve the sequential additions of IPP (isoprene, C_5) units to produce geranyl pyrophosphate (GPP, C_{10}), farnesyl pyrophosphate (FPP, C_{15}), and geranylgeranyl pyrophosphate (GGPP, C_{20}), carried out by prenyl transferases [4,5]. The first molecule with the characteristic aliphatic carotenoid-like structure is the colorless 15-*cis*-phytoene, consisting of a symmetrical polyene chain formed by the condensation of two GGPP units by phytoene synthase (Figure 1). The light-absorbing features of colored carotenoids are due to the presence of a chromophore, consisting of a series of conjugated double bonds generated by the activity of desaturase enzymes. Different subsequent chemical changes, usually the introduction of a cyclic end group (β and ε rings are the most common, while γ, κ, φ and χ rings are less frequent) on at least one of the ends of the molecule, and/or oxidative reactions (e.g., hydroxylation, epoxidation, carboxylation, esterification, etc), gives rise to the large diversity of known carotenoids [5].

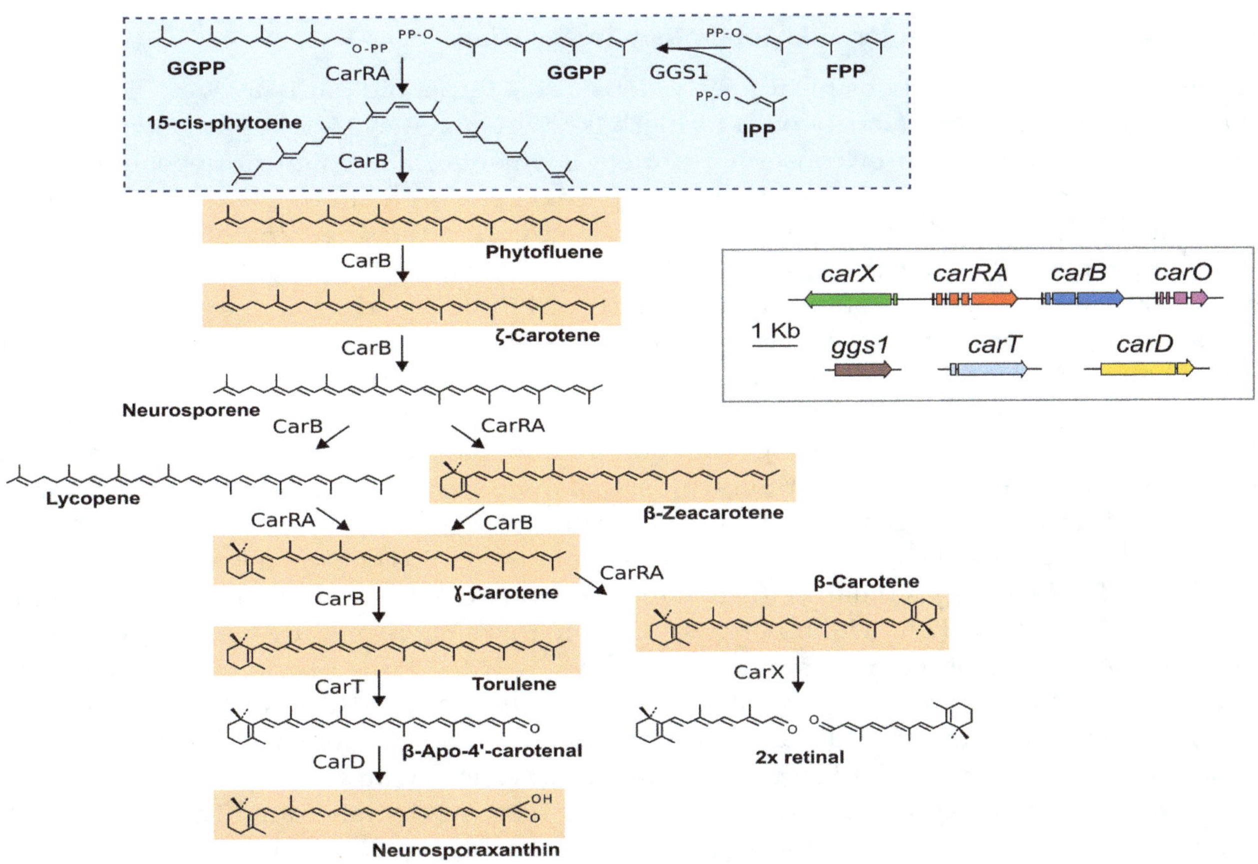

Figure 1. Carotenoid biosynthesis in *Fusarium*. Gene products responsible for the biosynthetic steps are indicated close to each arrow. Genomic organization of the genes is shown in the inserted box. Carotenoids detected in the High Performance Liquid Chromatography (HPLC) analyses described in Figure 4 are shaded in orange. Compounds shaded in blue (dotted box) are not detected in the chromatograms because of lack of absorption in the covered detection range (350–600 nm).

2. *Fusarium* Carotenoids

Early reports mentioned the occurrence of carotenoids in *Fusarium* species, e.g., *F. oxysporum* [3], but the first detailed biochemical analyses were described in *Fusarium aquaeductuum* [15]. Such analyses showed the occurrence of non-polar carotenoids and an acidic carotenoid fraction, which reminded the one assigned as the xanthophyll neurosporaxanthin (hereafter NX) in *Neurospora crassa* [16]. NX is a carboxylic C_{35} apocarotenoid, chemically defined as β-apo-4′-carotenoic acid [17]. NX is not a frequent carotenoid in fungi, and to date, besides the genera *Neurospora* and *Fusarium*, it has been only found in *Verticillium* [18,19] and *Podospora* [20].

Carotenoid analyses were refined in *F. aquaeductuum* in subsequent years, and confirmed the occurrence of NX together with precursor neutral carotenes, that included ζ-carotene, neurosporene, lycopene, γ-carotene, and torulene [21]. Later studies in *Fusarium fujikuroi*, and data presented in Section 5, corroborated the carotenoid composition of *F. aquaeductuum* with the exception of the detection of β-zeacarotene instead of lycopene. This indicates that the cyclization may be achieved either on neurosporene or on lycopene, depending on the species. Taken together, the chemical analyses are consistent with the biosynthetic pathway depicted in Figure 1. From phytoene, five desaturations and the cyclization of the non-saturated end yield the reddish torulene. The steps from torulene to NX were deduced from the enzymatic activities of the responsible enzymes, discussed below. In *F. fujikuroi* there are minor amounts of β-carotene [22], showing that the CarRA cyclase is also able to recognize γ-carotene as substrate. Because of its carboxylic group, NX may be a subject of esterification. In a marine *Fusarium* species, a fraction of the produced NX is accumulated as a glycosyl ester [23].

3. Genes and Enzymes of Carotenoid Metabolism in *Fusarium*

Some of the enzymes responsible for early steps of the terpenoid pathway were first identified in *F. fujikuroi* thanks to the similarity with formerly known enzymes. The typical cloning protocol consisted in the identification of conserved protein segments, the design of degenerate primers from such sequences to clone the corresponding internal gene segment by PCR, and the use of the amplified segment as a probe to clone the whole gene through the screening of a genomic library. Such methodology led to clone the genes for HMG-CoA reductase [24], producing mevalonate, and the prenyl transferases responsible for the synthesis of FPP [25] and its conversion to GGPP [26]. The mutants of these genes are expected to be non-viable because of their participation in the synthesis of essential terpenoids, and therefore their functions were inferred from their close relatedness with orthologs in other organisms. *F. fujikuroi* contains a second GGPP synthase located in the cluster for the synthesis of gibberellins, growth-inducing plant hormones of the terpenoid family. This second GGPP synthase gene is presumably specific for the synthesis of these secondary metabolites, as it indicates its coregulation with most of the genes from the gibberellin cluster [27].

The first gene specifically involved in the synthesis of carotenoids known in *Fusarium* was *carB* [28], identified by a PCR strategy similar to the one mentioned above through its sequence similarity to the *N. crassa* ortholog *al-1* [29]. The function of the gene was confirmed by complementation of a phytoene-accumulating albino mutant and by generation of the same phenotype by *carB* targeted mutation. The CarB enzyme is responsible for the five desaturation steps of the NX biosynthetic pathway, as it indicated the identification of a *carB* mutant allele specifically affected in the fifth desaturation [30]. This was in accordance with former observations with *al-1* of *N. crassa*: its expression in *Escherichia coli* or in vitro studies with purified AL-1 enzyme demonstrated its capacity to carry out the five desaturation steps for NX biosynthesis in this species [31]. Closely linked to *carB* it was found the gene *carRA* (Figure 1), orthologous of phytoene synthase *al-2* of *N. crassa* [32] and interpreted as the phytoene synthase of *Fusarium*, a conclusion consistent with the albino phenotype of their mutants [33]. In both species, a carboxylic domain with similarity to carotene cyclases precedes the phytoene synthase domain. The cyclase domain has a high conservation with that of the orthologous gene *crtYB* of the yeast *Xanthophyllomyces dendrorhous*, where the cyclase activity was biochemically demonstrated [34]. No mutants specifically affected in the cyclase activity have been described in

Fusarium, but the mutants of the cyclase domain of the *al-2* gene of *N. crassa* were shown to be defective in the cyclization step [35,36]. Taken together, the available data strongly suggest the same function for the amino domain of the CarRA protein of *F. fujikuroi*.

The reactions from torulene to NX include a carotenoid cleavage step, typically achieved by a family of enzymes known as carotenoid oxygenases [37]. A gene for an enzyme of this family, called *carX*, was found to be linked and transcribed divergently from the gene *carRA* (Figure 1). However, its targeted mutation did not impede the synthesis of NX [38], and the study of its enzymatic activity revealed that it cleaved β-carotene symmetrically to produce retinal [39]. Interestingly, the *carX/carRA/carB* cluster is linked to a rhodopsin gene, called *carO* because of its regulatory connections with the rest of the genes of the cluster (see next sections). The rhodopsins typically bind retinal as a light-absorbing prosthetic group, providing coherence to the gene organization of the cluster: the genes *carRA*, *carB*, and *carX* are necessary to produce retinal, which is presumably used by the rhodopsin encoded by the coregulated *carO* gene. Retinal might be also used for a second rhodopsin, encoded by the gene *opsA* [40]. In addition, retinal might be subject to further chemical reactions. The *Fusarium* genomes contain a gene for an aldehyde dehydrogenase highly similar to RALDH enzymes converting retinal to retinoic acid in mammals. This enzyme, that was called CarY, was investigated in *F. verticillioides* and found to exhibit such enzymatic activity in vitro [41]. Targeted mutation of the *carY* gene in *F. verticillioides* had no effect on carotenogenesis, but resulted in diverse developmental alterations. However, no retinoic acid could be detected in the control fungal cells and hence the biochemical function of the CarY enzyme in *Fusarium* has not been solidly established.

The *Fusarium* genes responsible for the conversion of torulene to NX have been recently described. The search for genes encoding other putative carotenoid oxygenases in the *Fusarium* genome databases led to identify the gene *carT*, whose function was revealed by the presence of a mutation in a torulene accumulating mutant and its genetic complementation with the wild-type *carT* allele [42]. As biochemical confirmation, purified CarT enzyme efficiently cleaved torulene in vitro to produce β-apo-4′-carotenal. As additional support, the targeted mutation of its ortholog in *N. crassa*, called *cao-2* from carotenoid oxygenase, resulted in the block of NX production and the accumulation of torulene [43]. These findings established a new enzymatic class in the carotenoid oxygenase family. The function of the gene *carT* in *Fusarium* carotenogenesis, together with those of the genes *carRA* and *carB*, was also confirmed by targeted gene disruption in *Gibberella zeae*, teleomorph of *Fusarium graminearum* [44].

The product of CarT, β-apo-4′-carotenal, requires a further oxidation to produce NX. The responsible gene was first discovered in *N. crassa* thanks to the study of a yellow mutant, called *ylo-1*. This mutant exhibited a puzzling biochemical phenotype, since it contained a complex carotenoid mixture that varied with the culture conditions and that did not include NX [45,46]. The gene *ylo-1* encodes an aldehyde dehydrogenase mediating the last step of NX biosynthesis, as demonstrated in its ability to complement the *ylo-1* mutation [47] and the capacity of purified YLO-1 protein to convert in vitro β-apo-4′-carotenal into NX. This finding allowed the identification of the orthologous gene in *F. fujikuroi*, that was called *carD* [48]. The function of *carD* was confirmed by targeted mutation, which resulted in the lack of NX and the accumulation of unusual carotenoids, interpreted as β-apo-4′-carotenal derivatives.

4. Regulation by Light

Early observations in *F. oxysporum* revealed that illumination promotes the accumulation of carotenoids in the mycelium [3]. First detailed studies on the effect of light were carried out with *F. aquaeductuum*. In an initial report, focused on the effect of temperature on the photoinduction process [49], the time course of the response to a transient light exposure revealed a gradual carotenoid accumulation that reached a maximum at about 12 h after illumination. However, the synthesis keeps increasing at a slower rate for at least three days if the culture is maintained under light [50].

Regarding the effect of temperature, the secondary reaction (formation of colored carotenoids) decreases at lower temperatures while the primary reaction to light, presumably a photochemical process, is independent of temperature in the range of 5–25 °C.

The photoinduction is ineffective under anaerobic conditions, as indicated by the lack of carotenoid accumulation if the illuminated mycelia are transferred to an oxygen-free atmosphere [51]. However, the photoinduced state is maintained, as indicated by the onset of carotenoid biosynthesis if the aerobic conditions are restored, even in the dark. Under aeration, the photoinduction requires protein synthesis, as shown by the lack of response if cycloheximide is added before or immediately after light exposure [50]. As happens with anaerobiosis, the removal of cycloheximide allows the start of carotenoid accumulation at any time in the dark, at least up to 30 h, pointing again to a high stability of the induced state of the photoreception system. In an ingenious experiment, when a *F. aquaeductuum* culture was illuminated, incubated in the dark to allow the start of enzyme production, and then exposed to anaerobiosis and to cycloheximide to block further enzyme production, there was no carotenoid production. However, if the culture was returned afterwards to aerobiosis, carotenoid biosynthesis started even in the presence of cycloheximide, indicating that the effect of anaerobiosis is at the level of enzyme activity, and that the enzymes are sufficiently stable [51].

In *F. aquaeductuum*, the effect of light may be partially replaced by the addition of the sulfhydryl oxidizing reagents p-chloro- and p-hydroxymercuribenzoate [50,52] or the oxidative reagent hydrogen peroxide [53], suggesting that oxidation of -SH groups plays a role in the light detection system. Accordingly, the photoinduction is abolished upon addition of reducing agents as dithionite and hydroxylamine, but the response is recuperated if the reducing agent is removed, indicating the recovery of the photoreceptor system [53]. However, while a short exposure to light is sufficient for a sustained photoinduction, p-hydroxymercuribenzoate must be all the time present to maintain its stimulating effect, and such stimulation is additive with that of light [54], indicating different mechanisms of action. Moreover, p-hydroxymercuribenzoate was ineffective in *F. fujikuroi* in the dark, while this species exhibits a similar photoinduction [55].

The study of light dependence of the photoinduction in *F. aquaeductuum* [56] showed that the amount of produced carotenoids depends on the incident light over a 100-fold range, with a reciprocity law holding true over a wide range of exposure times and light intensities. Action spectrum for NX photoinduction extends from 400 to 500 nm, with maxima at 375/380 nm and 450/455 nm and a shoulder at 430/440 nm, a shape consistent with the participation of a flavin photoreceptor. No induction is detected with wavelengths beyond 500 nm, and actually red light proved to be ineffective, discarding the participation of a phytochrome-like photoreceptor [57]. However, incubation of the fungus with methylene blue or toluidine blue allows it to respond to red light [58], suggesting that these chemicals may act as artificial photoreceptors. The shape of the action spectrum also discards the carotenoids as light-absorbing chromophores. As supporting evidence, the accumulation of phytoene in the albino *carB* mutant SG43 is still dependent on light [22], and the mutants of gene *carX* [38], involved in retinal formation, or those of the rhodopsin genes *carO* [59] and *opsA* [40] exhibit a full carotenoid photoinduction.

In *N. crassa*, a fungus with a similar action spectrum for light-induced carotenoid biosynthesis [60], the photoinduction is totally dependent on the White Collar complex, consisting of the photoreceptor WC-1 and its partner WC-2 (reviewed by [61]). However, the mutants of the *wc-1* orthologous genes of *F. fujikuroi* (*wcoA* [62]) and *F. oxysporum* (*wc1* [63]) conserve to different extents a detectable carotenoid photoinduction under continuous illumination. The targeted mutation of the orthologous *wc-1* and *wc-2* genes in *F. graminearum*, *fgwc-1* and *fgwc-2*, result in a paler pigmentation of the surface colonies under light, but the levels of carotenoids either in the wild type and the mutants were not chemically determined [64].

The analysis of carotenoid accumulation on surface cultures revealed a two-stage response in *F. fujikuroi*, with a first rapid increase dependent on WcoA and a slower subsequent carotenoid accumulation depending on another flavin photoreceptor, the DASH cryptochrome CryD [65].

The photoactivity of this flavin photoreceptor has been experimentally demonstrated [66], and its participation as a second photoreceptor explains the maintenance of photoinduced carotenoid accumulation in the *wcoA* mutants under constant illumination. The photoreceptors WcoA and CryD are not only involved in the regulation of carotenogenesis, as indicated by the alteration in the production of other pigments and secondary metabolites in their corresponding mutants [62,67], even in the dark in the case of WcoA.

The photoinduction of *Fusarium* carotenogenesis results from a rapid increase in the transcript levels of most of the structural genes. *Northern* blot experiments in *F. fujikuroi* showed similar induction kinetics for the four genes of the *car* cluster, *carRA*, *carB*, *carO* [59], and *carX* [38], as well as *carT* [42]. The transcripts reach maximal levels after one hour of exposure to light and decay afterwards, a down-regulating phenomenon known as photoadaptation. This photoinduction pattern has been confirmed by RT-qPCR approaches (Figure 2), and similar results have been obtained in *F. oxysporum* [68] and *F. verticillioides* [69]. The transcriptional photoresponse is similar to the one exhibited by the orthologous genes in *N. crassa* (reviewed by [70]), with the exception of the GGPP synthase gene *al-3*, that exhibits in this fungus a strong photoinduction while its *F. fujikuroi* ortholog *ggs1* is hardly affected by light [26]. Our recent RNA-seq data have revealed however a significant photoinduction of *ggs1* mRNA (Figure 3). A minor photoresponse was exhibited by the gene *carD* from *F. fujikuroi* [47], corroborated by the RNA-seq data (Figure 3), while no photoinduction was detected in the case of *ylo-1* from *N. crassa* [48], both genes encoding the ALDH enzymes responsible for the last reaction for NX production (Section 3). Therefore, in *F. fujikuroi*, the whole NX biosynthetic pathway is regulated by light.

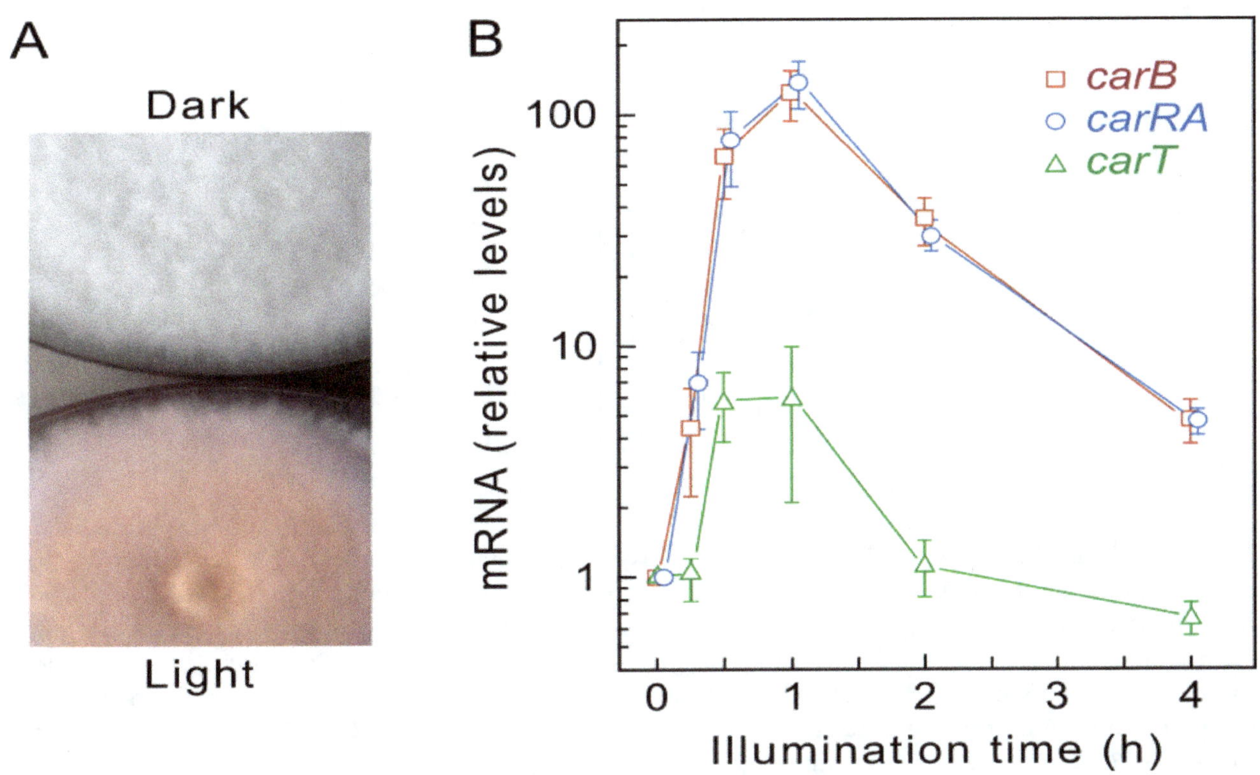

Figure 2. Photoinduction of carotenogenesis in *F. fujikuroi*. (**A**) Aspect of 7-day old colonies of the wild type IMI58289 grown on minimal medium in the dark or under continuous illumination; (**B**) Example of the kinetics of mRNA accumulation for the genes *carRA*, *carB*, and *carT* in the wild type FKMC1995. Transcript levels were determined by RT-qPCR (real time quantitative PCR) and referred to those of β-tubulin gene. Levels in the dark for each gene were taken as 1 (adapted from [65]).

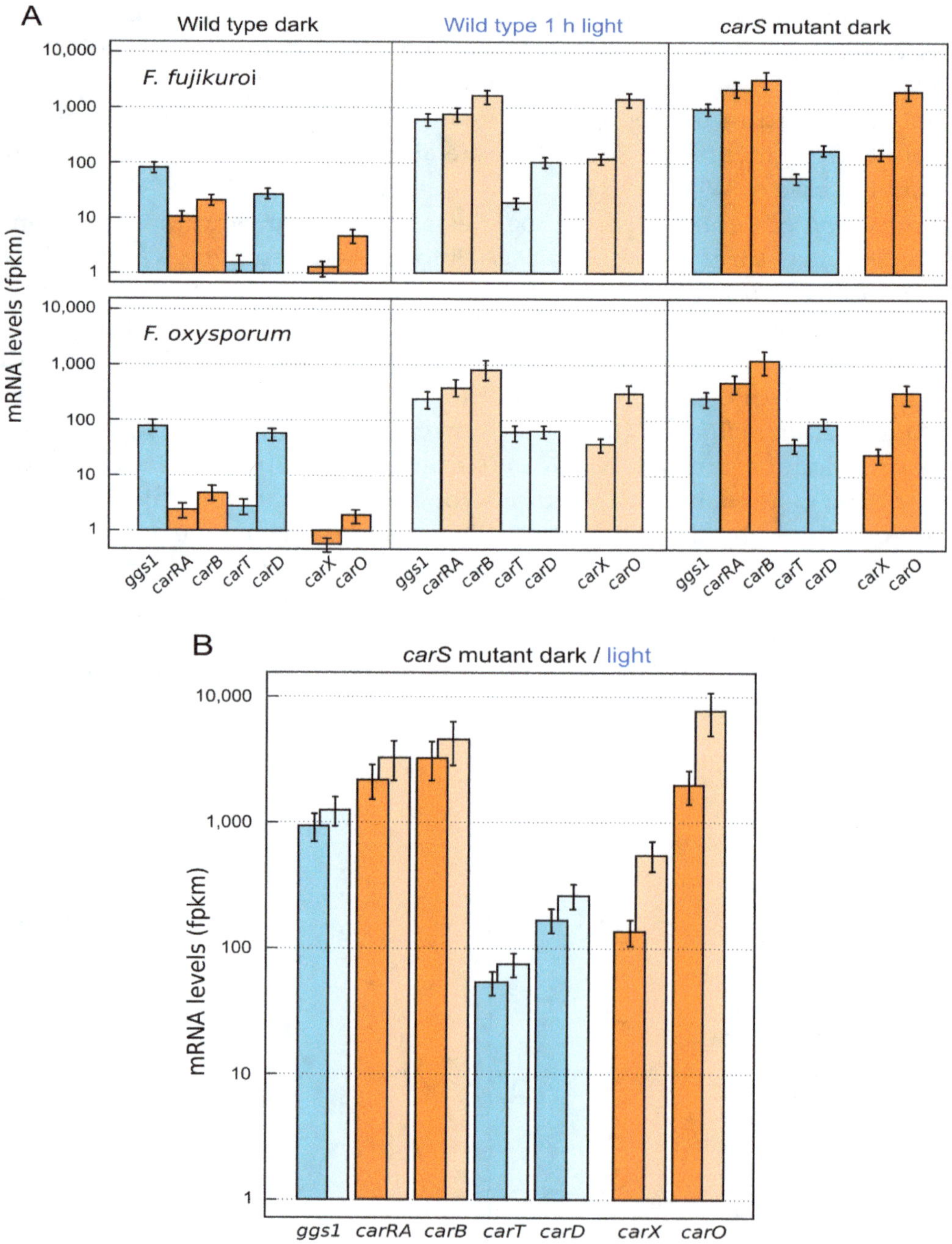

Figure 3. Effect of light and *carS* mutation on the expression levels for the *Fusarium* genes involved in carotenoid metabolism. The genes are grouped according to their functions: *ggs1*, *carRA*, *carB*, *carT*, and *carD* involved in NX biosynthesis and *carX* and *carO* involved in retinal and CarO rhodopsin production. Orange colors correspond to the genes organized as a cluster, and blue color to genes unlinked in the genome. Darker colors indicate cultures incubated in the dark and paler colors indicate cultures illuminated for one hour. (**A**) Above: wild-type strain IMI58289 of *F. fujikuroi* and its *carS* mutant SG39. Below: wild type strain 4287 of *F. oxysporum* (f. sp. *lycopersici*) and its *carS* mutant SX2; (**B**) Effect of light on the mRNA levels of the *carS* mutant SG39. RNA-seq data and culture conditions correspond to analyses already described [71]. In brief, all strains were cultured for three days in dark in minimal medium (DG medium for *F. fujikuroi* and DG with 3 g asparagine instead of $NaNO_3$ for *F. oxysporum*). In the cases indicated, the cultures were exposed to one hour of illumination. RNA samples were sequenced with Illumina technology and the resulting data were analyzed with software tools for read mapping (Bowtie and TopHat), transcriptome assembly (Cuffmerge and Cufflinks), and differential gene expression analysis (Cuffdiff and CummeRbund).

The study of the transcriptional photoinduction of the structural genes for carotenogenesis in the mutants of the *wcoA* and *cryD* genes is consistent with different mechanisms of action for the encoded photoreceptors [65]. After illumination, the *wcoA* mutants exhibit photoinduction of carotenogenesis, but they accumulate carotenoids more slowly and their total levels do not reach those of the wild type. However, both the carotenoid content and the mRNA levels for the structural *car* genes are much lower in the dark in the *wcoA* mutants than in the wild type. Even more, the transcriptional photoinduction of the *car* genes is basically absent in the *wcoA* mutants, with mRNA levels after illumination lower than those of the wild type in the dark. Therefore, the induction of carotenogenesis by the alternative photoreceptor, presumably CryD, must be done through a transcription-independent mechanism. The molecular basis for such a mechanism, that could involve mRNA stability or translation and enzymatic activities or stabilities, remains to be elucidated.

The decay of the mRNA levels after photoinduction has been investigated in detail in *N. crassa*. In this fungus, the small flavin prohotoreceptor VIVID (VVD), its name due to the higher carotenoid levels of the *vvd* mutants under light, plays a key role in deactivating the WC photoreceptor system after illumination [70]. In contrast, the mutation of the VIVID ortholog in *F. fujikuroi*, called *vvdA*, results in a lower carotenoid accumulation in this fungus [72]. The kinetics of carotenogenesis in the *vvdA* mutants show a faster carotenoid accumulation immediately after light exposure, consistent with an attenuating function of VvdA on the photoactivated WcoA, but a slower accumulation after more prolonged growth under light, suggesting an up-regulating role during the second stage of carotenoid photoinduction [65].

5. Down-Regulation of Carotenoid Production: The *CarS* Gene

Standard mutagenesis protocols allow the identification of *Fusarium* mutants affected in the synthesis of carotenoids, easy to detect because of their changes in pigmentation. Such mutants are more easily identified on minimal medium, such as the DG minimal agar [73], than on richer media, e.g., Potato Dextrose Agar (PDA), which frequently results in the accumulation of other unrelated pigments which may mask the color of carotenoids. Carotenoid mutants usually identified by visual inspection are either albino in the light, as those lacking an early enzymatic step of the carotenoid pathway, or deeply pigmented in the dark, affected in a down-regulating function. The deep-orange mutants, exhibiting a light-independent carotenoid production and generically called *carS*, have been described in *F. fujikuroi* [22] and *F. oxysporum* [68]. The *carS* mutants have been investigated in detail in *F. fujikuroi*, where they accumulate an increasing amount of carotenoids with aging [55] and contain large amounts of mRNA for the structural genes, from *carRA* to *carD* [38,42,48,59] (Figure 3). This up-regulation results in an increase of carotenogenic enzymes, as deduced from the study of in vitro carotenogenesis in a cell-free system of a *carS* mutant compared to the wild type [74]. Because of their high carotenoid content, the effect of light is less apparent in the *carS* mutants, but expression studies have found a detectable photoinduction of the transcript levels for the structural genes. This has been found in *F. fujikuroi* [42] and in *F. oxysporum* [68], and it has been recently corroborated in *F. fujikuroi* in RNA-seq analyses (Figure 3).

The qualitative mixture of carotenoids accumulated by two *carS* mutants of *F. fujikuroi* obtained in different genetic backgrounds is similar to that of their corresponding wild types under light, with minor differences, mainly a lower torulene content in the *carS* strains (Figure 4). Because of the large accumulation of carotenoids in the dark, these mutants are also useful to investigate possible damaging effects of light on the carotenoids. The comparison of the chromatograms from dark or light-grown cultures in two *carS* mutants reveal again only minor changes, with a decrease in torulene and an increase in β-carotene in the samples from the illuminated cultures (Figure 4), which could be due to different sensitivities of these carotenes to light. Because of their intense pigmentation, the *carS* mutants have been useful tools to detect mutants with qualitative alterations in the carotenoid mixtures, e.g., making possible the identification of torulene accumulating mutants [22] or mutants with alterations in substrate recognition by the CarB desaturase [30]. The *carS* mutants have been also an optimal background to check the effect of possible carotenoid inhibitors on *Fusarium* carotenogenesis, revealing the differential efficiency of some cyclase inhibitors that were fully active in other species [55].

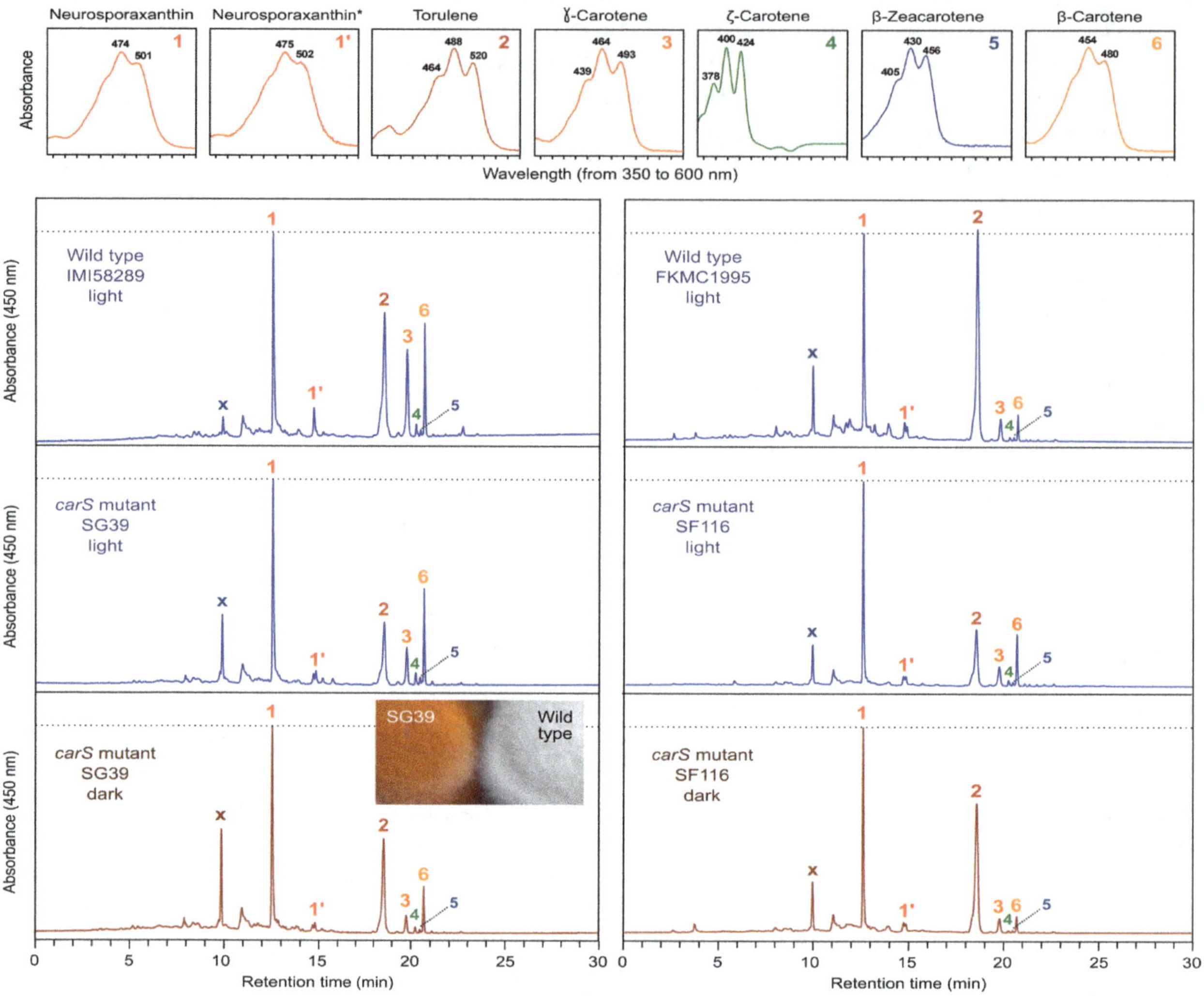

Figure 4. HPLC analysis of carotenoids produced by two different wild types of *F. fujikuroi* and the *carS* mutants SG39 [22] and SF116 [48]. The strains were grown for 7 days on DG minimal medium in the dark or under light (7 W m^{-2} white light). The method for HPLC separation has been described [75]. In brief, the samples were run in a reverse-phase C18 column with a binary-gradient elution (acetone : deionized water) at a flow rate of 1 mL/min. Detection was performed at 450 nm and online UV/Visible absorption spectra were acquired in the wavelength range 350–600 nm. The UV/Visible spectra of the identified carotenoids are shown above. The peak X corresponds to an unknown carotenoid with maximal absorbance at 480 nm, whose identity is under investigation. The peak 1′ is assigned as a neurosporaxanthin isomer. Phytofluene cannot be detected at 450 nm, but traces of this carotene were detected at 368 nm (data not shown). The inner picture shows the pigmentation of 7-day old colonies of the wild strain IMI58289 and the *carS* mutant SG39 grown on minimal DG medium in the dark.

The gene *carS* was found to code for a protein of the RING finger (RF) family. The assignation was based in the identification of relevant mutations in all the *carS* mutants analyzed so far, the deep-pigmented phenotype resulting from the targeted mutation of the gene in *F. oxysporum* [68] and the complementation of the *carS* mutation in *F. fujikuroi* [76]. The predicted CarS protein has sequence similarity with the protein CrgA of *M. circinelloides*, whose mutation results in a similar carotenoid overproducing pattern in this species [77,78]. The degree of similarity between the CarS and CrgA proteins is not very high, as expected for two taxonomically distant fungi, but it covers the most relevant CrgA domains. These include two amino-terminal RF domains, one of them initially disregarded in CarS because of a wrong intron assignation in the *F. fujikuroi* genome annotation [71], and a LON protease domain [79]. However, CrgA has two glutamine-rich regions and a carboxy-terminal isoprenylation site, the first one underrepresented and the second absent in CarS. In other proteins, the RF domains interact with E3 ligase-type enzymes that mediate ubiquitylation of target proteins, frequently as a label for their degradation. The function of CrgA of *M. circinelloides* has received considerable attention, and the available information may provide some clues on the hypothetical way of action of CarS in *Fusarium*. At least one of the RF domains of CrgA is essential for its regulatory function in carotenogenesis, suggesting that this protein might function as an E3 ubiquitin ligase [78]. However, CrgA interacts with one of the three WC proteins of *M. circinelloides*, MCWC-1b, to trigger its degradation through an ubiquitylation-independent mechanism [80]. There are other similarities between CarS and CrgA functions. First, as found for the *carS* mutants in *Fusarium*, carotenoid biosynthesis maintains a detectable photoinduction in the *crgA* mutants of *M. circinelloides* [77]. Second, both CrgA and CarS are involved in other processes in addition to carotenogenesis, pointing to wider regulatory functions. This is supported by the alterations in growth and sporulation by the *crgA* mutants of *M. circinelloides* [81,82], and in gibberellin and bikaverin productions by the *carS* mutants of *F. fujikuroi* [83,84].

Interestingly, *carS* mutants have not been described in other *Fusarium* species and they could not be found in mutagenesis screenings in *F. verticillioides* (J. García-Martínez and V. Díaz-Sánchez, unpublished observations). However, the *carS* gene of *F. fujikuroi* is able to restore a basal carotenoid production upon introduction in a *carS* mutant of *F. oxysporum*, although not totally at the level of expression of the structural genes, indicating a functional conservation in both species [68].

6. Regulation by Light-Independent Factors

Nitrogen availability controls the production of different *Fusarium* secondary metabolites. Well-known examples are the productions of gibberellins, bikaverin, and fusarubin, induced by the absence of nitrogen through a complex regulatory network [85]. However, other metabolites, as fusarins and fusaric acid, exhibit an opposite regulatory pattern, with a higher production under excess of nitrogen [86]. Experiments with immobilized mycelia of *F. fujikuroi* incubated under low nitrogen conditions, to which excess nitrogen was subsequently added, revealed a higher synthesis of carotenoids under nitrogen starvation [87]. The negative effect of nitrogen was confirmed with the exchange of media between immobilized mycelia samples with different nitrogen contents, resulting in enhanced or reduced synthesis depending on the pass to low or high nitrogen conditions, respectively, indicating that the repressing effect is reversible.

In a different approach, the effect of nitrogen was studied in shake cultures of wild type and *carS* mutants. The production of carotenoids was higher in low nitrogen medium (low N/C ratio) than in high nitrogen medium (high N/C) in either of the tested strains [84]. A different set of experiments, in which cultures grown under an excess of nitrogen were transferred to a nitrogen-free solution, showed an increase of transcript levels of the structural genes *carRA* and *carB* after the transfer, which was accompanied by an enhanced accumulation of carotenoids. As found for photoinduction (see Section 3), the transcriptional increase was transitory, with maximal levels found in this case after more than 10 h following nitrogen deprivation.

The stimulatory effect of nitrogen starvation on *F. fujikuroi* carotenogenesis, either at mRNA or carotenoid levels, is additive with the one produced by light [84], indicating different activating mechanisms. The regulation of carotenogenesis by nitrogen has been also described in *N. crassa*, where the mRNA levels for the *carRA* and *carB* orthologs *al-1* and *al-2* increase considerably under nitrogen limitation compared to standard nitrogen conditions, either in the wild type or in blind *wc* mutants [88]. The regulation by nitrogen might involve control of expression at the level of chromatin structure, as indicated by the differences in histone methylation found for the genes of the *car* cluster in a mutant of the methyltransferase KMT6 in *Fusarium graminearum* [89]. In *F. fujikuroi*, the expression of several nitrogen-regulated clusters for secondary metabolite production correlate with H3K9 acetylation [90]. This suggests a similar mechanism in the regulation of the structural *car* genes, which is currently under investigation.

Other regulatory circuits may also influence the control of carotenoid biosynthesis. A major regulatory pathway is the one involving cAMP, produced by adenylate cyclase under the stimulation of a Gα unit from a heterotrimeric G protein. cAMP acts on other proteins, including protein kinases that phosphorylate target proteins to module their activity. The mutation of the adenylate cyclase gene *acyA* in *F. fujikuroi* results in higher levels of carotenoids in the dark but in a reduced photoinduction [91]. The mutation affects other phenotypes, as the growth pattern or the production of other secondary metabolites. Another study in the same species, that extended the analysis to the effect of mutations of two adenylyl cyclase-stimulating Gα subunits and two cAMP-dependent PKs, confirmed different alterations in secondary metabolism and development [92], but in this case the effect on carotenogenesis was not investigated. Carotenogenesis may also have a regulatory connection with sexual development, as suggested by the lower carotenoid photoinduction of the mutants of the MAT1-2-1 mating-type gene of *F. verticillioides* [69].

In *N. crassa*, besides the regulation by light, the synthesis of carotenoids is coupled to conidiation in a light-independent manner [70]. Actually, conidiation is stimulated by light and aerial growth in this fungus, leading to a massive conidia production that provides a typical orange pigmentation in slant cultures. Conidiation is less abundant in *Fusarium*, and the synthesis of carotenoids in the conidia has been usually disregarded. A comparison of the effect of light on the amount of carotenoids in mycelia and conidia from surface cultures in a wild type and a *carS* mutant of *F. oxysporum* showed the conservation in the conidia of the regulation by light and the deregulation by the *carS* mutation [93] (Figure 5). The low amount of carotenoids in wild type conidia in the dark indicates the lack of a developmental induction as the one described in *N. crassa*.

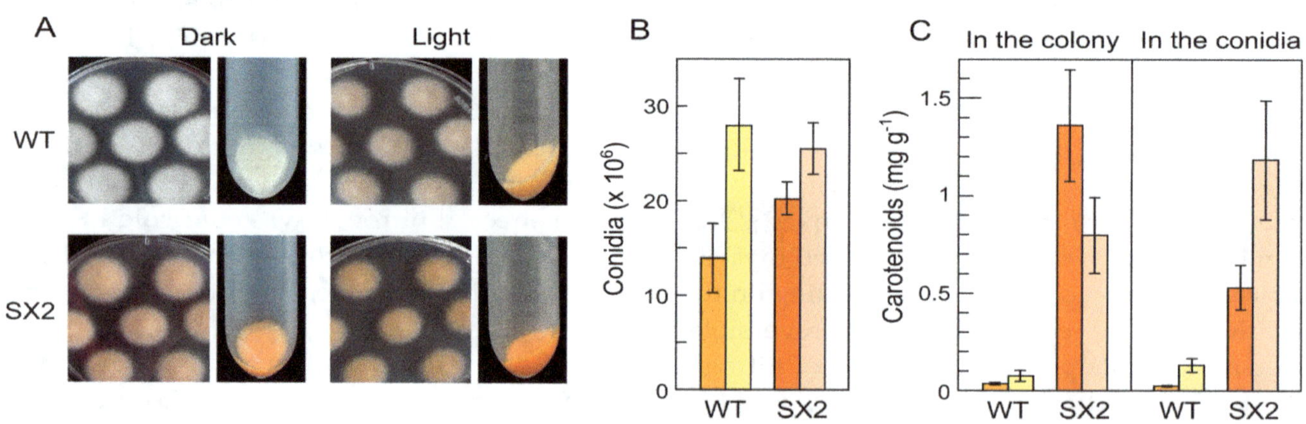

Figure 5. Conidiation and carotenoid biosynthesis in *F. oxysporum*. (**A**) Aspect of colonies of the wild type strain 4287 of *F. oxysporum* (f. sp. *lycopersici*) and its *carS* mutant SX2 grown on DG minimal medium at 30 °C for 7 days in the dark or under 5 w m^{-2} white light. Pictures on the right show the aspect of precipitated conidia obtained from the colonies as described [59]; (**B**) total amounts of conidia per Petri dish produced by the same strains under the indicated culture conditions; (**C**) total amounts of carotenoids accumulated in the mycelia and in the conidia mentioned above. The strains were grown in the dark (left darker bars) or under light (right paler bars). Data adapted from [93].

The cellular location of carotenoid biosynthesis is a regulatory aspect that has received little attention. Because of their hydrophobic nature, the carotenoids are assumed to interact with membranes, but their subcellular distribution in *Fusarium* is unknown. A biochemical approach based on the specific labeling of different terpenoids from ^{14}C-labeled mevalonate allowed to establish that the carotenoids are produced in different cell compartments than those where the gibberellins and the sterols are synthesized [94]. Experiments to visualize the physical location of the enzymes for carotenoid biosynthesis in the fungal cells, based in their fusion to fluorescent proteins, are currently underway.

7. Conclusions and Future Prospects

Following the early studies in *F. aquaeductuum*, the fungus *F. fujikuroi*, and more recently also *F. oxysporum*, have become reference models in the research of fungal carotenogenesis and a wealth of information has been accumulated on the biochemistry and genetics of their NX production. All the genes and enzymes of the biosynthetic pathway are known and current research work is dedicated to elucidate the molecular mechanisms of the regulation. The available information on the control by light has revealed an unexpected complexity in relation to the model system *N. crassa*, in which the White Collar complex is the only transcriptional activator of the structural genes for neurosporaxanthin biosynthesis. At least a second photoreceptor, the DASH cryptochrome CryD, is also involved in the photoresponse in *Fusarium*. Our current research efforts are centered in a better understanding of the regulation by light and by the CarS protein, and their possible connections. Studies on the mechanism of action of CarS are focused in the alteration of its expression and in the identification of interacting proteins. Efforts are also centered in the identification of regulatory proteins able to bind the promoters of the structural *car* genes, which might include CarS targets. The regulation may include epigenetic mechanisms, as those related with chromatin structure, also under study. While the biochemistry of the carotenoid pathway has many similarities between different producing fungi, the regulatory mechanisms have been the object of a higher diversification, allowing refined adaptations to respond more efficiently to the different needs in their respective natural habitats. The studies on the regulatory processes governing carotenoid biosynthesis in *Fusarium* are a promising area of research, which predictably may lead to the discovery of novel molecular mechanisms.

Acknowledgments: We thank the Spanish Government (projects BIO2012-39716, BIO2015-69613-R, AGL2014-53195R and CaRed Network BIO2015-71703-REDT) and Junta de Andalucía (project CTS-6638) for financial support.

References

1. Booth, C. *The Genus Fusarium*; Commonwealth Agricultural Bureaux for the Commonwealth Mycological Institute: Farnham Royal, UK, 1971.
2. Leslie, J.F.; Summerell, B.A. *The Fusarium Laboratory Manual*; Blackwell Professional: Ames, IA, USA, 2006.
3. Carlile, M.J. A study of the factors influencing non-genetic variation in a strain of *Fusarium oxysporum*. *J. Gen. Microbiol.* **1956**, *14*, 643–654. [CrossRef] [PubMed]
4. Britton, G.; Liaaen-Jensen, S.; Pfander, H. *Carotenoids*; Birkhäuser Verlag: Basel, Switzerland, 1998; Volumes 1 and 2.
5. Britton, G.; Liaaen-Jensen, S.; Pfander, H. *Carotenoids: Handbook*; Birkhauser Verlag: Basel, Switzerland, 2004.
6. Sandmann, G.; Misawa, N. Fungal carotenoids. In *The Mycota X. Industrial Applications*; Osiewacz, H.D., Ed.; Springer: Berlin, Germany, 2002; pp. 247–262.
7. Avalos, J.; Díaz-Sánchez, V.; García-Martínez, J.; Castrillo, M.; Ruger-Herreros, M.; Limón, M.C. Carotenoids. In *Biosynthesis and Molecular Genetics of Fungal Secondary Metabolites*; Martín, J.F., García-Estrada, C., Zeilinger, S., Eds.; Springer: New York, NY, USA, 2014; pp. 149–185.
8. Phadwal, K. Carotenoid biosynthetic pathway: Molecular phylogenies and evolutionary behavior of *crt* genes in eubacteria. *Gene* **2005**, *345*, 35–43. [CrossRef] [PubMed]
9. Boronat, A.; Rodríguez-Concepción, M. Terpenoid biosynthesis in prokaryotes. *Adv. Biochem. Eng. Biotechnol.* **2015**, *148*, 3–18. [PubMed]
10. Mandrioli, M.; Rivi, V.; Nardelli, A.; Manicardi, G.C. Genomic and cytogenetic localization of the carotenoid genes in the aphid genome. *Cytogenet. Genome Res.* **2016**, *149*, 207–217. [CrossRef] [PubMed]
11. Domonkos, I.; Kis, M.; Gombos, Z.; Ughy, B. Carotenoids, versatile components of oxygenic photosynthesis. *Prog. Lipid Res.* **2013**, *52*, 539–561. [CrossRef] [PubMed]

12. Blomhoff, R.; Blomhoff, H.K. Overview of retinoid metabolism and function. *J. Neurobiol.* **2006**, *66*, 606–630. [CrossRef] [PubMed]
13. Avalos, J.; Limón, M.C. Biological roles of fungal carotenoids. *Curr. Genet.* **2015**, *61*, 309–324. [CrossRef] [PubMed]
14. Rohdich, F.; Kis, K.; Bacher, A.; Eisenreich, W. The non-mevalonate pathway of isoprenoids: Genes, enzymes and intermediates. *Curr. Opin. Chem. Biol.* **2001**, *5*, 535–540. [CrossRef]
15. Rau, W.; Zehender, C. Die Carotinoide von *Fusarium aquaeductuum* Lagh. *Arch. Mikrobiol.* **1959**, *32*, 423–428. [CrossRef] [PubMed]
16. Zalokar, M. Isolation of an acidic pigment in *Neurospora*. *Arch. Biochem. Biophys.* **1957**, *70*, 568–571. [CrossRef]
17. Aasen, A.J.; Jensen, S.L. Fungal carotenoids II. The structure of the carotenoid acid neurosporaxanthin. *Acta Chem. Scand.* **1965**, *19*, 1843–1853. [CrossRef] [PubMed]
18. Valadon, L.R.G.; Mummery, R.S. Biosynthesis of neurosporaxanthin. *Microbios* **1969**, *1A*, 3–8.
19. Valadon, L.R.G.; Osman, M.; Mummery, R.S.; Jerebzoff-Quintin, S.; Jerebzoff, S. The effect of monochromatic radiation in the range 350 to 750 nm on the carotenogenesis in *Verticillium agaricinum*. *Physiol. Plant.* **1982**, *56*, 199–203. [CrossRef]
20. Strobel, I.; Breitenbach, J.; Scheckhuber, C.Q.; Osiewacz, H.D.; Sandmann, G. Carotenoids and carotenogenic genes in *Podospora anserina*: Engineering of the carotenoid composition extends the life span of the mycelium. *Curr. Genet.* **2009**, *55*, 175–184. [CrossRef] [PubMed]
21. Bindl, E.; Lang, W.; Rau, W. Untersuchungen über die lichtabhängige Carotinoidsynthese. VI. Zeitlicher Verlauf der Synthese der einzelnen Carotinoide bei *Fusarium aquaeductuum* unter verschiedenen Induktionsbedingungen. *Planta* **1970**, *94*, 156–174. [CrossRef] [PubMed]
22. Avalos, J.; Cerdá-Olmedo, E. Carotenoid mutants of *Gibberella fujikuroi*. *Curr. Genet.* **1987**, *25*, 1837–1841. [CrossRef]
23. Sakaki, H.; Kaneno, H.; Sumiya, Y.; Tsushima, M.; Miki, W.; Kishimoto, N.; Fujita, T.; Matsumoto, S.; Komemushi, S.; Sawabe, A. A new carotenoid glycosyl ester isolated from a marine microorganism, *Fusarium* strain T-1. *J. Nat. Prod.* **2002**, *65*, 1683–1684. [CrossRef] [PubMed]
24. Woitek, S.; Unkles, S.E.; Kinghorn, J.R.; Tudzynski, B. 3-Hydroxy-3-methylglutaryl-CoA reductase gene of *Gibberella fujikuroi*: Isolation and characterization. *Curr. Genet.* **1997**, *31*, 38–47. [CrossRef] [PubMed]
25. Homann, V.; Mende, K.; Arntz, C.; Ilardi, V.; Macino, G.; Morelli, G.; Bose, G.; Tudzynski, B. The isoprenoid pathway: Cloning and characterization of fungal FPPS genes. *Curr. Genet.* **1996**, *30*, 232–239. [CrossRef] [PubMed]
26. Mende, K.; Homann, V.; Tudzynski, B. The geranylgeranyl diphosphate synthase gene of *Gibberella fujikuroi*: Isolation and expression. *Mol. Gen. Genet.* **1997**, *255*, 96–105. [CrossRef] [PubMed]
27. Tudzynski, B. Gibberellin biosynthesis in fungi: Genes, enzymes, evolution, and impact on biotechnology. *Appl. Microbiol. Biotechnol.* **2005**, *66*, 597–611. [CrossRef] [PubMed]
28. Fernández-Martín, R.; Cerdá-Olmedo, E.; Avalos, J. Homologous recombination and allele replacement in transformants of *Fusarium fujikuroi*. *Mol. Gen. Genet.* **2000**, *263*, 838–845. [CrossRef] [PubMed]
29. Schmidhauser, T.J.; Lauter, F.R.; Russo, V.E.; Yanofsky, C. Cloning, sequence, and photoregulation of *al-1*, a carotenoid biosynthetic gene of *Neurospora crassa*. *Mol. Cell. Biol.* **1990**, *10*, 5064–5070. [CrossRef] [PubMed]
30. Prado-Cabrero, A.; Schaub, P.; Díaz-Sánchez, V.; Estrada, A.F.; Al-Babili, S.; Avalos, J. Deviation of the neurosporaxanthin pathway towards β-carotene biosynthesis in *Fusarium fujikuroi* by a point mutation in the phytoene desaturase gene. *FEBS J.* **2009**, *276*, 4582–4597. [CrossRef] [PubMed]
31. Hausmann, A.; Sandmann, G. A single five-step desaturase is involved in the carotenoid biosynthesis pathway to β-carotene and torulene in *Neurospora crassa*. *Fungal Genet. Biol.* **2000**, *30*, 147–153. [CrossRef] [PubMed]
32. Schmidhauser, T.J.; Lauter, F.R.; Schumacher, M.; Zhou, W.; Russo, V.E.; Yanofsky, C. Characterization of *al-2*, the phytoene synthase gene of *Neurospora crassa*. Cloning, sequence analysis, and photoregulation. *J. Biol. Chem.* **1994**, *269*, 12060–12066. [PubMed]
33. Linnemannstöns, P.; Prado, M.M.; Fernández-Martín, R.; Tudzynski, B.; Avalos, J. A carotenoid biosynthesis gene cluster in *Fusarium fujikuroi*: the genes *carB* and *carRA*. *Mol. Genet. Genomics* **2002**, *267*, 593–602. [CrossRef] [PubMed]

34. Verdoes, J.C.; Krubasik, P.; Sandmann, G.; van Ooyen, A.J.J. Isolation and functional characterisation of a novel type of carotenoid biosynthetic gene from *Xanthophyllomyces dendrorhous*. *Mol. Gen. Genet.* **1999**, *262*, 453–461. [CrossRef] [PubMed]
35. Arrach, N.; Schmidhauser, T.J.; Avalos, J. Mutants of the carotene cyclase domain of *al-2* from *Neurospora crassa*. *Mol. Genet. Genom.* **2002**, *266*, 914–921.
36. Díaz-Sánchez, V.; Estrada, A.F.; Trautmann, D.; Limón, M.C.; Al-Babili, S.; Avalos, J. Analysis of *al-2* mutations in *Neurospora*. *PLoS ONE* **2011**, *6*, e21948. [CrossRef] [PubMed]
37. Sui, X.; Kiser, P.D.; von Lintig, J.; Palczewski, K. Structural basis of carotenoid cleavage: From bacteria to mammals. *Arch. Biochem. Biophys.* **2013**, *539*, 203–213. [CrossRef] [PubMed]
38. Thewes, S.; Prado-Cabrero, A.; Prado, M.M.; Tudzynski, B.; Avalos, J. Characterization of a gene in the car cluster of *Fusarium fujikuroi* that codes for a protein of the carotenoid oxygenase family. *Mol. Genet. Genom.* **2005**, *274*, 217–228. [CrossRef] [PubMed]
39. Prado-Cabrero, A.; Scherzinger, D.; Avalos, J.; Al-Babili, S. Retinal biosynthesis in fungi: Characterization of the carotenoid oxygenase CarX from *Fusarium fujikuroi*. *Eukaryot. Cell* **2007**, *6*, 650–657. [CrossRef] [PubMed]
40. Estrada, A.F.; Avalos, J. Regulation and targeted mutation of *opsA*, coding for the NOP-1 opsin orthologue in *Fusarium fujikuroi*. *J. Mol. Biol.* **2009**, *387*, 59–73. [CrossRef] [PubMed]
41. Díaz-Sánchez, V.; Limón, M.C.; Schaub, P.; Al-Babili, S.; Avalos, J. A RALDH-like enzyme involved in *Fusarium verticillioides* development. *Fungal Genet. Biol.* **2016**, *86*, 20–32. [CrossRef] [PubMed]
42. Prado-Cabrero, A.; Estrada, A.F.; Al-Babili, S.; Avalos, J. Identification and biochemical characterization of a novel carotenoid oxygenase: Elucidation of the cleavage step in the *Fusarium* carotenoid pathway. *Mol. Microbiol.* **2007**, *64*, 448–460. [CrossRef] [PubMed]
43. Saelices, L.; Youssar, L.; Holdermann, I.; Al-Babili, S.; Avalos, J. Identification of the gene responsible for torulene cleavage in the *Neurospora* carotenoid pathway. *Mol. Genet. Genom.* **2007**, *278*, 527–537. [CrossRef] [PubMed]
44. Jin, J.M.; Lee, J.; Lee, Y.W. Characterization of carotenoid biosynthetic genes in the ascomycete *Gibberella zeae*. *FEMS Microbiol. Lett.* **2010**, *302*, 197–202. [CrossRef] [PubMed]
45. Goldie, A.H.; Subden, R.E. The neutral carotenoids of wild-type and mutant strains of *Neurospora crassa*. *Biochem. Genet.* **1973**, *10*, 275–284. [CrossRef] [PubMed]
46. Sandmann, G. Photoregulation of carotenoid biosynthesis in mutants of *Neurospora crassa*: Activities of enzymes involved in the synthesis and conversion of phytoene. *Z. Naturforsch.* **1993**, *48c*, 570–574.
47. Estrada, A.F.; Youssar, L.; Scherzinger, D.; Al-Babili, S.; Avalos, J. The *ylo-1* gene encodes an aldehyde dehydrogenase responsible for the last reaction in the *Neurospora* carotenoid pathway. *Mol. Microbiol.* **2008**, *69*, 1207–1220. [CrossRef] [PubMed]
48. Díaz-Sánchez, V.; Estrada, A.F.; Trautmann, D.; Al-Babili, S.; Avalos, J. The gene *carD* encodes the aldehyde dehydrogenase responsible for neurosporaxanthin biosynthesis in *Fusarium fujikuroi*. *FEBS J.* **2011**, *278*, 3164–3176. [CrossRef] [PubMed]
49. Rau, W. Über den Einfluss der Temperatur auf die lichtabhängige Carotinoidbildung von *Fusarium aquaeductuum*. *Planta* **1962**, *59*, 123–137. [CrossRef]
50. Rau, W. Untersuchungen über die lichtabhängige Carotinoidsynthese. II. Ersatz der Lichtinduktion durch Mercuribenzoat. *Planta* **1967**, *74*, 263–277. [CrossRef] [PubMed]
51. Rau, W. Untersuchungen über die lichtabhängige Carotinoidsynthese. VII. Reversible Unterbrechung der Reaktionskette durch Cycloheximid und anaerobe Bedingungen. *Planta* **1971**, *101*, 251–264. [CrossRef] [PubMed]
52. Rau, W.; Feuser, B.; Rau-Hund, A. Substitution of p-chloro- or p-hydroxymercuribenzoate for light in carotenoid synthesis by *Fusarium aquaeductuum*. *Biochim. Biophys. Acta* **1967**, *136*, 589–590. [CrossRef]
53. Theimer, R.R.; Rau, W. Untersuchungen über die lichtabhängige Carotinoidsynthese V. Aufhebung der Lichtinduktion dutch Reduktionsmittel und Ersatz des Lichts durch Wasserstoffperoxid. *Planta* **1970**, *92*, 129–137. [CrossRef] [PubMed]
54. Theimer, R.R.; Rau, W. Untersuchungen über die lichtabhängige Carotinoidsynthese VIII. Die unterschiedlichen Wirkungsmechanismen von Licht und Mercuribenzoat. *Planta* **1972**, *106*, 331–343. [CrossRef] [PubMed]
55. Avalos, J.; Cerdá-Olmedo, E. Chemical modification of carotenogenesis in *Gibberella fujikuroi*. *Phytochemistry* **1986**, *25*, 1837–1841. [CrossRef]

56. Rau, W. Untersuchungen über die lichtabhängige Carotinoidsynthese. I. Das Wirkungsspektrum von *Fusarium aquaeductuum*. *Planta* **1967**, *72*, 14–28. [CrossRef] [PubMed]
57. Schrott, E.L.; Huber-Willer, A.; Rau, W. Is phytochrome involved in the light-mediated carotenogenesis in *Fusarium aquaeductuum* and *Neurospora crassa*? *Photochem. Photobiol.* **1982**, *35*, 213–216. [CrossRef]
58. Lang-Feulner, J.; Rau, W. Redox dyes as artificial photoreceptors in light-dependent carotenoid synthesis. *Photochem. Photobiol.* **1975**, *21*, 179–183. [CrossRef] [PubMed]
59. Prado, M.M.; Prado-Cabrero, A.; Fernández-Martín, R.; Avalos, J. A gene of the opsin family in the carotenoid gene cluster of *Fusarium fujikuroi*. *Curr. Genet.* **2004**, *46*, 47–58. [CrossRef] [PubMed]
60. De Fabo, E.C.; Harding, R.W.; Shropshire, W., Jr. Action spectrum between 260 and 800 nanometers for the photoinduction of carotenoid biosynthesis in *Neurospora crassa*. *Plant Physiol.* **1976**, *57*, 440–445. [CrossRef] [PubMed]
61. Fischer, R.; Aguirre, J.; Herrera-Estrella, A.; Corrochano, L.M. The complexity of fungal vision. *Microbiol. Spectr.* **2016**, *4*, 1–22.
62. Estrada, A.F.; Avalos, J. The White Collar protein WcoA of *Fusarium fujikuroi* is not essential for photocarotenogenesis, but is involved in the regulation of secondary metabolism and conidiation. *Fungal Genet. Biol.* **2008**, *45*, 705–718. [CrossRef] [PubMed]
63. Ruiz-Roldán, M.C.; Garre, V.; Guarro, J.; Mariné, M.; Roncero, M.I. Role of the white collar 1 photoreceptor in carotenogenesis, UV resistance, hydrophobicity, and virulence of *Fusarium oxysporum*. *Eukaryot. Cell* **2008**, *7*, 1227–1230. [CrossRef] [PubMed]
64. Kim, H.; Son, H.; Lee, Y.-W. Effects of light on secondary metabolism and fungal development of *Fusarium graminearum*. *J. Appl. Microbiol.* **2014**, *116*, 380–389. [CrossRef] [PubMed]
65. Castrillo, M.; Avalos, J. The flavoproteins CryD and VvdA cooperate with the white collar protein WcoA in the control of photocarotenogenesis in *Fusarium fujikuroi*. *PLoS ONE* **2015**, *10*, e0119785. [CrossRef] [PubMed]
66. Castrillo, M.; Bernhardt, A.; Avalos, J.; Batschauer, A.; Pokorny, R. Biochemical characterization of the DASH-type cryptochrome CryD from *Fusarium fujikuroi*. *Photochem. Photobiol.* **2015**, *91*, 1356–1367. [CrossRef] [PubMed]
67. Castrillo, M.; García-Martínez, J.; Avalos, J. Light-dependent functions of the *Fusarium fujikuroi* CryD DASH cryptochrome in development and secondary metabolism. *Appl. Environ. Microbiol.* **2013**, *79*, 2777–2788. [CrossRef] [PubMed]
68. Rodríguez-Ortiz, R.; Michielse, C.; Rep, M.; Limón, M.C.; Avalos, J. Genetic basis of carotenoid overproduction in *Fusarium oxysporum*. *Fungal Genet. Biol.* **2012**, *49*, 684–696. [CrossRef] [PubMed]
69. Ádám, A.L.; García-Martínez, J.; Szücs, E.P.; Avalos, J.; Hornok, L. The MAT1–2-1 mating-type gene upregulates photo-inducible carotenoid biosynthesis in *Fusarium verticillioides*. *FEMS Microbiol. Lett.* **2011**, *318*, 76–83. [CrossRef] [PubMed]
70. Avalos, J.; Corrochano, L.M. Carotenoid biosynthesis in *Neurospora*. In *Neurospora: Genomics and Molecular Biology*; Kasbekar, D.P., McCluskey, K., Eds.; Caister Academic Press: Norfolk, UK, 2013; pp. 227–241.
71. Ruger-Herreros, M. Participación de la Proteína CarS en la Regulación de la Carotenogénesis y el Estrés en *Fusarium fujikuroi*. Ph.D. Thesis, Universidad de Sevilla, Seville, Spain, 2016.
72. Castrillo, M.; Avalos, J. Light-mediated participation of the VIVID-like protein of *Fusarium fujikuroi* VvdA in pigmentation and development. *Fungal Genet. Biol.* **2014**, *71*, 9–20. [CrossRef] [PubMed]
73. Avalos, J.; Casadesús, J.; Cerdá-Olmedo, E. *Gibberella fujikuroi* mutants obtained with UV radiation and N-methyl-N′-nitro-N-nitrosoguanidine. *Appl. Environ. Microbiol.* **1985**, *49*, 187–191. [PubMed]
74. Avalos, J.; Mackenzie, A.; Nelki, D.S.; Bramley, P.M. Terpenoid biosynthesis in cell-extracts of wild type and mutant strains of *Gibberella fujikuroi*. *Biochim. Biophys. Acta* **1988**, *966*, 257–265. [CrossRef]
75. Delgado-Pelayo, R.; Hornero-Méndez, D. Identification and quantitative analysis of carotenoids and their esters from sarsaparrilla (*Smilax aspera* L.) berries. *J. Agric. Food Chem.* **2012**, *60*, 8225–8232. [CrossRef] [PubMed]
76. Rodríguez-Ortiz, R.; Limón, M.C.; Avalos, J. Functional analysis of the *carS* gene of *Fusarium fujikuroi*. *Mol. Genet. Genom.* **2013**, *288*, 157–173. [CrossRef] [PubMed]
77. Navarro, E.; Lorca-Pascual, J.M.; Quiles-Rosillo, M.D.; Nicolás, F.E.; Garre, V.; Torres-Martínez, S.; Ruiz-Vázquez, R.M. A negative regulator of light-inducible carotenogenesis in *Mucor circinelloides*. *Mol. Genet. Genom.* **2001**, *266*, 463–470.

78. Lorca-Pascual, J.M.; Murcia-Flores, L.; Garre, V.; Torres-Martínez, S.; Ruiz-Vázquez, R.M. The RING-finger domain of the fungal repressor *crgA* is essential for accurate light regulation of carotenogenesis. *Mol. Microbiol.* **2004**, *52*, 1463–1474. [CrossRef] [PubMed]
79. Navarro, E.; Ruiz-Pérez, V.L.; Torres-Martínez, S. Overexpression of the *crgA* gene abolishes light requirement for carotenoid biosynthesis in *Mucor circinelloides*. *Eur. J. Biochem.* **2000**, *267*, 800–807. [CrossRef] [PubMed]
80. Silva, F.; Navarro, E.; Peñaranda, A.; Murcia-Flores, L.; Torres-Martínez, S.; Garre, V. A RING-finger protein regulates carotenogenesis via proteolysis-independent ubiquitylation of a White Collar-1-like activator. *Mol. Microbiol.* **2008**, *70*, 1026–1036. [CrossRef] [PubMed]
81. Quiles-Rosillo, M.D.; Torres-Martínez, S.; Garre, V. *Ciga*, a light-inducible gene involved in vegetative growth in *Mucor circinelloides* is regulated by the carotenogenic repressor *crgA*. *Fungal Genet. Biol.* **2003**, *38*, 122–132. [CrossRef]
82. Murcia-Flores, L.; Lorca-Pascual, J.M.; Garre, V.; Torres-Martínez, S.; Ruiz-Vázquez, R.M. Non-AUG translation initiation of a fungal RING finger repressor involved in photocarotenogenesis. *J. Biol. Chem.* **2007**, *282*, 15394–15403. [CrossRef] [PubMed]
83. Candau, R.; Avalos, J.; Cerdá-Olmedo, E. Gibberellins and carotenoids in the wild type and mutants of *Gibberella fujikuroi*. *Appl. Environ. Microbiol.* **1991**, *57*, 3378–3382. [PubMed]
84. Rodríguez-Ortiz, R.; Limón, M.C.; Avalos, J. Regulation of carotenogenesis and secondary metabolism by nitrogen in wild-type *Fusarium fujikuroi* and carotenoid-overproducing mutants. *Appl. Environ. Microbiol.* **2009**, *75*, 405–413. [CrossRef] [PubMed]
85. Studt, L.; Tudzynski, B. Gibberellins and the red pigments bikaverin and fusarubin. In *Biosynthesis and Molecular Genetics of Fungal Secondary Metabolites*; Martín, J.-F., García-Estrada, C., Zeilinger, S., Eds.; Springer: New York, NY, USA, 2014; pp. 209–238, ISBN 978-1-4939-1190-5.
86. Niehaus, E.-M.; Díaz-Sánchez, V.; von Bargen, K.W.; Kleigrewe, K.; Humpf, H.-U.; Limón, M.C.; Tudzynski, B. Fusarins and Fusaric Acid in Fusaria. In *Biosynthesis and Molecular Genetics of Fungal Secondary Metabolites*; Martín, J.-F., García-Estrada, C., Zeilinger, S., Eds.; Springer: New York, NY, USA, 2014; pp. 239–262, ISBN 978-1-4939-1190-5.
87. Garbayo, I.; Vílchez, C.; Nava-Saucedo, J.E.; Barbotin, J.N. Nitrogen, carbon and light-mediated regulation studies of carotenoid biosynthesis in immobilized mycelia of *Gibberella fujikuroi*. *Enzyme Microb. Technol.* **2003**, *33*, 629–634. [CrossRef]
88. Sokolovsky, V.Y.; Lauter, F.R.; Müller-Röber, B.; Ricci, M.; Schmidhauser, T.J.; Russo, V.E.A. Nitrogen regulation of blue light-inducible genes in *Neurospora crassa*. *J. Gen. Microbiol.* **1992**, *138*, 2045–2049. [CrossRef]
89. Connolly, L.R.; Smith, K.M.; Freitag, M. The *Fusarium graminearum* histone H3 K27 methyltransferase KMT6 regulates development and expression of secondary metabolite gene clusters. *PLoS Genet.* **2013**, *9*, e1003916. [CrossRef] [PubMed]
90. Wiemann, P.; Sieber, C.M.; von Bargen, K.W.; Studt, L.; Niehaus, E.M.; Espino, J.J.; Huss, K.; Michielse, C.B.; Albermann, S.; Wagner, D.; et al. Deciphering the cryptic genome: Genome-wide analyses of the rice pathogen *Fusarium fujikuroi* reveal complex regulation of secondary metabolism and novel metabolites. *PLoS Pathog.* **2013**, *9*, e1003475. [CrossRef] [PubMed]
91. García-Martínez, J.; Ádám, A.L.; Avalos, J. Adenylyl cyclase plays a regulatory role in development, stress resistance and secondary metabolism in *Fusarium fujikuroi*. *PLoS ONE* **2012**, *7*, e28849. [CrossRef] [PubMed]
92. Studt, L.; Humpf, H.-U.; Tudzynski, B. Signaling governed by G proteins and cAMP is crucial for growth, secondary metabolism and sexual development in *Fusarium fujikuroi*. *PLoS ONE* **2013**, *8*, e58185. [CrossRef] [PubMed]
93. Rodríguez-Ortiz, R. Análisis Genético y Molecular del Fenotipo CarS en *Fusarium*. Ph.D. Thesis, Universidad de Sevilla, Seville, Spain, 2012.
94. Domenech, C.E.; Giordano, W.; Avalos, J.; Cerdá-Olmedo, E. Separate compartments for the production of sterols, carotenoids and gibberellins in *Gibberella fujikuroi*. *Eur. J. Biochem.* **1996**, *239*, 720–725. [CrossRef] [PubMed]

10

Fungal Diversity in the Phyllosphere of *Pinus heldreichii* H. Christ—An Endemic and High-Altitude Pine of the Mediterranean Region

Jelena Lazarević [1,*] and Audrius Menkis [2]

1 Biotechnical faculty, University of Montenegro, Mihaila Lalića 15, 81 000 Podgorica, Montenegro
2 Department of Forest Mycology and Plant Pathology, Uppsala BioCenter, Swedish University of Agricultural Sciences, P.O. Box 7026, SE-75007 Uppsala, Sweden; audrius.menkis@slu.se
* Correspondence: enalazarevic@ac.me

Abstract: *Pinus heldreichii* is a high-altitude coniferous tree species naturaly occurring in small and disjuncted populations in the Balkans and southern Italy. The aim of this study was to assess diversity and composition of fungal communities in living needles of *P. heldreichii* specifically focusing on fungal pathogens. Sampling was carried out at six different sites in Montenegro, where 2–4 year-old living needles of *P. heldreichii* were collected. Following DNA isolation, it was amplified using ITS2 rDNA as a marker and subjected to high-throughput sequencing. Sequencing resulted in 31,831 high quality reads, which after assembly were found to represent 375 fungal taxa. The detected fungi were 295 (78.7%) Ascomycota, 79 (21.0%) Basidiomycota and 1 (0.2%) Mortierellomycotina. The most common fungi were *Lophodermium pinastri* (12.5% of all high-quality sequences), *L. conigenum* (10.9%), *Sydowia polyspora* (8.8%), *Cyclaneusma niveum* (5.5%), Unidentified sp. 2814_1 (5.4%) and *Phaeosphaeria punctiformis* (4.4%). The community composition varied among different sites, but in this respect two sites at higher altitudes (harsh growing conditions) were separated from three sites at lower altitudes (milder growing conditions), suggesting that environmental conditions were among major determinants of fungal communities associated with needles of *P. heldreichii*. Trees on one study site were attacked by bark beetles, leading to discolouration and frequent dieback of needles, thereby strongly affecting the fungal community structure. Among all functional groups of fungi, pathogens appeared to be an important component of fungal communities in the phyllosphere of *P. heldreichii*, especially in those trees under strong abiotic and biotic stress.

Keywords: needle pathogens; high altitude forests; DNA metabarcoding; Montenegro

1. Introduction

Pinus heldreichii is a high-altitude (grows at ca. 1200–2000 m) conifer tree species with a discontinuous and restricted distribution in the Mediterranean region. Its forests are naturaly regenerated and consists of small and disjuncted populations located in high mountain areas influenced by the Mediterranean climate in the Balkans and southern Italy. Although in the past *P. heldreichii* formed a continuous forest belt in the Balkans, currently its forests are scattered and largely isolated. Growing primary on shallow calcareous soils, it inhabits typical tree line habitats, often on steep ridges, mountain sides and screens. Such habitats are nutrient poor, exposed, dry and cold during the winter [1,2]. Larger *P. heldreichii* forests can still be found on mountain plateaus or in valleys situated at altitudes of ca. 1200–1300 m that are characterised by more developed soils (cambisols). Those forests are considered to be climazonal. i.e., characterised by permanent and stable vegetation forests of *P. heldreichii*, that are growing at ecological optimum for the species [3]. Being a tertiary relic, *P. heldreichii* evolved to survive severe winter frosts, and short, dry and warm summers with intensive sun radiation. Due to the short growing season, *P. heldreichii* grows very slowly, reaching ca. 15 m height after about 150 years [1]. It develops thick and branched roots, which penetrate deep in cracks of calcareous stones [2]. In the last few centuries, *P. heldreichii* forests were affected by intensive livestock grazing, exploitation and forest fires [4]. Intensive grazing, as well as the short and dry growing season have also resulted in very limited natural regeneration [2–4], though recent observations show some regeneration in the abandoned mountain areas [5].

Pinus heldreichii is a protected species both in Balkan countries and in Italy due to the key importance of nature conservation, protection against gravitational natural hazards, landscape conservation and recreation. Hence, *P. heldreichii* requires special attention, i.e., the development and application of conservation measures [1–4,6].

Fungi represent the largest microbial component associated with forest trees. They play key roles in forest ecosystems, especially in pines forest that are obligatory mutualistic, and are important contributors to the primary production and carbon, nutrient and water cycling [7–9]. Pathogenic fungi may negatively affect health and growth of forest trees [10,11], while fungal endophytes and epiphytes support ecological adaptations of host plants and constitute an important component of microbial biodiversity [12,13]. However, information about fungal communities associated with *P. heldreichii* is limited. Previous studies on needle pathogens of *P. heldreichii* have focused on either those that were affecting natural regeneration [14–16], or potentially invasive ones such as *Dothistroma septosporum* [17]. Information about the occurrence of *D. septosporum* in stands of *P. heldreichii* was not known until recently [17], what indicated the need for a wider assessment of fungal diversity, including fungal pathogens associated with the phyllosphere of this tree species.

The aim of this study was to assess the diversity and composition of fungal communities in living needles of *P. heldreichii* specifically focusing on fungal pathogens. Needles were sampled across the natural distribution range of *P. heldreichii* in Montenegro, including sites under different environmental conditions. This was expected to demonstrate potential site-specific effects of environmental conditions on fungal diversity and community composition. By using high-throughput sequencing of fungal ITS2 rDNA, we examined fungal communities in 2–4 year-old living needles of *P. heldreichii* from six sites situated in four mountain areas in Montenegro.

2. Materials and Methods

2.1. Study Sites

Samples were taken at six sites that were situated in four mountain areas, representing the *P. heldreichii* distribution range in Montenegro (Figure 1). The sampling sites were Orjen (ORJ), Prekornica (PRE), Kučka Korita (KKO), Kučka korita North (KKN), Kuči MT (KMT) and Prokletije (PRO) (Figure 1, Table 1).

The sampling sites differed in altitude, morphology of terrain and soil properties (Table 1, Figure 2). Sites at ORJ and KMT were on altitudes of ca. 1800 m, with trees growing on stone ridges, high inclinations, and lithosols. Sites at PRE, PRO and KKO were on an altitude of ca. 1250 m, on flat terrain and more developed soil that was leptic cambisol. A site at KKN was in close proximity to the KKO, but it was on a slope with a thinner soil layer (molic leptosol). Trees at the KKN were in groups heavily infested by *Tomicus* sp. bark beetles (Scolitinae). Trees at ORJ and KMT were growing at the upper tree line, and thus, these sites were considered as sites with harsh growing conditions, while other sites had moderate growing conditions.

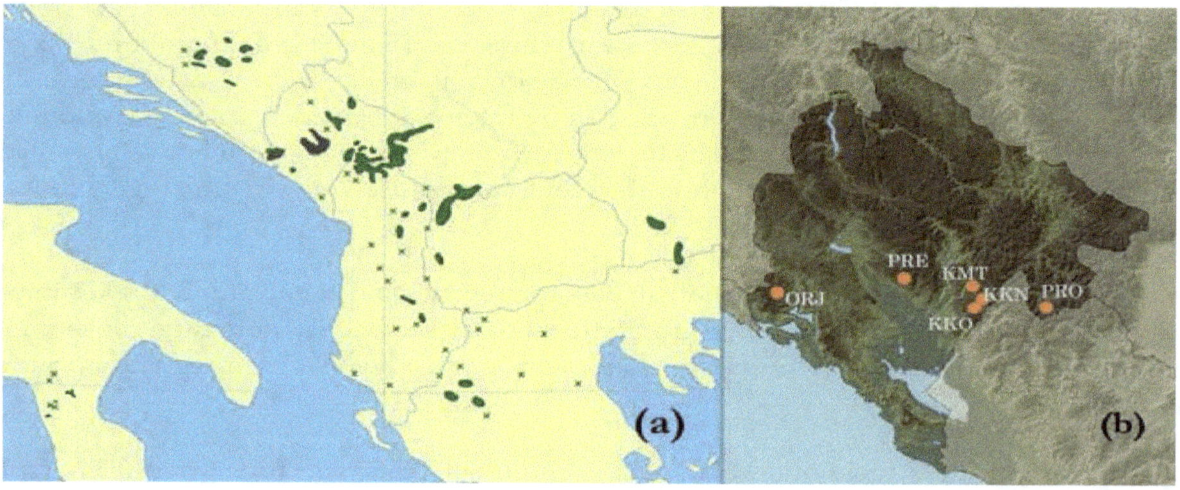

Figure 1. A map showing (**a**) distribution of *Pinus heldreichii* in the Mediterranean region; a grey area was added and represents an additional *P. heldreichii* site; the map is adapted from [18] and (**b**) sampling sites in Montenegro (ORJ-Orjen; PRE-Prekornica; KKO-Kuči Mt, Kučka korita; KKN-Kuči Mt., Kučka korita North, KMT- Kuči mountain, Momonjevo, PRO- Prokletije); a map was adapted from [19].

Table 1. Characteristics of *Pinus heldreichii* sampling sites.

Site	GPS Coordinates	Altitude (m)	Soil Type [a]	Climate [b]	Health [c]
Orjen, Reovačka greda (ORJ)	42°35′ N 18°35′ E	1700	Lithosol	Cfsb	Healthy-looking
Kuči Mt., Momonjevo (KMT)	42°36′ N 19°32′ E	1800	Lithosol	Cfb	Healthy-looking
Prekornica, Studeno (PRE)	42°38′ N 19°12′ E	1200	Leptic cambisol	Cfsb	Healthy-looking
Prokletije, Ropojana (PRO)	42°29′ N 19°48′ E	1250	Leptic cambisol	Cfb	Healthy-looking
Kučka korita (KKO)	42°29′ N 19°30′ E	1300	Leptic cambisol	Cfs″b	Healthy-looking
Kučka korita North (KKN)	42°30′ N 19°32′ E	1450	Molic leptosol	Cfs″b	Attacked by insects

[a] WRB soil clasification system [20,21], [b] Koopen climate classification [22,23], [c] Health status of the trees.

Figure 2. (**a**) *Pinus heldreichii* shoots with needles; trees growing at (**b**) Kuči Mountains, Momonjevo (KMT), and (**c**) Kuči Mountains, Kučka korita (KKO) sites.

Climate at the study sites is characterised as a humid warm temperate climate type (Cf), which is represented by subtypes Cfsb, Cfs″b and Cfb [22,23]. Summers are short, dry and chilly, and winters are cold and windy. At different study sites, the mean annual air temperature is between 1 °C (at PRO) and 6 °C (at ORJ). The mean daily summer maximum is between 7 and 11 °C, and the mean daily winter minimum is between −5 and 0 °C. The absolute minimum is in winter at ca. −30 °C and the maximum is in summer, at ca. 30 °C. The mean annual precipitation is highest at ORJ with ca. 3800 mm, and lowest at PRO with ca. 2000 mm. The rainfall reaches maximum in late autumn and early winter, while the minimum is during the summer months that is often followed by 40–70 day-long periods of droughts. The mean summer precipitation is 300 mm at ORJ and ca. 220 mm on other sites, making ca. 10% of the total annual precipitation [22].

The bedrock at the study sites is solid chalk limestone, which contains a small proportion of insoluble residues. In general, soils are poorly developed, very water porous, skeletal leptosols (rendzina), having only A horizon, characterized by accumulation of humus (A-R profile). According to the classification of texture, the soil is a sandy loam and the structure is powdery. Soil is faintly differentiated down in the soil profile. Soil is rich in humus content (10–25%), poor in calcium-carbonate and has a weak acid reaction (pH ca. 6). On more flat terrains, leptosol develops into the leptic cambisol, which contains initial B horizon. Brown B horizon improves the water retention capacity of soil [20,24].

2.2. Experimental Design and Sampling

At the sampling sites, stands of *P. heldreichii* were healthy-looking (Figure 2a). An exception was the KKN site, where trees were damaged by *Tomicus* sp. bark beetles, resulting in discolouration and fungal infection of needles. At all sites, the shoot dieback was occasional, but with higher rates at the KKN site.

Sampling was carried out in May 2015, i.e., before the beginning of the growing season. At each site, experimental design included sampling of five twigs with needles from five different trees. Hence, at each site, five mature *P. heldreichii* trees situated at a distance of ca. 50 m from each other were selected and five twigs of up to 15 cm long and up to 2 m from the ground were sampled from different parts of the crown using secateurs. This sampling approach was used in order (i) to get a joint representative sample per each site (all samples per site were amplified with the same barcode, see below); (ii) to compare fungal communities among different sites, but not within the same site. Collected twigs were placed in plastic bags and transported to the laboratory.

In the laboratory, twigs and needles were assessed for the presence of disease symptoms. The current-year needles were typically green and without disease symptoms. Symptomatic needles were generally two years old or older and symptoms included changed colour that was yellow, orange, pale green or yellowish with red bands. Other symptoms were dying needle tips and a necrotic base of needles. In the case of insect attack (KKN site), defoliation and necrotic needles were common. In order to sample the potentially entire fungal community associated with needles of *P. heldreichii*, both healthy looking needles and needles with disease symptoms were selected. No surface sterilization was carried out. Following morphological examination, up to 5 representative needles that were 2–4 years old, were selected per twig. Selected needles were cut into 0.5–1 cm long segments, containing a random mixture of both asymptomatic and symptomatic parts of needles, placed in 2 mL screw cap tubes, and stored frozen at −20 °C before DNA extraction.

2.3. DNA Isolation, Amplificationand Sequencing

DNA extractions were done from 150 samples (6 sites × 5 trees × 5 needle samples), which previously were freeze-dried for 48 h. For isolation of total DNA, needles were homogenised in a Fastprep machine (Precellys, Montigny-le-Bretonneux, France). The extraction was completed using CTAB protocol [25]. After extraction, DNA samples were purified using a JetQuick DNA purification kit (Genomed GmbH, Leinfelden, Germany). The DNA concentration of each sample was determined using a NanoDrop™ One spectrophotometer (Thermo Scientific, Rodchester, NY, USA) and adjusted to 1–10 ng/μL. Amplification by PCR of the ITS2 rDNA region was done using barcoded fungal-specific primer gITS7 [26] and barcoded universal primer ITS4 [27]. All 25 samples from the same site were amplified using primers with the same barcode, resulting in six different barcodes representing each site. Amplification of several samples with the same barcode was done to get a broader representativeness of fungal communities per site. Amplifications were performed using the Applied Biosystems 2720 thermal cycler (Foster City, CA, USA). An initial denaturation step started at 95 °C for 2 min, followed by 27 amplification cycles of denaturation at 95 °C for 30 s, annealing at 55 °C for 30 s, and extension at 72 °C for 60 s. The thermal cycling was ended by a final extension step at 72 °C for 7 min. The PCR products were analyzed using gel electrophoresis on 1% agarose gels stained with Nancy-520 (Sigma-Aldrich, Sweden). PCR products were purified using a sodium acetate protocol [28]. Purified PCR products were quantified using a Qubit fluorometer 4.0 (Thermo Fisher Scientific, Waltham, MA, USA), and an equimolar mix of all PCR products with six barcodes (one per each site) was used for high-throughput sequencing using a Pacific Biosciences platform (Menlo Park, CA, USA) at the SciLifeLab (Uppsala, Sweden).

2.4. Bioinformatics

Principles of bioinformatics followed that specified in [29]. The sequences obtained from the six samples that represented six sampling sites (Table 2), were subjected to quality control and clustering in the SCATA NGS sequencing pipeline [30]. The initial procedure started with quality filtering of the sequences that included the removal of sequences shorter than 200 bps, sequences with low read quality, primer dimers and homopolymers, which were collapsed to 3 bps before clustering. Only sequences containing a barcode and primer were retained. Then, the primer and sample barcodes were removed from the sequence, but information on the sample and sequence association was stored as meta-data. A single-linkage clustering based on 98% similarity was used to cluster sequences into different taxa. For each cluster, the sequence of the most common genotype was used for taxonomic identification. For clusters containing only two sequences, a consensus sequence was produced. The taxa were taxonomically identified using the GenBank database and the Blastn algorithm [31]. The following criteria were used for identification: sequence coverage >80%; 94–97% similarity to genus level and >98% similarity to species level. Sequences deviating from these criteria were identified only to a high taxonomic rank and were given unique names as shown in Table 3 and Table S1. Representative sequences of fungal nonsingletons are available from GenBank under accession numbers MT241905–MT242268.

Table 2. Generated high-quality ITS2 rDNA fungal sequences and detected diversity of fungal taxa at different sampling sites of *Pinus heldreichii*. Within the column *No. of Fungal Taxa*, values followed by the same letter in chi-square test do not differ significantly at $p > 0.05$.

Site	No. of Fungal Sequences	No. of Fungal Taxa	Shannon Diversity Index
Orjen –ORJ	4830	139 a	3.4
Kuči Mt.-KMT	4910	119 ab	2.9
Prekornica-PRE	5828	192 a	3.1
Prokletije -PRO	3679	127 a	3.2
Kučka korita -KKO	8748	167 b	2.9
Kučka korita North –KKN	3834	104 ab	2.7
All	31,829	375	-

Table 3. Relative abundance of the 25 most common fungal taxa detected in needles of *Pinus heldreichii* at six sites in Montenegro. Sites are as in Table 1.

Fungal Taxon	Phylum *	GenBank Accession No.	Compared (bp)	Similarity (%)	Sites						All
					ORJ	KMT	PRE	PRO	KKO	KKN	
Lophodermium pinastri	A	KC608049	361/361	100	4.7	22.1	14.3	14.2	14.9	0.4	12.5
Lophodermium conigenum	A	HM060650	453/453	100	-	0.1	32.4	5.5	14.5	2.8	10.9
Sydowia polyspora	A	KU516591	473/473	100	6.8	16.0	2.1	8.2	0.4	32.2	8.8
Cyclaneusma niveum	A	AF013223	442/442	99	1.3	1.0	4.5	5.7	12.9	0.9	5.5
Unidentified sp. 2814_1	A	MF976656	238/245	97	20.2	-	0.5	-	8.2	-	5.4
Phaeosphaeria pontiformis	A	KT000144	442/442	98	-	-	3.7	11.1	8.9	-	4.4
Unidentified sp. 2814_11	A	KU062806	203/252	81	-	-	1.3	9.0	10.3	-	4.1
Unidentified sp. 2814_10	A	KX220267	257/257	100	3.3	-	1.5	4.1	5.4	5.4	3.4
Phaeomoniella sp. 2814_15	A	GQ153187	257/259	99	7.2	5.9	0.2	1.2	2.9	0.5	3.0
Neocatenulostroma germanicum	A	KR995100	242/242	100	1.3	-	0.4	1.0	3.4	7.9	2.3
Microsphaeropsis olivacea	A	MH871969	249/249	100	0.7	0.1	0.3	0.1	-	16.8	2.2
Allantophomopsiella pseudotsugae	A	MH857222	240/240	100	6.4	7.1	-	-	-	-	2.1
Lachnellula calyciformis	A	MH858771	239/239	100	4.7	8.1	-	-	-	-	2.0
Unidentified sp. 2814_23	A	KF983527	209/244	86	1.4	-	1.5	3.7	0.3	4.3	1.5
Ramoconidiophora euphorbiae	A	MG592740	232/239	97	3.2	6.0	-	-	-	0.6	1.5
Geastrumia sp. 2814_26	A	FJ438389	220/235	94	2.7	2.6	0.4	0.6	0.5	2.3	1.4
Unidentified sp. 2814_16	A	KP892077	240/242	99	4.8	0.1	0.6	1.9	0.3	0.5	1.2
Unidentified sp. 2814_27	A	KT244857	209/248	84	-	-	3.5	5.0	-	-	1.2
Collophora sp. 2814_30	A	NR_137726	229/241	95	3.2	2.2	0.4	-	-	0.3	0.9
Unidentified sp. 2814_29	A	MF976139	209/240	87	1.4	-	0.4	4.1	0.4	-	0.9
Rhytismataceae sp. 2814_22	A	KR266446	234/237	99	0.4	-	1.0	1.4	1.6	-	0.8
Dothideomycetes sp. 2814_13	A	KP991484	253/257	98	1.1	3.5	-	0.1	-	0.4	0.8
Mollisia ligni	A	MF161301	237/241	98	-	4.7	-	-	-	-	0.7
Cenangium acuum	A	MG597445	239/239	100	-	-	3.9	-	-	-	0.7
Athelia acrospora	B	KP814375	296/299	99	-	-	1.8	0.1	1.3	-	0.7
All of 25 taxa					74.9	79.9	74.6	76.6	86.4	75.4	79.0

* A—Ascomycota, B—Basidiomycota.

2.5. Statistical Analyses

Rarefaction analysis was carried out using Analytical Rarefaction v.1.3 [32]. Differences in richness of fungal taxa in different study sites of *P. heldreichii* were compared by nonparametric chi-square testing [33]. As each of the datasets was subjected to multiple comparisons, confidence limits for *p*-values of chi-square tests were reduced the corresponding number of times as required by the Bonferroni correction [34]. The Shannon diversity index, qualitative Sorensen similarity index and nonmetric multidimensional scaling (NMDS) in Canoco 5 [33,35,36] were used to characterize the diversity and composition of fungal communities.

3. Results

The phyllosphere fungi of *P. heldreichii* were examined from four different mountain regions, represented by six sampling sites across the distribution range of *P. heldreichii* in Montenegro (Figure 1). Amplification of fungal ITS2 rDNA from 150 needle samples, PacBio sequencing and quality filtering resulted in 31,829 high quality reads. Sequence assembly and BLASTn analyses showed that the fungal community in the phyllosphere of *P. heldreichii* was composed of 375 fungal taxa (Table 2, Table S1). Nonfungal taxa and singletons were excluded. The detected fungi were 295 (78.7%) Ascomycota, 79 (21.0%) Basidiomycota and 1 (0.2%) Mortierellomycotina. Identification at least to genus level was successful for 254 (67.7%) out of 375 fungal taxa (Table S1), and those represented 78.5% of all fungal sequences.

In different study sites, the number of fungal taxa varied between 104 and 192 (Table 2). The chi-square test showed that the largest difference in richness of fungal taxa was between KKO and the remaining sites (Table 2). Rarefaction showed that fungal taxa detected in all sites did not reach the species saturation (Figure 3).

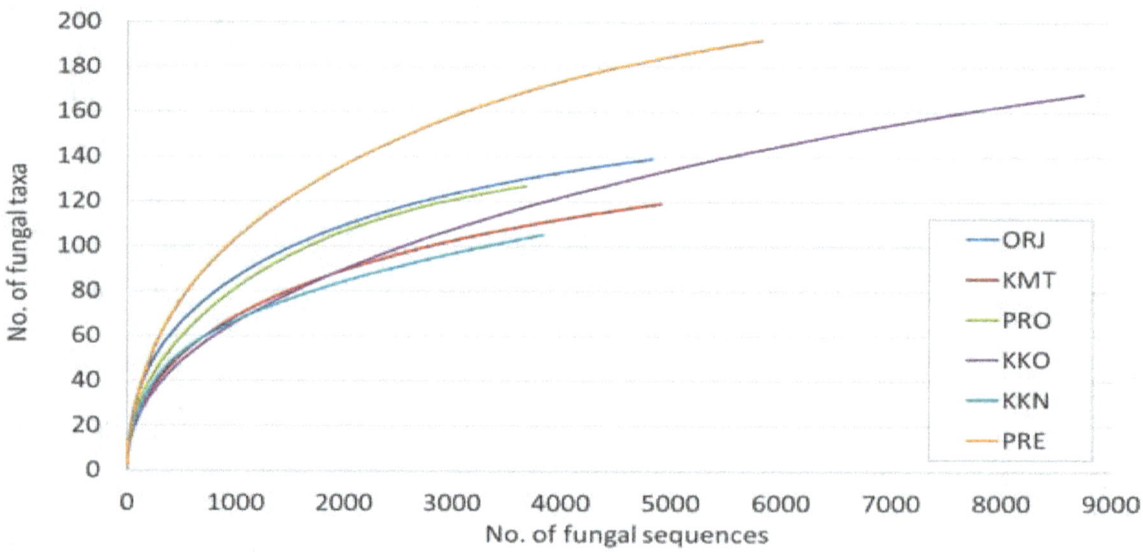

Figure 3. Rarefaction curves showing the relationship between the cumulative number of fungal taxa and the number of ITS2 rDNA sequences from needles of *Pinus heldreichii* from six different sites.

Information on the 25 most common fungal taxa representing 79.0% of all fungal sequences is in Table 3. The most common fungi in the phyllosphere of *P. heldreichii* were *Lophodermium pinastri* (12.5% of all fungal sequences), *Lophodermium conigenum* (10.9%), *Sydowia polyspora* (8.8%), *Cyclaneusma niveum* (5.5%) and Unidentified sp. 2814_1 (5.4%) (Table 3). The most common fungal pathogens of pine needles were *S. polyspora* (8.9%), *C. niveum* (5.5%), *Neocatenulostroma germanicum* (2.3%), *Allantophomopsiella pseudotsugae* (2.1%) and *Cenangium acuum* (0.7%) (Table 3). A pine needle pathogen, *Dothistroma septosporum* (0.25%) was also detected, but at lower relative abundance (Table S1). The other detected pathogenic species that are mainly known as wound pathogens of deciduous trees and/or agricultural crops, were *Phaeomoniella* 2814_15 (3.0%), *Ramoconidiophora euphorbiae* (1.5%), *Collophorina sp.* (0.9%), *Geastrumia sp.* (1.4%) and *Athelia acrospora* (0.7%). The detected fungal endophytes that are known to produce antimicrobial metabolites were *Phaeosphaeria pontiformis* (4.4%), *Microsphaeropsis olivacea* (2.2%), *Lachnellula calyciformis* (2.0%) and *Mollisia ligni* (0.7%). Ubiquitous saprotrophs were rare (Table S1).

The community composition of the phyllosphere fungi varied among different sites (Figure 4). Dothideomycetes dominated fungal communities at PRO and KKN sites, while Leotiomycetes dominated at KMT and PRE sites (Figure 4). At ORJ and KKO sites, the relative abundance of these two classes was similar (Figure 4). Sordariomycetes showed higher relative abundance at ORJ (22.6%), while all other fungal classes in different sampling sites were less abundant (Figure 4). Agaricomycetes (Basidiomycotina) were rare and their relative abundance at different sites varied between 0.03% and 6.3% (Figure 4).

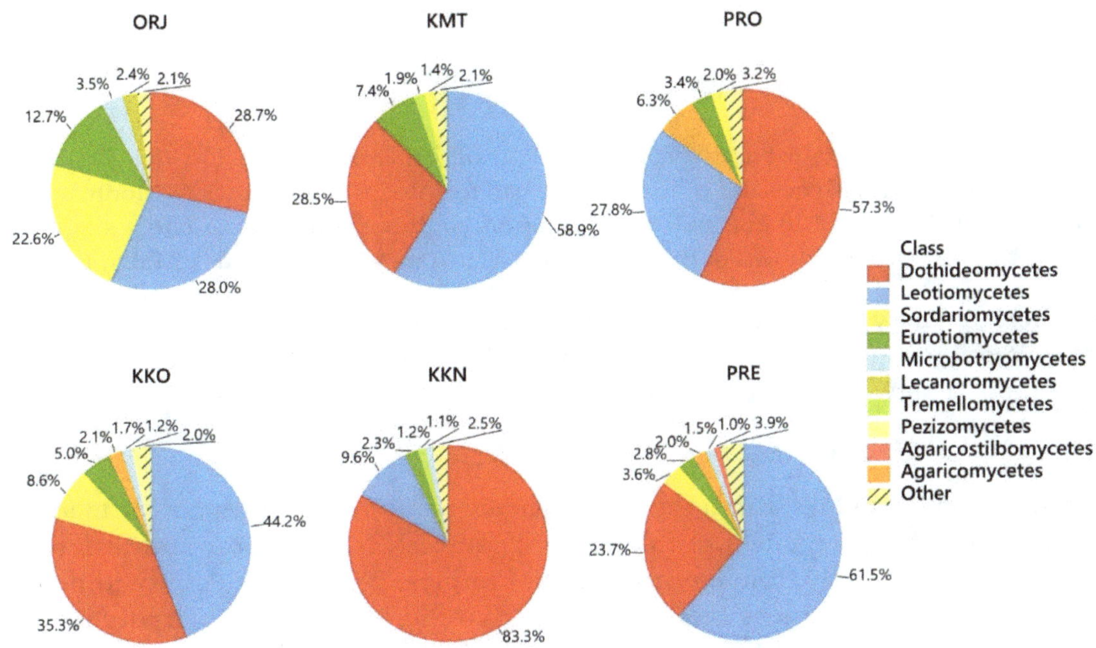

Figure 4. Relative abundance of different fungal classes in needles of *Pinus heldreichii* from six different sites in Montenegro. Classes comprising <1.0% of relative abundance for a given set of sequences were combined and shown as other. Sites are as in Table 1.

The NMDS of fungal communities associated with needles of *P. heldreichii* (Figure 5) showed that KKO, PRO and PRE sites clustered together and were separated from the remaining sites along axis 1. Fungal communities at the ORJ and KMT showed a closer proximity (Figure 5), while fungal communities at the KKN site differed from all remaining sites and were separated along both axis 1 and 2.

At ORJ and KMT sites that are characterized by harsh growing conditions, the fungal community was dominated by *Lophodermium pinastri*, *Sydowia polyspora*, *Phaeomoniella* sp., *Allanthopomopsiella pseudotsugae*, *Lachnellula calyciformis*, *Ramoconidiophora euphorbiae*, *Collophorina* sp., *Geastrumia* sp. and *Mollisia ligni* (Table 3). At PRE, PRO and KKO sites that are characterized by moderate growing conditions, fungal community was dominated by *Lophodermium conigenum*, *L. pinastri*, *Cyclaneusma niveum*, *Phaeosphaeria punctiformis*, Unidentified sp. 2814_11, *Cenangium acuum* and *Dothistroma septosporum*. At the KKN, with insect attacks and partly necrotic needles, fungal community was dominated by *S. polyspora*, *Microsphaeropsis olivacea*, *Neocatelunostroma germanicum*, Uncultured sp. 2814_10, *Chaetothyriales* 2814_18 and *Geastrumia* sp. (Table S1).

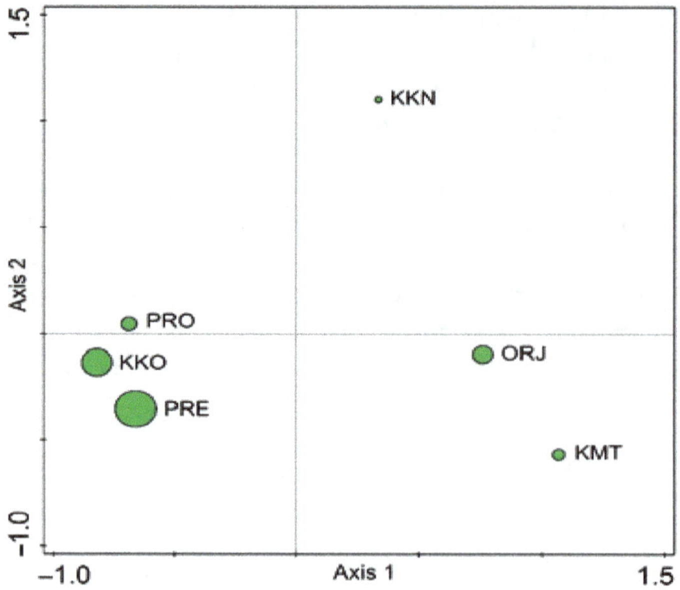

Figure 5. Ordination diagram based on nonmetric multidimensional scaling (NMDS) of fungal communities associated with needles of *Pinus heldreichii* from six different sites in Montenegro, with 56.5% variation on Axis 1 and 28.2% on Axis 2. Each point in the diagram represents a single site and the size of the point reflects richness of the fungal taxa as in Table 2. Sites are as in Table 1.

The Shannon diversity index of fungal communities in different study sites was between 2.7 and 3.4 (Table 2). Among different sites, Sørensen similarity index of fungal communities was low to moderate as it ranged between 0.28 and 0.54 (Table S2).

4. Discussion

A recent development of high-throughput sequencing methods provides powerful tools to explore fungal diversity. Such tools enable identification of complex fungal communities and individual community components directly from environmental samples. Besides, while providing detailed and semi-quantitative information, these tools enable studying the effects of different factors on fungal diversity and community composition [29,37,38]. By using PacBio sequencing, we detected nearly 400 fungal taxa, which were associated with needles of *P. heldreichii* including fungi present in very small abundances (Table S1). Many fungal taxa remained unidentified, which remains a major challenge in fungal taxonomy [39]. However, while using high-throughput sequencing, we need be aware of potential limitations that can include methodological biases, limitations of markers and bioinformatics challenges [40,41].

The observed diversity of the phyllosphere fungi can be considered being high and comparable with similar studies on different *Pinus* species. For example, there were 446 and 260 fungal taxa recorded during studies of needle-associated fungi of *Pinus sylvestris* in Sweden and Poland, respectively [39,42]. Furthermore, by using fungal culturing, there were 118 fungal taxa associated with needles of *Pinus taeda* in USA [43], 35 taxa associated with needles of *Pinus halepensis* [44] and *P. sylvestris* in Spain [45]. According to [45], the species diversity of fungal endophytes varied from 21 in *Pinus monticola* to 49 in *Pinus nigra*. At different study sites, the growing conditions of *P. heldreichii* had in general a minor effect of the absolute richness of fungal taxa, which was similar among different sites (Table 2). However, with respect to the number of sequences obtained from samples of each site, lower richness of fungal taxa was at the KKO (Table 2), what was likely due to the dominance of several fungal taxa such as, e.g., *Lophodermium* species, *Cylaneusma niveum*, Unidentified sp. 2814_11, *Phaeosphaeria pontiformis* and Unidentified sp. 2814_1. Several fungal taxa of the present study were recorded for the first time in the Balkan region such as, e.g., pathogens *Neocatelunostroma germanicum* and *Allantophomopsiella pseudotsugae*. The study also revealed a number of fungal pathogens that were previously known from agricultural crops as, e.g., *Phaeomoniella, Ramoconidiophora, Geastrumia* and *Athelia.* These could have been overlooked in other studies due to latent occurrence [46]. As *P. heldreichii* belongs to a group of forest trees that grows under harsh enviromental conditions, the possibility should not be excluded that both the host and the specific environment influence the composition of associated fungal communities. Studies on plant–endophyte associations in high stress habitats have revealed that at least some fungal endophytes can contribute to stress tolerance of host plants [12]. Indeed, the results of this study revealed site-specific differences in fungal communities associated with needles of *P. heldreichii*. For example, fungal communities were more similar among sites situated at lower altitudes, i.e., with moderate growing conditions (PRE, PRO, KKO) as compared to those sites at higher altitudes, i.e., with harsh growing conditions (ORJ, KMT) (Table 1, Figure 5). Further, the fungal community at the KKN site was largely different from the remaining sites, what was likely due to insect damage to trees, resulting in mainly symptomatic and partly necrotic needles. These observations demonstrate that the composition of fungal communities and their succession in needles of *P. heldreichii* can be determined by different abiotic and biotic factors. Different environmental factors have earlier been shown to play an important role in shaping fungal communities both in the phyllosphere and in the rhizosphere of forest trees [5,42,47,48].

Fungal pathogens represent an important biotic factor that may negatively affect health and growth of forest trees [11]. In the present study, a fungal pathogen *Sydowia polyspora* was among the most dominant fungi (Table 3). *Sydowia polyspora* has a wide geographical range [48] and is common in Europe. The pathogenicity of *S. polyspora* to young conifers (genera *Thuja, Abies, Tsuga, Larix, Picea* and *Pinus*) was previously reported [49]. In the Balkan region, it has been reported in forest plantations of *P. nigra* and *P. sylvestris* in Serbia, occurring on needles damaged by drought or frost [50]. It was also detected on needles of *P. halepensis* in Italy and Spain [44,51], *P. sylvestris* in Poland and Lithuania [29,39] and on *P. ponderosa* in North America [52]. *Sydowia polyspora* has recently been reported as a pathogen that dominates fungal communities vectored by bark beetles associated with *P. radiata, P. nigra* subsp. *salzmannii* and *P. sylvestris* in Spain [53] and *Pinus yunnanensis* in China [54]. *Sydowia* symptoms include needle necrosis and shoot dieback. The fungus is favoured by a warm climate, especially if the host is stressed by summer drought or insect or mite attack [53]. Indeed, in the present study the highest relative abundance of *S. polyspora* was at the KKN (Table 2), the site that was subjected to insect attack. Besides, it appears that *S. polyspora* had a negative effect on the relative abundance of *L. pinastri* (Table 2), which is in agreement with Behnke-Borowczyk et al. [39], who have observed the similar pattern on *P. sylvestris*. *Cyclaneusma niveum* was also among the dominant fungi (Table 3). It is one of the two fungal pathogens causing Cyclaneusma needle cast, which is an important needle disease reported from many pine species including *P. nigra* and *P. sylvestris* in the Balkan region and in Crimea [50,55,56], and reported from *P. halepensis* in Spain [44]. It was suggested that *C. niveum* is more frequently associated with *P. nigra* under warmer climate conitions, while *Cyclaneusma minus* is favoured by wet, humid, above-freezing conditions, and thus, more commonly infects *P. sylvestris* grown in central and northern Europe [39,50,57]. Our study is in agreement, as on *P. heldreichii, C. niveum* dominated the fungal community, while *C. minus* was recorded at a low frequency (Table S1). Interestingly, *C. niveum* was more common on sites with prevailing moderate than harsh enviromental conditions. *Neocatenulostroma germanicum* is a recently identified fungal pathogen causing needle blight on *P. mugo, P. sylvestris* and *P. nigra* subsp. *pallsiana* in Lithuania, Ukraine and Poland, where it was found to be commonly associated with *Dothistroma, Lecanosticta acicola* and Cyclaneusma needle cast infections [39,58]. In this study, it was present in the needles of *P. heldreichii* from five sites, with considerably higher relative abundance on symptomatic needles at the KKN site and the neighbouring KKO site as compared to the remaining sites. *Allantophomopsiella pseudotsugae* (syn. *Phomopsis pseudotsugae*) was a pathogen commonly detected on needles of *P. heldreichii* grown on sites under harsh growth conditions (ORJ and KMT) (Table 3). While being a pathogen of conifers, that is, mainly infecting pines, it develops on young shoots [59] and has been reported in several European countries [60]. The present finding of the fungus represents a new record for this part of Europe, indicating that *A. pseudotsugae* could be among important pathogens of *P. heldreichii* growing under harsh conditions. *Cenangium accum* was recorded on *P. heldreichii* needles at the PRE site (Table 3). Previously it was demonstrated as a pathogen of weakened *P. nigra* and *P. sylvestris* in the Balkan region [50,55]. It has also been shown to be associated with Cyclaneusma needle cast in *P. sylvestris* in western Poland [39]. *Cenangium accum* develops predominantly on needles damaged by frosts or drought and its development is favoured by a high humidity. Similarly, *Dothistroma septosporum*, was recorded at a low relative abundance and only on the PRE (Table S1). Investigation of *Dothistroma septosporum* accomplished using PCR and species-specific primers, has demonstrated the presence of this potentially invasive pathogen across the *P. heldreichii* distribution range in Montenegro [17]. It was suggested that enviromental conditions present at *P. heldreichii* sites suppress the development of Dothistroma needle blight and that *P. heldreichii* is only slightly susceptible to *D. septosporum*. However, the infection level by *D. septosporum* may vary in different years, especially after rainy periods [10,17].

Among the other fungi detected, there were fungal endophytes that are known to be commonly associated with different tissues of forest trees [61,62]. The relationship between endophytic fungi and plants is not clearly understood and may change depending on the health status of the plant. However, for at least part of their life, they colonise plant tissues asymptomatically [12,63,64]. In the present study, *L. pinastri* and *L. conigenum* were the most commonly detected fungi on *P. heldreichii* needles. They are globaly distributed and commonly associated with pines [42,65]. It was shown recently that *L. pinastri* colonises healthy needles latently as an endophyte, initiates active growth at the begining of needle senescence and sporulates after needle fall. It is a dominant coloniser of dying needles and a saprotroph contributing to a needle decomposition [39]. In the Balkan region, *L. pinastri* was frequently reported from *P. nigra* and *P. sylvestris* grown in forests, forest nurseries and plantations [50,55]. *Lophodermium conigenum* is known as a coloniser of damaged needles and can also form fruitbodies on broken branches [65]. In comparison to *L. pinastri*, it is less frequently reported on both *P. nigra* and *P. sylvestris* in Serbia [50] and on *P. sylvestris* in Spain [45] and the UK [66]. In the present study, *L. pinastri* was present on all investigated sites, while *L. conigenum* showed a higher relative abundance on sites characterised by moderate growth conditions (PRO, KKO and PRE). The relative abundance of *L. pinastri* was at least 10 times lower on the KKN site than on any other site, what was likely due to its exclusion by pathogenic fungi which dominated on this site.

Some fungal endophytes of conifers can produce bioactive secondary metabolites, supporting ecological adaptations of host plants owing strong antimicrobial activities [63,67]. *Microsphaeropsis olivacea* (syn. *Coniothyrium olivaceum*) is a fungal endophyte, which was shown to produce bioactive compounds that are considered as promising antibacterial and antifungal agents [68]. *Microsphaeropsis olivacea* was among the most abundant species on the KKN, where *Sydowia polyspora* and *Neocatenulostroma germanicum* were also abundant. *Mollisia ligni* has been shown to produce mollisin, a compound that is known for a strong fungicidal activity against *Sydowia polyspora* [69]. A high frequency of *Mollisia ligni* may result in decreased frequency of *S. polyspora* on *P. sylvestris* [39]. On *P. heldreichii*, *N. germanicum* was absent in the cases where *M. ligni* was present (KMT).

The study has also detected a number of fungi previously known as pathogens of deciduous trees and/or of agricultural crops including *Phaeomoniella* 2814_15, *Ramoconidiophora euphorbiae* and *Geastrumia sp.*, which were for the first time recorded on conifers, possibly as latent endophytes. Fungi from genus *Phaeomoniella* are important pathogens of grapevines, causing grapevine trunk disease [70], *Ramoconidiophora* and *Collophorina* species have been reported from necrotic and symptomless wood and leaves of *Prunus, Castanea, Vitis,* and from roots of *Calluna* [71], while *Geastrumia sp.* is known as one of the species involved in disease complex of sooty blotch and flyspeck of apple (*Malus domestica*) [72]. *Phaeomoniella sp., Ramoconidiophora, Collophorina* and *Geastrumia* were more abundant on sites with harsh growth conditions (ORJ and KMT). Among other pathogens, *Lachnellula calyciformis* was abundant on sites under harsh growing conditions, while *Phaeosphaeria pontiformis* was on sites characterised by moderate growing conditions. Corticoid fungus from genus *Athelia*, which typically occurs as saprotroph [73,74] was present on sites with moderate growth conditions.

Fungi may modulate stress tolerance, enhance growth and increase reproduction [7,47,75]. Habitats subjected to a high abiotic stress can be inhabited by a species-specific fungal community [76,77]. Our results demonstrated that needles of *P. heldreichii* constitute a habitat for a species-rich community of fungi, the composition of which was found to be largely driven by environmental conditions and/or health status of host trees. Indeed, similar patterns were also observed for fungal communities in roots of *P. heldreichii* [5], indicating that without the habitat-adapted fungal symbionts, plants are hardly capable of surviving in high stress habitats.

Fungal communities evolve together with host plants [12], suggesting that relic and endemic pine species such as *P. heldreichii* [5,54], can be associated with a specific fungal community. For example among the 50 mushroom species, which were recorded in *P. heldreichii* forests to date [78–81], many were rarely observed and exclusively recorded in *P. heldreichii* forests. Hence, *Chalciporus ammarelus*, *Geastrum minimum*, *Hygroporus gliocyclus*, *Hygroporus hypothejus*, *Rhizopogon roseolus* and *Morchella esculenta* s.l. are listed on the preliminary red list [79], and thus, are protected [82,83]. Furthermore, recently described species *Erioscyphella curvispora* [84] was discovered in *P. heldreichii* needle litter and a newly described genus *Perzia*, accommodated by a type species *Perzia triseptata*, gen. nov. was discovered on the xeric bark of *P. heldreichii* [85]. Other findings include newly described species *Velutarina bertinscensis* [86], *Peziza montrivicola* [87], *Cenangiopsis ragvanii*, *C. junipericola* [88] and rare species *Trichophaea flavobrunea* [89] that were determined on other tree species growing on *P. heldreichii* sites. Those findings represents a special value to science, and together with overall recorded fungal diversity demonstrate a high value of *P. heldreichii* habitats as being unique and biodiversity hotspots in high-altitude mountain areas.

In summary, the results demonstrated that needles of *P. heldreichii* were associated with a species-rich community of fungi, the composition of which was found to be largely dependent on environmental conditions and/or health status of host trees. Needle pathogens appeared to be an important component of fungal communities associated with needles of *P. heldreichii*, but caused limited damage likely due to interaction with other fungal species and/or due to plant defence mechanisms. In order not to become a threat to the health and growth of forest trees, the present and potentially new invasive pathogens should be monitored regularly. Further research is also needed to understand patterns of different fungal species coexistence.

Author Contributions: Conceptualisation, J.L. and A.M.; methodology, J.L. and A.M.; formal analysis, J.L. and A.M.; resources, J.L.; data curation, J.L. and A.M.; writing—original draft preparation, J.L.; writing—review and editing, J.L. and A.M.; funding acquisition, J.L. All authors have read and agreed to the published version of the manuscript.

References

1. Janković, M.M. Betrachtungen uber gegenseitigen Bezichungen der Molika (*Pinus peuce*) und Panzerkiefer (*Pinus heldreichii*) sowie auch Uber ihre okologishen Eigenschaften, besonders in Bezug auf ihre geologishe Grundlage. *Bull. L'Inst. Jard. Bot. L'univ. Beogr.* **1960**, *1*, 141–181.
2. Jovanović, B. *Dendrology*, 6th ed.; Faculty of Forestry, University of Belgrade: Belgrade, Serbia, 2007; p. 536.
3. Blečić, V.; Lakušić, R. Forests of Pinius heldreichii Christ at Štitovo and Bjelasica in Montenegro. *Bull. Repub. Inst. Prot. Nat. Mus. Nat. Hist. Titogr.* **1969**, *4*, 5–10.
4. Stevanović, V.; Jovanović, S.; Lakušić, D. Diversity of vegetation of Yugoslavia. In *Biodiversity of Yugoslavia with List of Species of Special Importance*; Radović, I., Angelus, J., Eds.; Yugoslavia: Ecolibri, Beograd, 1995; pp. 219–241.

5. Lazarević, J.; Menkis, A. Fungi inhabiting fine roots of *Pinus heldreichii* in the Montenegrin montane forests. *Symbiosis* **2018**, *74*, 189–197. [CrossRef]
6. Vendramin, G.G.; Fineschi, S.; Fady, B. *EUFORGEN Technical Guidelines for Genetic Conservation and Use for Bosnian Pine (Pinus heldreichii)*; Bioversity International: Rome, Italy, 2008; ISBN 978-92-9043-789-5.
7. Baldrian, P. Forest microbiome: Diversity, complexity and dynamics. *Fems Mycrobiol. Rev.* **2017**, *41*, 109–130. [CrossRef]
8. Stewart, J.E.; Kim, M.S.; Klopfenstein, N.B. Molecular genetic approaches toward understanding forest-associated fungi and their interactive roles within forest ecosystems. *Curr. For. Rep.* **2018**, *4*, 72–84. [CrossRef]
9. Zanne, A.E.; Abarenkov, K.; Afkhami, M.E.; Aguilar-Trigueros, C.A.; Bates, S.; Bhatnagar, J.M.; Busby, P.E.; Christian, N.; Cornwell, W.K.; Crowther, T.W.; et al. Fungal functional ecology: Bringing a trait-based approach to plant-associated fungi. *Biol. Rev.* **2020**, *95*, 409–433. [CrossRef]
10. Drenkhan, R.; Tomešová-Haataja, V.; Fraser, S.; Bradshaw, R.; Vahalik, P.; Mullett, M.; Martín-García, J.; Bulman, L.; Wingfield, M.; Kirisits, T.; et al. Global geographic distribution and host range of *Dothistroma* species: A comprehensive review. *For. Pathol.* **2016**, *46*, 408–442. [CrossRef]
11. Elvira-Recuenco, M.; Cacciola, S.O.; Sanz-Ros, A.V.; Garbelotto, M.; Aguayo, J.; Solla, A.; Mullett, M.; Drenkhan, T.; Oskay, F.; Aday Kaya, A.G.; et al. Potential Interactions between Invasive *Fusarium circinatum* and Other Pine Pathogens in Europe. *Forests* **2020**, *11*, 7. [CrossRef]
12. Rodriguez, R.J.; Redman, R.S. More than 400 million years of evolution and some plants still can't make it on their own: Plant stress tolerance via fungal symbiosis. *J. Exp. Bot.* **2008**, *59*, 1109–1114. [CrossRef]
13. Terhonen, E.; Blumenstein, K.; Kovalchuk, A.; Asiegbu, F.O. Forest tree microbiomes and associated fungal endophytes: Functional roles and impact on forest health. *Forests* **2019**, *10*, 42. [CrossRef]
14. Lazarević, J. Contribution to the study of physiological characteristics of the pathogenic fungus *Herpotrichia juniperi*. *Agric. For.* **2003**, *49*, 95–109.
15. Lazarević, J. *Pinus heldreichii* as the host of brown felt blight (*Herpotrichia juniperi*) in Montenegro. *Mycol. Monten.* **2004**, *7*, 77–89.
16. Lazarević, J. Contribution to the study of te morphological characteristics of the pathogenic fungus *Herpotrichia juniperi*. *Mycol. Monten.* **2005**, *8*, 137–147.
17. Lazarević, J.; Davydenko, K.; Millberg, H. Dothistroma Needle Blight on High Altitude Pine Forests in Montenegro. *Balt. For.* **2017**, *23*, 294–302.
18. Caudullo, G.; Welk, E.; San-Miguel-Ayanz, J. Chorological maps for the main European woody species. *Data Brief* **2017**, *12*, 662–666. [CrossRef]
19. Dinaric_Alps_map-fr.svg: Sémhur, Relief Map of Montenegro svg. Available online: https://commons.wikimedia.org/wiki/File:Relief_Map_of_Montenegro.svg (accessed on 10 January 2020).
20. Fuštić, B.; Đuretić, G. *Soils of Montenegro*; Biotechnical Institute, University of Montenegro: Podgorica, Montenegro, 2000; pp. 156–172.
21. IUSS Working Group WRB. Descriptions, distribution, use and management of Reference Soil Groups. In *World Reference Base for Soil Resources. International Soil Classification System for Naming Soils and Creating Legends for Soil Maps, Udate 2015*; World Soil Resource Reports No. 106; Food and Agriculture Organization of the United Nations: Rome, Italy, 2015; pp. 144–180.
22. Burić, M.; Micev, B.; Mitrović, L. *Atlas of Climate of Montenegro*; Montenegrin Academy of Science and Arts: Podgorica, Montenegro, 2012; pp. 18–62, 86–102, 126–129.
23. Burić, D.; Ducić, V.; Mihajlović, J. The climate of Montenegro: Modificators and types-part two. *Bull. Serb. Geogr. Soc.* **2014**, *94*, 73–90. [CrossRef]
24. Janković, M.M.; Stefanović, K. Ecological relations between relict (sub)endemic pine species *Pinus heldreichii* and the character of geological substrate and soil in Jugoslavia. *Ekologija* **1971**, *6*, 49–61.
25. Menkis, A.; Ihrmark, K.; Stenlid, J.; Vasaitis, R. Root-associated fungi of Rosa rugosa grown on the frontal dunes of the Baltic Sea coast in Lithuania. *Microb. Ecol.* **2014**, *67*, 769–774. [CrossRef]
26. Ihrmark, K.; Bodeker, I.T.M.; Cruz-Martinez, K.; Friberg, H.; Kubartova, A.; Schenck, J.; Strid, Y.; Stenlid, J.; Brandstrom-Durling, M.; Clemmensen, K.E.; et al. New primers to amplify the fungal ITS2 region—Evaluation by 454-sequencing of artificial and natural communities. *FEMS Microbiol. Ecol.* **2012**, *82*, 666–677. [CrossRef]

27. White, T.J.; Bruns, T.; Lee, S.; Taylor, J. Amplification and direct sequencing of fungal ribosomal RNA genes for phylogenetics. In *PCR Protocols: A Guide to Methods and Applications*; Innis, M.A., Gelfand, D.H., Sninsky, J.J., White, T.J., Eds.; Academic Press, Inc.: San Diego, CA, USA, 1990; pp. 315–322.
28. Menkis, A.; Marčiulynas, A.; Gedminas, A.; Lynikienė, J.; Povilaitienė, A. High-throughput sequencing reveals drastic changes in fungal communities in the phyllosphere of Norway spruce (*Picea abies*) following invasion of the spruce bud scale (*Physokermes piceae*). *Microb. Ecol.* **2015**, *70*, 904–911. [CrossRef]
29. Lynikienė, J.; Marčiulynienė, D.; Marčiulynas, A.; Gedminas, A.; Vaičiukynė, M.; Menkis, A. Managed and unmanaged *Pinus sylvestris* forest stands harbour similar diversity and composition of the phyllosphere and soil fungi. *Microorganisms* **2020**, *8*, 259. [CrossRef]
30. Brandström Durling, M.; Clemmensen, K.; Stenlid, J.; Lindahl, B. SCATA—An Efficient Bioinformatic Pipeline for Species Identification and Quantification after High-Throughput Sequencing of Tagged Amplicons (Submitted). 2011. Available online: https://www.scata.mykopat.slu.se (accessed on 12 November 2019).
31. Altschul, S.F.; Madden, T.L.; Schäffer, A.A.; Zhang, J.; Zhang, Z.; Miller, W.; Lipman, D.J. Gapped BLAST and PSI-BLAST: A new generation of protein database search programs. *Nucleic Acids Res.* **1997**, *25*, 3389–3402. [PubMed]
32. Analytic Rarefaction, UGA Stratigraphy Lab. Available online: http://www.uga.edu/strata/software/index.html (accessed on 10 January 2020).
33. Magurran, A.E. *Ecological Diversity and Its Measurement*; Princeton University Press: Princeton, NJ, USA, 1988; p. 192.
34. Sokal, R.R.; Rohlf, F.J. *Biometry: The Principles and Practice of Statistics in Biological Research*, 4th ed.; H. Freeman & Company: New York, NY, USA, 2005; p. 896.
35. Shannon, C.E. A mathematical theory of communication. *Bell Syst. Tech. J.* **1948**, *27*, 379–423.
36. Ter Braak, C.J.F.; Smilauer, P. *Canoco Reference Manual and User's Guide to Canoco for Windows: Software for Canonical Community Ordination, Version 4*; Microcomputer Power: Ithaca, NY, USA, 1998; p. 351.
37. Bringel, F.; Couee, I. Pivotal roles of phyllosphere microorganisms at the interface between plant functioning and atmospheric trace gas dynamics. *Front. Microbiol.* **2015**, *6*, 486. [CrossRef] [PubMed]
38. Porazinska, D.L.; Sung, W.; Giblin-Davis, R.M.; Thomas, W.K. Reproducibility of read numbers in high-throughput sequencing analysis of nematode community composition and structure. *Mol. Ecol. Resour.* **2010**, *10*, 666–676. [CrossRef] [PubMed]
39. Behnke-Borowczyk, J.; Kwaśna, H.; Kulawinek, B. Fungi associated with *Cyclaneusma* needle cast in Scots pine in the west of Poland. *For. Pathol.* **2019**, *49*, e12487. [CrossRef]
40. Lindahl, B.D.; Nilsson, R.H.; Tedersoo, L.; Abarenkov, K.; Carlsen, T.; Kjøller, R.; Kõljalg, U.; Pennanen, T.; Rosendahl, S.; Stenlid, J.; et al. Fungal community analysis by high-throughput sequencing of amplified markers—A user's guide. *New Phytol.* **2013**, *199*, 288–299.
41. Tedersoo, L.; Tooming-Klunderud, A.; Anslan, S. PacBio metabarcoding of Fungi and other eukaryotes: Errors, biases and perspectives. *New Phytol.* **2018**, *217*, 1370–1385.
42. Millberg, H.; Boberg, J.; Stenid, J. Changes in fungal community of Scots pine (*Pinus sylvestris*) needles along a latitudinal gradient in Sweden. *Fungal Ecol.* **2015**, *17*, 126–139. [CrossRef]
43. Oono, R.; Lefèvre, E.; Simha, A.; Lutzoni, F. A comparison of the community diversity of foliar fungal endophytes between seedling and adult loblolly pines (*Pinus taeda*). *Fungal Biol.* **2015**, *119*, 917–928. [CrossRef]
44. Botella, L.; Santamaria, O.; Diez, J.J. Fungi associated with decline of *Pinus halepensis* in Spain. *Fungal Divers.* **2010**, *40*, 1–11. [CrossRef]
45. Sanz-Ros, A.V.; Muller, M.M.; San Martin, R.; Diez, J.J. Fungal endophytic communities on twigs of fast and slow growing Scots pine (*Pinus sylvestris* L.) in northern Spain. *Fungal Biol.* **2015**, *119*, 870–883. [CrossRef]
46. Parfitt, D.; Hunt, J.; Dockrell, D.; Rogers, H.J.; Boddy, L. Do all trees carry the seeds of their own destruction? PCR reveals numerous wood decay fungi latently present in sapwood of a wide range of angiosperm trees. *Fungal Ecol.* **2010**, *3*, 338–346. [CrossRef]
47. Bowman, E.A.; Arnold, A.E. Distribution of ectomicorrhizal and foliar endophytic fungal communities associated with *Pinus ponderosa* along a spatially constrained elevation gradient. *Am. J. Bot.* **2018**, *105*, 687–699. [CrossRef]
48. Botella, L.; Diez, J.J. Phylogenetic diversity of fungal endophytes in Spanish stands of *Pinus halepensis*. *Fungal Divers.* **2011**, *47*, 9–18.

49. Talgo, V.; Chastanger, G.; Thomsen, I.M.; Cech, T.; Riley, K.; Lang, K.; Klemsdal, S.S. *Sydowia polyspora* associated with current season needle necrosis (CSNN) on true fir (*Abies* spp.). *Fungal Biol.* **2010**, *114*, 545. [CrossRef] [PubMed]
50. Karadžić, D.; Milijašević, T. The most important parasitic and saprophytic fungi in Austrian pine and Scots pine plantations in Serbia. *Bull. Fac. For.* **2008**, *97*, 147–170.
51. Tinivella, F.; Dani, E.; Minuto, G.; Minuto, A. First Report of *Sydowia polyspora* on Aleppo Pine (*Pinus halepensis*) in Italy. *Plant Dis.* **2014**, *98*, 2. [CrossRef]
52. Ridout, M.; Newcombe, G. *Sydowia polyspora* is both a Foliar Endophyte and a Preemergent Seed Pathogen in *Pinus Ponderosa*. *Plant Dis.* **2018**, *118*, 3. [CrossRef]
53. Muñoz-Adalia, E.J.; Sanz-Ros, A.V.; Flores-Pacheco, J.A.; Hantula, J.; Diez, J.J.; Vainio, E.J.; Fernández, M. *Sydowia polyspora* Dominates Fungal Communities Carried by Two *Tomicus* Species in Pine Plantations Threatened by *Fusarium circinatum*. *Forests* **2017**, *8*, 127. [CrossRef]
54. Pan, Y.; Ye, H.; Lu, J.; Chen, P.; Zhou, X.D.; Qiao, M.; Yu, Z.-F. Isolation and identification of *Sydowia polyspora* and its pathogenicity on *Pinus yunnanensis* in Southwestern China. *J. Phytopatol.* **2018**, *166*, 386–395. [CrossRef]
55. Dobreva, M.; Georgieva, M.; Dermedzhiev, P.; Nachev, R.; Velinov, V.; Terziev, P.; Georgiev, G. Fungal pathogens associated with *Pinus species* in the region of forest protection station Plovdiv in the period 2013–2016. *For. Sci.* **2016**, 1–2.
56. Minter, D.W.; Dudka, I.O. *Fungi of Ukraine. A Preliminary Checklist*; International Mycological Institute: Egham, Surrey, UK; M.G. Kholodny Institute of Botany: Kiev, Ukraine, 1996; pp. 1–361.
57. Drenkhan, R.; Hanso, M. Recent invasion of foliage fungi of pines (*Pinus* spp.) to the Northern Baltics. *For. Stud.* **2009**, *51*, 49–64. [CrossRef]
58. Markovskaja, S.; Kačergius, A.; Davydenko, K.; Fraser, S. First record of *Neocatenulostroma germanicum* on pines in Lithuania and Ukraine and its co- occurrence with *Dothistroma* spp. and other pathogens. *Path* **2016**, *46*, 522–533. [CrossRef]
59. Wajihi, A.H.; Lee, S.; Das, K.; Eom, A.; Jung, H. First Report of *Allantophomopsiella pseudotsugae* Isolated from Soil in Korea. *Korean J. Mycol.* **2019**, *47*, 29–34. [CrossRef]
60. Crous, P.W.; Quaedvlieg, W.; Hansen, K.; Hawksworth, D.L.; Groenewald, J.Z. *Phacidium* and *Ceuthospora* (Phacidiaceae) are congeneric: Taxonomic and nomenclatural implications. *Ima Fungus* **2014**, *5*, 173–193. [CrossRef]
61. Sieber, T.N. Endophytic fungi in forest trees: Are they mutualists? *Fungal Biol. Rev.* **2007**, *21*, 75–89. [CrossRef]
62. Kogel, K.-H.; Franken, P.; Huckelhoven, R. Endophyte or parasite-what decides. *Curr. Opin. Plant Biol.* **2006**, *9*, 358–369. [CrossRef]
63. Soltani, J.; Hosseyni Moghaddam, M.S. Fungal Endophyte Diversity and Bioactivity in the Mediterranean Cypress *Cupressus Sempervirens*. *Curr. Microbiol.* **2015**, *70*, 580–586.
64. Jeewon, R.; Yeung, Q.S.; Wannasinghe, D.N.; Rampadarath, S.; Puchooa, D.; Wang, H.K.; Hyde, K.D. Hidden mycota of pine needles: Molecular signatures from PCRDGGE and Ribosomal DNA phylogenetic characterization of novel phylotypes. *Sci. Rep.* **2018**, *8*, 18053. [CrossRef]
65. Minter, D.W.; Staley, J.M.; Millar, C.S. Four species of *Lophodermium* on *Pinus sylvestris*. *Trans. Br. Mycol. Soc.* **1978**, *71*, 295–301.
66. Reignoux, S.A.; Green, S.; Ennos, R.A. Molecular identification and relative abundance of cryptic *Lophodermium* species in natural populations of Scots pine, *Pinus sylvestris* L. *Fungal Biol.* **2014**, *118*, 835–845. [CrossRef] [PubMed]
67. Stierle, A.A.; Stierle, D.B. Bioactive secondary metabolites produced by fungal Endophytes of conifers. *Nat. Prod. Commun.* **2015**, *10*, 1671–1682. [PubMed]
68. Wang, L.W.; Xu, B.G.; Wang, J.Y.; Su, Z.Z.; Lin, F.C.; Zhang, C.L.; Kubicek, C.P. Bioactive metabolites from Phoma species, an endophytic fungus from the Chinese medicinal plant Arisaema erubescens. *Appl. Microbiol. Biotechnol.* **2012**, *93*, 1231–1239. [CrossRef] [PubMed]
69. Johne, S.; Watzke, R.; Meusel, W.; Mollmann, V.; Harti, A.; Dahse, H.M.; Matthes, B.; Seifert, K. Biotechnological production and bioactivities of mollisin and two new, structuraly related fungal naphthequinone metabolites. *Chem. Biodivers.* **2005**, *2*, 1109–1115. [PubMed]

70. Bertscha, C.; Ramırez-Sueroa, M.; Magnin-Robertb, M.; Larignonc, P.; Chonga, J.; Abou-Mansourd, E.; Spagnolob, A.; Clementb, C.; Fontaineb, F. Grapevine trunk diseases: Complex and still poorly understood. *Plant Pathol.* **2013**, *62*, 243–265. [CrossRef]
71. Bien, S.; Kraus, C.; Damm, U. Novel *Collophorina*-like genera and species from *Prunus* trees and vineyards in Germany. *Persoonia* **2020**, *45*, 46–67. [CrossRef]
72. Piperkova, N.; Yonkova, I. Symptoms, Etiology and Control of Sooty Blotch and Flyspeck in Bulgaria. *Turk. J. Agric. Nat. Sci.* **2014**, *1*, 817–822.
73. Epidemiology of the Fungus *Athelia arachnoidea* in Epiphytic Communities of Broadleaved Forests under Strong Anthropogenic Impact. Available online: http://www.elibrary.lt/resursai/LMA/Ekologija/0504_07_Eko.pdf (accessed on 5 March 2020).
74. Adams, G.C.; Kropp, B.R. *Athelia arachnoidea*, the sexual state of *Rhizoctonia carotae*, a pathogen of carrot in cold storage. *Mycologia* **1996**, *88*, 459–472. [CrossRef]
75. Menkis, A.; Vasiliauskas, R.; Taylor, A.F.S.; Stenlid, J.; Finlay, R. Fungal communities in mycorrhizal roots of conifer seedlings in forest nurseries under different cultivation systems, assessed by morphotyping, direct sequencing and mycelial isolation. *Mycorrhiza* **2005**, *16*, 33–41.
76. Gehring, C.A.; Theimer, T.C.; Whitham, G.; Keim, P. Ectomycorrhizal fungal community structure of pinyon pines growing in two environmental extremes. *Ecology* **1998**, *79*, 1562–1572.
77. Rosa, L.H.; Vaz, A.B.M.; Caligiorne, R.B.; Campolina, S.; Rosa, C.A. Endophytic fungi associaed with Antarctic grass *Deschampsia antarctica* Devs (Poaceae). *Polar Biol.* **2009**, *32*, 161–167. [CrossRef]
78. Perić, B.; Perić, O. Diversity of macromycetes in Montenegro. *MASA* **1997**, *11*, 45–142.
79. Perić, B.; Perić, O. Preliminary red list of Macromycetes of Montenegro 2. *Mycol. Monten.* **2004**, *7*, 7–33.
80. Lazarević, J.; Perić, O.; Perić, B. Ectomycorrhizal fungi in Montenegro—Diversity and distribution. *Mycol. Monten.* **2011**, *14*, 85–115.
81. Lazarević, J. Forests and biodiversity of Kuči mountains. In *Katun of the Kuči Mountains*; Laković, I., Ed.; University of Montenegro: Podgorica, Montenegro, 2017; pp. 64–78.
82. Official Gazette of Montenegro, No 054/16, 15.08.2016. Law of Nature Protection of Montenegro. Available online: http://www.sluzbenilist.me/pregled-dokumenta/?id=\protect\T1\textbraceleft0C659042-9DD7-43FB-8E8C-B7DF445298A1\protect\T1\textbraceright (accessed on 23 March 2020).
83. Official Gazette of Montenegro, No 76/06, 12.12.2006, Act on Protection of Certain Plant and Animal Species. 2006. Available online: http://www.sluzbenilist.me/pregled-dokumenta/?id=\protect\T1\textbraceleft631C3E5D-4129-4985-B55B-CE4D3703CA2E\protect\T1\textbraceright (accessed on 23 March 2020).
84. Perić, B.; Baral, H.-O. *Erioscyphella curvispora*, spec. nov. from Montenegro. *Mycol. Monten.* **2015**, *17*, 89–104.
85. Baral, H.-O.; Perić, B. *Perzia triseptata* Gen. et sp. nov. (Ascomycota, Insertae sedis) from xeric bark in France and Montenegro. *Mycol. Monten.* **2016**, *19*, 7–20.
86. Baral, H.-O.; Perić, B. *Velutarina bertiscensis* and *V. alpestris* spp. nov., with a redescription of *V. rufoolivacea* and discussion of its synonyms. *Mycol. Monten.* **2015**, *17*, 17–52.
87. Perić, B.; Grebenc, T. Une espèce nouvelle du genre *Peziza* (Pezizales): *P. ontirivicola* spec. Nov. *Ascomycete. Org.* **2015**, *7*, 347–356.
88. Perić, B.; Baral, H.-O.; Partel, K. *Cenangiopsis raghavanii* and *C. junipericola* spp. nov (Cenangiaceae, Helotiales) collected in Montenegro, with redescription of a recent collection of *C. quercicola*. *Mycol. Monten.* **2015**, *17*, 7–40.
89. Perić, B.; van Vooren, N.; Healy, R.; Lazarević, J. Une *Trichophaea* rare récoltée en France et au Monténégro: *T. Flavobrunnea Comb. Nov. (Pezizales). Mycol. Monten.* **2014**, *16*, 65–87.

11

Production and New Extraction Method of Polyketide Red Pigments Produced by Ascomycetous Fungi from Terrestrial and Marine Habitats

Juliana Lebeau [1], Mekala Venkatachalam [1], Mireille Fouillaud [1], Thomas Petit [2], Francesco Vinale [3], Laurent Dufossé [1] and Yanis Caro [1,*]

[1] Laboratoire de Chimie des Substances Naturelles et des Sciences des Aliments (LCSNSA), Université de la Réunion, F-97490 Sainte-Clotilde, Ile de la Réunion, France; juliana.lebeau@gmail.com (J.L.); mekalavenkat@gmail.com (M.V.); mireille.fouillaud@univ-reunion.fr (M.F.); laurent.dufosse@univ-reunion.fr (L.D.)

[2] UMR QualiSud, Université de la Réunion, IUT, F-97410 Saint-Pierre, Ile de la Réunion, France; thomas.petit@univ-reunion.fr

[3] Istituto per la Protezione Sostenibile delle Piante (IPSP-CNR) and Dipartimento di Agraria, Università degli Studi di Napoli Federico II, 80055 Portici, NA, Italy; francesco.vinale@ipsp.cnr.it

* Correspondence: yanis.caro@univ-reunion.fr.

Abstract: The use of ascomycetous fungi as pigment producers opens the way to an alternative to synthetic dyes, especially in the red-dye industries, which have very few natural pigment alternatives. The present paper aimed to bio-prospect and screen out 15 selected ascomycetous fungal strains, originating from terrestrial and marine habitats belonging to seven different genera (*Penicillium*, *Talaromyces*, *Fusarium*, *Aspergillus*, *Trichoderma*, *Dreschlera*, and *Paecilomyces*). We identified four strains, *Penicillium purpurogenum rubisclerotium*, *Fusarium oxysporum*, marine strains identified as *Talaromyces* spp., and *Trichoderma atroviride*, as potential red pigment producers. The extraction of the pigments is a crucial step, whereby the qualitative and quantitative compositions of each fungal extract need to be respected for reliable identification, as well as preserving bioactivity. Furthermore, there is a growing demand for more sustainable and cost-effective extraction methods. Therefore, a pressurized liquid extraction technique was carried out in this study, allowing a greener and faster extraction step of the pigments, while preserving their chemical structures and bioactivities in comparison to conventional extraction processes. The protocol was illustrated with the production of pigment extracts from *P. purpurogenum rubisclerotium* and *Talaromyces* spp. Extracts were analyzed by high-performance liquid-chromatography combined with photodiode array-detection (HPLC-DAD) and high-resolution mass spectrometry (UHPLC-HRMS). The more promising strain was the isolate *Talaromyces* spp. of marine origin. The main polyketide pigment produced by this strain has been characterized as *N*-threoninerubropunctamine, a non-toxic red *Monascus*-like azaphilone pigment.

Keywords: red pigment; fungal pigment; *Talaromyces*; *Penicillium*; marine fungi; ascomycetous; *N*-threoninerubropunctamine

1. Introduction

Natural colorants are widely used in the world in many industries such as food, cosmetics, pharmaceuticals, and textiles. The majority of authorized natural food colorants in the market are of either a plant or vegetable origin, and have numerous drawbacks such as instability against light, heat, or adverse pH, and a low water solubility [1]. The dye industry is currently suffering from the cost increase of feedstock associated with the higher demands of eco-friendly pigments for replacing synthetic dyes (like azo dyes). This is even more the case in the red-dye industries, which have no, or very few, natural red pigment alternatives for food processes. For instance, there is a strong need for red colours other than plant-originated anthocyanins, which cannot be used over the whole pH range. As of now, red pigments used in foods are mainly from insects (carmine). Carmine (or carminic acid, cochineal extract) is produced in Peru, Bolivia, Mexico, Chile, and Spain (Canary islands) from the dried bodies of female cochineal insects (*Dactylopius coccus*), primarily grown on *Opuntia* cacti [2]. Carmine is considered as one of the most stable natural food red colorants in terms of light and heat. Between 2004 and 2009, a 76% increase in new European food product launches listing carmine as an ingredient was observed, an increase also linked to the consequences of the "Southampton six" study, which promoted a warning for child hyperactivity related to the occurrence of six artificial colorants in food, including three sulphonated mono azo red dyes (E122 carmoisine/azorubine, E124 Ponceau 4R, and E129 Allura Red AC). However, carmine holds ethical issues for some social groups, and another drawback of carmine products is that from a stable level of 15 USD per kg, it surged in 2010–2011 up to 120 USD per kg and decreased again to 15 USD per kg. As a conclusion, "Dr Jekyll's" (positive) aspect of carmine is its excellent stability in food formulations, whereas the "Mr Hyde" (negative) ones are: (i) it cannot be used by vegans-vegetarians-kosher-halal, (ii) its price versatility, and (iii) allergenicity in some cases [3]. The world's largest food color company, Chr. Hansen, which sources one third of global carmine production, decided in 2011 to explore whether it would be commercially viable to produce carmine with a controlled fermentation process (proof of concept test).

Thus, there is an increasing interest from the academic world and industrial sector about the readily available natural sources of red pigments. Among non-conventional sources, ascomycetous fungi are known to produce an extraordinary range of red polyketide pigments that are often more stable and soluble than plant pigments [4–6]. So, fungal red polyketides, such as azaphilone, naphtoquinone, and hydroxyanthraquinone red compounds, are most promising in this respect, even if unusual microbial red carotenoids should be investigated. The development of such a fungal-based pigments industry and its sustainability rely on the selection of adequate strains regarding the three following parameters: (i) profitable yields, (ii) pigment purity and stability, (iii) and the total absence of toxic compounds in the fungal pigment extract. Furthermore, fungal pigments are of interest due to the broad spectrum of their biological activities and their potential applications in designing new pharmaceutical products [7].

Nowadays, some fermentative natural colorants from filamentous fungi like *Blakeslea trispora*, *Ashbya gossypii*, *Penicillium oxalicum*, and *Monascus* sp., are available for replacing the yellow, orange, and red synthetic dyes [6–9]. Over the past five years, very few reports have been published on the *Monascus*-like azaphilone red pigments produced by non-mycotoxigenic strains of *Talaromyces* species [6–9]. In the literature, this biosynthetic potential has been linked to species such as *Talaromyces purpurogenus*, *T. albobiverticillius*, *T. marneffei*, and *T. minioluteus*, often known under their previous *Penicillium* names. For example, in 2012, a European patent was granted for a submerged cultivation method for some of the non-mycotoxigenic strains of *Talaromyces* sp., whereby the concentration of pigments was significantly enhanced, with the polyketide azaphilone purple pigment PP-V [(10Z)-12-carboxyl-monascorubramine] being the major compound [10–12]. *N*-glutarylmonascorubramine and *N*-glutarylrubropunctamine were the water-soluble *Monascus*-like polyketide azaphilone red pigments discovered in the extracellular pigment extract obtained from the liquid medium of *Penicillium purpurogenum* [13]. Recently, Frisvad et al. [14] concluded that the isolate of *T. atroroseus* sp. nov., which produces *Monascus*-like azaphilone red pigments and mitorubrins, without being accompanied by mycotoxin synthesis, can be used industrially for red pigment production (patent application EP2262862 B1 [12]). However, they indicated that isolates identified as *T. purpurogenus* may not be recommended for the industrial production of red pigments due to their potential coproduction of mycotoxins, such as rubratoxin A and B, and luteoskyrin, in addition to potential toxic extrolites, such as spiculisporic acid and rugulovasine A and B.

Few reports have been published on the following polyketide naphthoquinone red pigments produced by *Fusarium* species: aurofusarin in *Fusarium graminearum* [15] and bikaverin and its minor coproduct nor-bikaverin in *Fusarium fujikuroi* [6,16]. Along similar lines, some species of the genus *Aspergillus* were found to produce known polyketide hydroxyantraquinone red pigments, such as erythroglaucin, catenarin, and rubrocristin [2,6,17]. Some strains of *Trichoderma* such as *T. aureoviride*, *T. harzianum*, and *T. polysporum* are found to produce the hydroxyanthraquinone orange-red pigment chrysophanol [6]. The hydroxyanthraquinone red pigments catenarin and erythroglaucin have also been isolated from cultures of strains among *Drechslera* species and from a culture of *Curvularia lunata* [6].

The present paper aimed to bio-prospect and screen out 15 selected ascomycetous fungal strains, belonging to seven different genera (*Penicillium*, *Talaromyces*, *Fusarium*, *Aspergillus*, *Trichoderma*, *Dreschlera*, and *Paecilomyces*) originating from terrestrial and marine habitats. Recent literature abundantly reports the interest in marine microorganisms with respect to the production of new molecules and, among them, new pigments [7]. The biotechnological properties of the 15 strains for the production of extracellular water-soluble pigments and intracellular polyketide pigments were investigated in submerged cultures.

The choice of extraction protocol is crucial, as the extraction solvents and conditions can drastically influence the final composition, quality, and efficiency of the process. Indeed, extended extraction times, and the exposure to organic solvents and a higher temperature, can result in a tremendous loss of bioactive substances due to hydrolysis, oxygen- and light-oxidation, as well as ionization. In order to preserve as much as possible of the qualitative and quantitative compositions of the pigmented molecules, the use of a recent extraction method, known as Pressurized Liquid Extraction, was investigated, resulting in the development of a six-stage pressurized liquid extraction protocol (PLE) for advanced mycelial pigment extraction. PLE has been mostly applied on environmental samples (recovery) [18], as well as food and biological samples [19], as an analytical method. Only a few applications have been reported on the extraction of bioactive phyto-compounds. Thus, to our knowledge, such methods, which use less and non-toxic solvents, have not been used on fungal matrixes thus far.

2. Materials and Methods

2.1. Fungal Strains

Eleven fungal strains used in this study originating from terrestrial environments were bought from the fungal culture collection of the Museum d'Histoire Naturelle de Paris (Paris, France): *Penicillium purpurogenum* LCP4890, *P. purpurogenum rubisclerotium* LCP4464, *P. erythromellis* LCP3684, *P. oxalicum* LCP4158, *Fusarium oxysporum* LCP531, *Aspergillus repens* LCP5511, *Paecilomyces farinosus* LCP3391, *Trichoderma harzianum* LCP3404, *T. polysporum* LCP3531, and *Dreschlera cynodontis* LCP2226. *Trichoderma harzianum* strain T22 is a commercial biological control strain. The four fungal isolates of marine origin investigated in this study and identified as *Talaromyces* spp. (code: 305_70), *Talaromyces verruculosus* (code: PA9), *Trichoderma atroviride* (code: 305_55), and *Aspergillus sydowii* (code: B34) were isolated by Mireille Fouillaud from samples collected in the back reef-flat and on the external slope of the coral reef on the west coast of La Reunion island. The fungal collection was stored at −80 °C at the LCSNSA laboratory (Reunion island).

2.2. Fermentation and Biomass Production

For inoculum preparation, 0.15 g of conidia and mycelium mixture was sampled from a seven days-old preculture on a potato dextrose agar (PDA) plate, and transferred into a microcentrifuge tube containing 1 mL of nutrient broth supplemented with 0.05 $g \cdot L^{-1}$ of Tween® 80 (Sigma-Aldrich Co, Saint Louis, MO, USA). The mycelium was crushed and the suspension was used to inoculate 250-mL flasks containing 100 mL of liquid media: (i) potato dextrose broth medium (PDB: composed of 4 $g \cdot L^{-1}$ potato infusion solids and 20 $g \cdot L^{-1}$ glucose; Sigma-Aldrich); (ii) defined minimal dextrose broth medium (DMD: composed of 1 $g \cdot L^{-1}$ ammonium sulfate, 30 $g \cdot L^{-1}$ glucose, 0.5 $g \cdot L^{-1}$ $MgSO_4$, 1.4 $g \cdot L^{-1}$

K_2HPO_4, 0.6 g·L^{-1} KH_2PO_4, 0.8 mg·L^{-1} $ZnSO_4$, 0.8 mg·L^{-1} $FeSO_4$, 0.8 mg·L^{-1} $CuSO_4$, 0.8 mg·L^{-1} NaH_2PO_4 and 0.4 mg·L^{-1} $MnSO_4$; Fisher Scientific UK Limited, Loughborough, Leicestershire LE, UK) based on Velmurugan et al. [20]; and (iii) yeast casamino dextrose broth (YCD: composed of 1 g·L^{-1} yeast extract (Becton, Dickinson and Co., Sparks, MD, USA), 5 g·L^{-1} casamino acids (BD Bacto), 20 g·L^{-1} glucose, 5 g·L^{-1} sodium chloride and 1 g·L^{-1} KH_2PO_4 (Fisher Scientific UK Limited, Loughborough, Leicestershire LE, UK) based on Guyomarc'h et al. [21]. The pH was adjusted to 6.0. Flasks were incubated at 26 °C and agitated at 150 rpm. For cultures performed in total darkness, flasks were wrapped in aluminum foil. After seven days of fermentation, all the contents of each flask were collected and centrifuged at 10,000 rpm for 10 min; the resulting supernatant was filtered through a Whatman filter paper (GF/C) at a reduced pressure using a Büchner funnel to obtain the culture filtrate. The mycelial biomass was washed with deionized water. After freezing at −84 °C in an ultra-low-temperature freezer (Sanyo, Guangzhou, China) for at least 2 h, the samples were quickly transferred to a LABCONCO FreeZone 2.5 lyophilizer (LABCONCO, Kansas City, MO, USA) and lyophilized for 24 h. During freezing, the condenser temperature and vacuum pressure were maintained at −47 °C and 200 mbar, respectively. Then, dried cells were weighed to estimate the biomass. All experiments were conducted in duplicate.

2.3. *Quantitative Colorimetric Analysis of Extracellular Extracts*

The colorimetric characterization of extracellular extracts was assessed from the pigmented culture filtrate after seven days of cultivation. Measurements were performed in the CIE L*a*b* (L*a*b* colorimetric system of the Commission Internationale de l'Eclairage) using a spectrocolorimeter CM 3500 with the SpectraMagic™ software v1.9 (Konica Minolta, Mahwah, NJ, USA). The so-called CIELab colorimetric system is based on the fact that light reflected from any colored surface can be visually matched by an additive mixture of the three primary colors: red, green, and blue [22,23]. To characterize a color in the CIE L*a*b* color system, three colorimetric coordinates are obtained from the spectrocolorimeter. L* defines the lightness (ranges from 0% to 100%, dark to light), a* value indicates the red/green value (from −60 to +60, green to red), and b* value denotes the blue/yellow value (from −60 to +60, blue to yellow). The attributes of color, C* and h°, describe the chroma (vividness or dullness) and the hue angle (or tone) of the color, respectively. The value of chroma C* is 0 at the center and increases according to the distance from the center. The hue angle h° is defined as starting at the +a* axis and is expressed in degrees: 0° would be +a* (red), 90° would be +b* (yellow), 180° would be −a* (green), and 270° would be −b* (blue). Hue values correspond to the angle of the a*/b* coordinates of the points. The "Y red chroma" value used in this study to link the positive a*-value (red color) and the others coordinates and attributes of the color of the pigment extract (such as b*, C* and h* values) was calculated as follow:

$$\text{“Y red chroma” value} = f(a^* \text{ value}) = b^* \text{ value} \times C^* \text{ chroma value} \times 1/h^* \quad (1)$$

2.4. *UV-Visible Spectrophotometry and Extracellular Polyketide Metabolites Quantification*

The culture filtrate was diluted in deionized water with a dilution factor (d) ranging from four to 10 with respect to the concentration of extrolites in the filtrate. The solution was used to investigate the sample absorption profile in the wavelength range of 200–800 nm. Then, the concentrations of polyketide metabolites in the culture filtrate were determined through the absorption values read at 276 nm using a UV-visible spectrophotometer (UV-1800, Shimadzu Corporation, Tokyo, Japan). Polyketide red carmine aqueous solutions were used as standards, and their maximum absorption values experimentally measured at 276 nm were taken as the reference for polyketide concentration determination (Figure S1). Carmine (hydrosoluble carmine in powder, kindly provided by Pronex, Lima, Peru) used as the standard was dissolved in pure water. The diluted uninoculated liquid broth (that is PDB, DMD, or YCD broth) was used as a blank before any absorbance measurements were taken. The coefficient of proportionality (ε), which links the absorbance at 276 nm with the extracellular polyketide metabolites concentration, was obtained by linear regression (Figure S1). The concentration of extracellular polyketide metabolites in culture filtrate was expressed as milli-equivalents of polyketide carmine per liter of liquid broth.

2.5. *New Extraction Method of Intracellular Polyketide Pigments*

The extraction of intracellular polyketide pigments from the mycelial biomass was performed using a new pressurized liquid extraction (PLE) process. The weighed sample (lyophilized biomass) was transferred to a 10-mL stainless steel extraction cell equipped with two cellulose filters on the bottom and containing glass balls (diam. 0.25–0.50 mm). PLE extraction was performed on a Dionex ASE system (ASE^{TM} 350 apparatus, Dionex, Germering, Germany). The ASE conditions were: temperature: 90 °C, pressure: 1500 psi, heating time: 5 min, static time: 18 min, flush: 100%, and purge: 5 min. The lyophilized biomass was subjected to a six-stage liquid solvent extraction under pressure as an attempt to entirely extract the intracellular pigments from the mycelium. The following six-stage PLE sequence was performed: purified water was used as the first extraction solvent, followed by 50% methanol, then 50% ethanol, >99.9% methanol, and MeOH:EtOH (1/1, *v*/*v*), and then, mycelium was depleted with >99.9% ethanol as the extraction solvent (Figure S2). The sequence of solvents was set to show a decreasing polarity profile. Solvents (methanol and ethanol, 99.9%-HPLC quality) were obtained from Carlo Erba (Val de Reuil, France). Purified water was obtained from a Milli-Q system (EMD Millipore Co., Billerica, MA, USA). A part of each color extract was used for chromatographic analysis and the rest was used for absorbance analysis.

2.6. *Absorbance and HPLC-DAD Analyses*

Each intracellular extract obtained in a collection bottle at the end of the six-stage pressurized liquid solvent extraction sequence was filtered onto a 0.2-μm poresized hydrophylic Millex-GV membrane (Millipore, Carrigtwohill, Ireland) and stored at −20 °C in amber glass vials (2 mL) with Teflon-lined caps, until further analysis. The total polyketide secondary metabolite content extracted from the mycelial biomass was first analyzed by measuring the absorbance of each extract by spectral analysis at 276 nm. The results obtained are expressed in terms of meqv. of carmine per g dry cell mass, a value proportional to the polyketide metabolite concentration extracted from the mycelia. Then, reverse phase high-resolution liquid-chromatography combined with photodiode array-detection (HPLC-DAD) analysis was performed on each extract (25 μL injection) on a Dionex HPLC-DAD system (Ultimate 3000 apparatus, Dionex, Germering, Germany) and using a Hypersil Gold^{TM} column (150 mm × 2.1 mm i.d., 5 μm; Thermo Scientific Inc., Waltham, MA, USA) maintained at 30 °C. The HPLC-DAD system was operated using a (A) purified water-(B) acetonitrile-(C) aqueous formic acid 1% (*v*/*v*) solution gradient system starting from a ratio of 45%(A)–45%(B)–10%(C) for 8 min and increasing to 95%(B)–5%(C) in 55 min, at which point it was maintained for 20 min. The flow rate was 0.4 mL·min^{-1}. Monitoring, data recording, and processing were led with the Chromeleon v.6.80 software (Dionex). Solvents (acetonitrile and methanol, 99.9%-HPLC quality) and formic acid (purity 99%) were obtained from Carlo Erba (Val de Reuil, France).

2.7. *UHPLC-HRMS Analyses*

The intracellular extracts were analyzed by ultra high performance liquid chromatography high-resolution mass spectrometry (UHPLC-HRMS) according to Klitgaard et al. [24]. Liquid chromatography was performed on an Agilent 1290 Infinity LC system with a DAD detector, coupled to an Agilent 6550 iFunnel Q-TOF with an electrospray ionization source (Agilent Technologies, Santa Clara, CA, USA). The separation was performed on a 2.1 mm × 250 mm, 2.7 μm Poroshell 120 Phenyl-Hexyl column (Agilent) at 60 °C with a water-acetonitrile gradient (both buffered with 20 mM formic acid) going from 10 % (*v*/*v*) to 100 % acetonitrile in 15 min, followed by 3 min with 100 % acetonitrile. The flow rate was kept constant at 0.35 mL/min throughout the run. The injection volume, depending on the sample concentration, typically varied between 0.1 and 1 μL. Mass spectra were recorded as centroid data for m/z 85–1700 in MS mode and m/z 30–1700 in MS/MS mode, with an acquisition rate of 10 spectra/s. The lock mass solution in 95% acetonitrile was infused in the second sprayer using an extra LC pump at a flow of 10–50 µL/min, and the solution contained 1 µM tributyle amine (Sigma-Aldrich), 10 µM Hexakis(2,2,3,3-tetrafluoropropoxy) phosphazene (Apollo Scientific Ltd., Cheshire, UK), and 1 µM trifluoroacetic acid (Sigma-Aldrich) as lock masses.

3. Results

3.1. Biomass and Polyketide Extrolites Production Capacities under Various Growth Conditions

Our results indicated that the 15 fungal strains investigated in the present study could be divided into two main color categories. The reddish polyketide pigment producers include the four following strains, ranked from the darker to the paler red-pigment producers: *P. purpurogenum rubisclerotium*, the local marine isolate identified as *Talaromyces* spp., followed by the strain *F. oxysporum*, and the marine isolate *Trichoderma atroviride*. The yellowish pigment producers encompass the following strains: *P. erythromellis*, *P. oxalicum*, *Talaromyces verruculosus*, *Trichoderma harzianum*, *Paecilomyces farinosus*, *Aspergillus repens*, and *A. sydowii* (Figure S3). The dried biomass concentration produced by the 15 ascomycetous strains is reported in Figure 1. Three different culture broths were investigated: (i) defined minimal dextrose medium (DMD), (ii) potato dextrose broth medium (PDB), and (iii) yeast casamino dextrose broth (YCD).

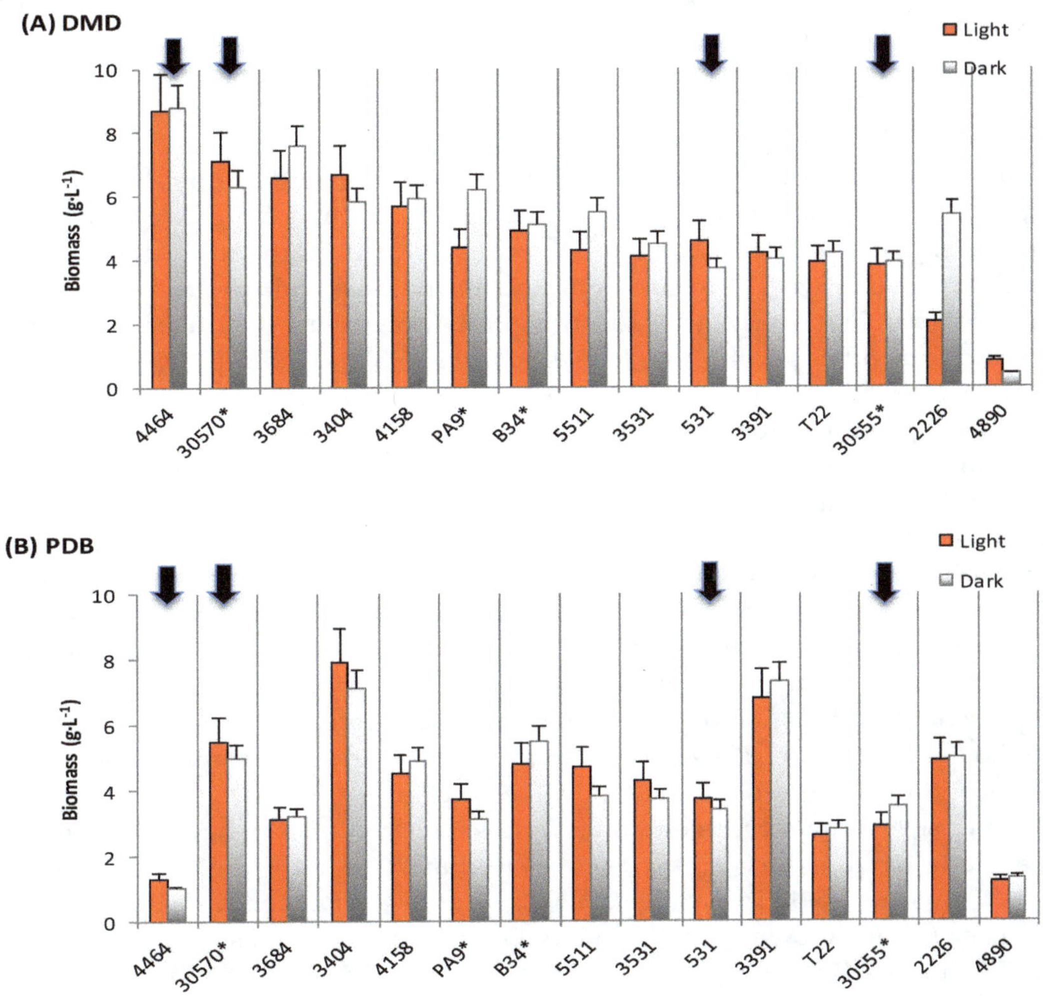

Figure 1. *Cont.*

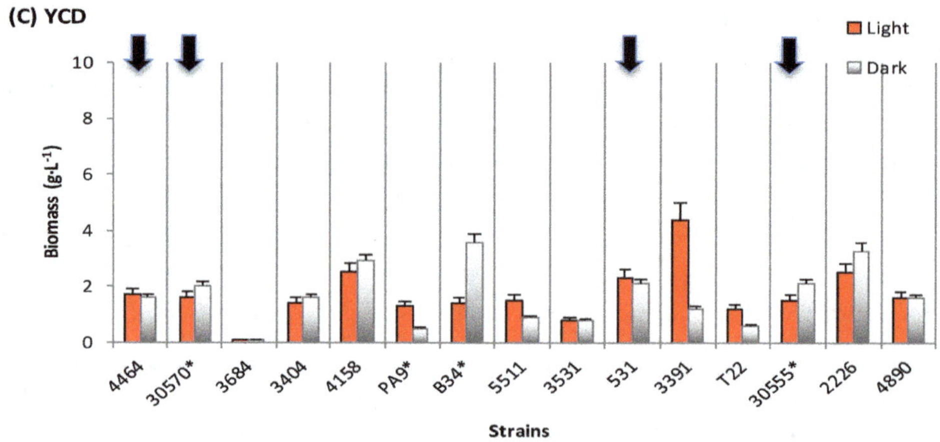

Figure 1. Fungal biomass production (mean in g·L^{-1} ± s.d.) in a submerged culture of the ascomycetous fungi. (**A**) Biomass production obtained in DMD (Defined Minimal Dextrose broth) submerged culture; (**B**) in PDB (Potato Dextrose Broth) submerged culture; (**C**) in YCD (Yeast Casamino Dextrose broth) submerged culture; Culture conditions: under illumination (▮: red) and in the dark (▮: grey); s.d.: standard deviation; Strain identification: *4464 =* ❶ *Penicillium purpurogenum rubisclerotium; 30570 =* ❷ *Talaromyces* spp. *(marine isolate)*; 531 =* ❸ *Fusarium oxysporum; 30555 =* ❹ *Trichoderma atroviride*; 4890= Penicillium purpurogenum; 2226 = Dreschlera cynodontis; 3684 = Penicillium erythromellis; T22 = Trichoderma harzianum; 4158 = Penicillium oxalicum; 5511= Aspergillus repens; PA9= Talaromyces verruculosus*; 3404= Trichoderma harzianum; 3391= Paecylomyces farinosus; B34 = Aspergillus sydowii*; 3531 = Trichoderma polysporum;* (* strains collected from marine biotopes of La Reunion island's reef flat).

Our results indicated that the DMD medium (Figure 1A) is the more favorable liquid broth for biomass production. Six fungal strains producing more than 5.5 g·L^{-1} of dry biomass were observed: *P. purpurogenum rubisclerotium*, i.e., the first reddish pigment producer described above, which produced 8.4 and 8.5 g·L^{-1} in the light and in the dark culture, respectively; followed by the marine isolate *Talaromyces* spp. (the second reddish pigments producer: 7.1 and 6.2 g·L^{-1} of dry biomass); and four other strains, i.e., *P. erythromellis* (6.5 and 7.5 g·L^{-1}), *Trichoderma harzianum* (6.7 and 5.8 g·L^{-1}), *P. oxalicum* LCP4158 (5.7 and 5.9 g·L^{-1}), and *T. verruculosus* (4.4 and 6.1 g·L^{-1}). The two other reddish pigment producers, i.e., the strain *F. oxysporum* and the marine isolate *T. atroviride*, produced less than 4.5 g·L^{-1} of biomass in this DMD medium. In PDB medium (Figure 1B), considering the four fungal strains identified above as potential reddish pigment producers, only the marine strain *Talaromyces* spp. produced more than 5.5 g·L^{-1} of dry biomass. Surprisingly, whereas the strain *P. purpurogenum rubisclerotium* showed the highest biomass productions in the DMD medium, it was the less productive strain in PDB medium (1.3 and 1.0 g dry biomass·L^{-1} in the light and in the dark culture, respectively). In YCD medium (Figure 1C), the biomass contents remained relatively lower (<2 g·L^{-1}) for most of the cultured fungal strains.

Figure 2 presents an estimation of the volumetric productions of polyketide extrolites secreted by each strain in the culture filtrate. The concentrations were expressed as milli-equivalents of carmine per liter of liquid broth. The color of the "light" histograms of Figure 2 indicates the shade of the culture filtrate after seven days of fermentation: reddish and yellowish shades were mainly observed, but an orange shade (obtained for *A. repens* cultivated in YCD under light conditions), pink shade (*T. verruculosus* grown in DMD in darkness, or *P. purpurogenum* grown in PDB and YCD), brown shade (*Dreschlera cynodontis* grown in DMD and YCD broths), and green shade (*Trichoderma polysporum* cultivated in PDB and DMD broths) were also noticed (S3). The lowest extracellular productivities (<200 meqv g·L^{-1}) were observed on "minimal" DMD medium for all strains investigated in this study (Figure 2A). Thus, the presence of a simple source of carbon (30 g·L^{-1} of glucose) and simple source of nitrogen (ammonium sulfate) in this DMD medium favored the production of mycelial biomass to the detriment of the production and excretion of extracellular diffusible polyketides. In contrast, the best polyketide extrolite productions were obtained in PDB liquid culture medium, reaching an average extracellular volumetric production ranging from 135 to 1118 meqv·L^{-1} for eight strains out of the 15 (Figure 2B). For example, the highest extrolite production was obtained in culture filtrate of the strain *P. purpurogenum rubisclerotium* grown in PDB liquid broth (1118 and 685 meqv·L^{-1} in the light and dark culture, respectively), while its biomass concentration was very low in this medium (only 1.0–1.3 g·L^{-1} of dry biomass).

The marine isolate *Talaromyces* spp. presents a different behavior towards the medium nutrient composition. It simultaneously showed a relatively high mycelial biomass content (above 5.5 g·L^{-1} of dry biomass) in PDB and a strong content in extracellular polyketide-derived metabolites (278 and 266 meqv·L^{-1} in the light and dark culture, respectively). The two other strains belonging to the reddish pigment producers, i.e., *F. oxysporum* and *T. atroviride* (marine isolate), presented extracellular polyketide productivities of 166 and 373 meqv·L^{-1}, respectively, but also low contents in the mycelial biomass in submerged cultures in PDB (< 3.5 g·L^{-1} of dry biomass).

The YCD medium that also contained a complex source of nitrogen (f.i. amino acids and proteins from yeast extract) had an intermediate position regarding extracellular polyketide extrolite production, exhibiting very heterogeneous productivities (Figure 2C).

The culture filtrates were then analyzed by quantitative colorimetric analysis in the CIE L*a*b* colorimetric system. It is worth noticing that for colored culture filtrates, the greater the positive red a*-value is, the greater the positive b*-value, the greater the positive C*-chroma, and the smaller the h*-value. Thus, a "Y chroma" value has been calculated in this study to link the positive a*-value (red color) and the other b*, C*, and h* values of the colored culture filtrates. The results are presented in Figure 3. In PDB medium with both complex sources of carbon and nitrogen, the four fungal strains belonging to the reddish pigment producers could produce diffusible polyketide red pigments: the culture filtrates exhibited an intense purple, red, or orange-red color, in the darkness, as well as under light conditions (Figure 3A), which is consistent with visual observations of the culture filtrate and the volumetric production of extracellular pigments noticed in these fungal cultures grown in PDB. The highest positive a*-value and positive "Y chroma" value were obtained for the culture filtrate of the marine isolate *Talaromyces* spp. (purple-red color; a*-value up to +63.8) and the strain *P. purpurogenum rubisclerotium* (purple-red color; a*-value up to +51.5). The two other strains, *F. oxysporum* and *T. atroviride*, exhibited lower positive a*-values: up to +35.2 (red color) and +18.7 (orange-red color), respectively.

Furthermore, it is worth noticing that in the "minimal" DMD liquid broth (with only glucose and ammonium sulfate, combined with some metal ions and inorganic salts), only the marine isolate *Talaromyces* spp. and the strain *F. oxysporum* can produce water-soluble reddish pigments: with a corresponding red a*-value from +10 to +20 (Figure 3B). The strain *P. purpurogenum rubisclerotium* grown in this DMD broth produced pinkish pigments exclusively under light conditions, and a pale yellowish pigmentation was noticed for the marine isolate *T. atroviride* grown in DMD. The results confirmed that the same fungal strain did not show the same metabolism of extracellular polyketide pigments in submerged cultures with respect to the medium composition used. Finally, in YCD medium (with glucose, yeast extract, and casamino acids as the complex nitrogen source), only one strain, that is *P. purpurogenum rubisclerotium*, can produce reddish pigments in liquid broth (red a*-value of +42.9 and +43.5 for light and dark culture, respectively) (Figure 3C). Interestingly, the marine isolate *T. atroviride* cultivated in YCD broth produced an orange-red pigmentation exclusively under light conditions, indicating that the production of orange-red pigments is a light-inducible metabolism.

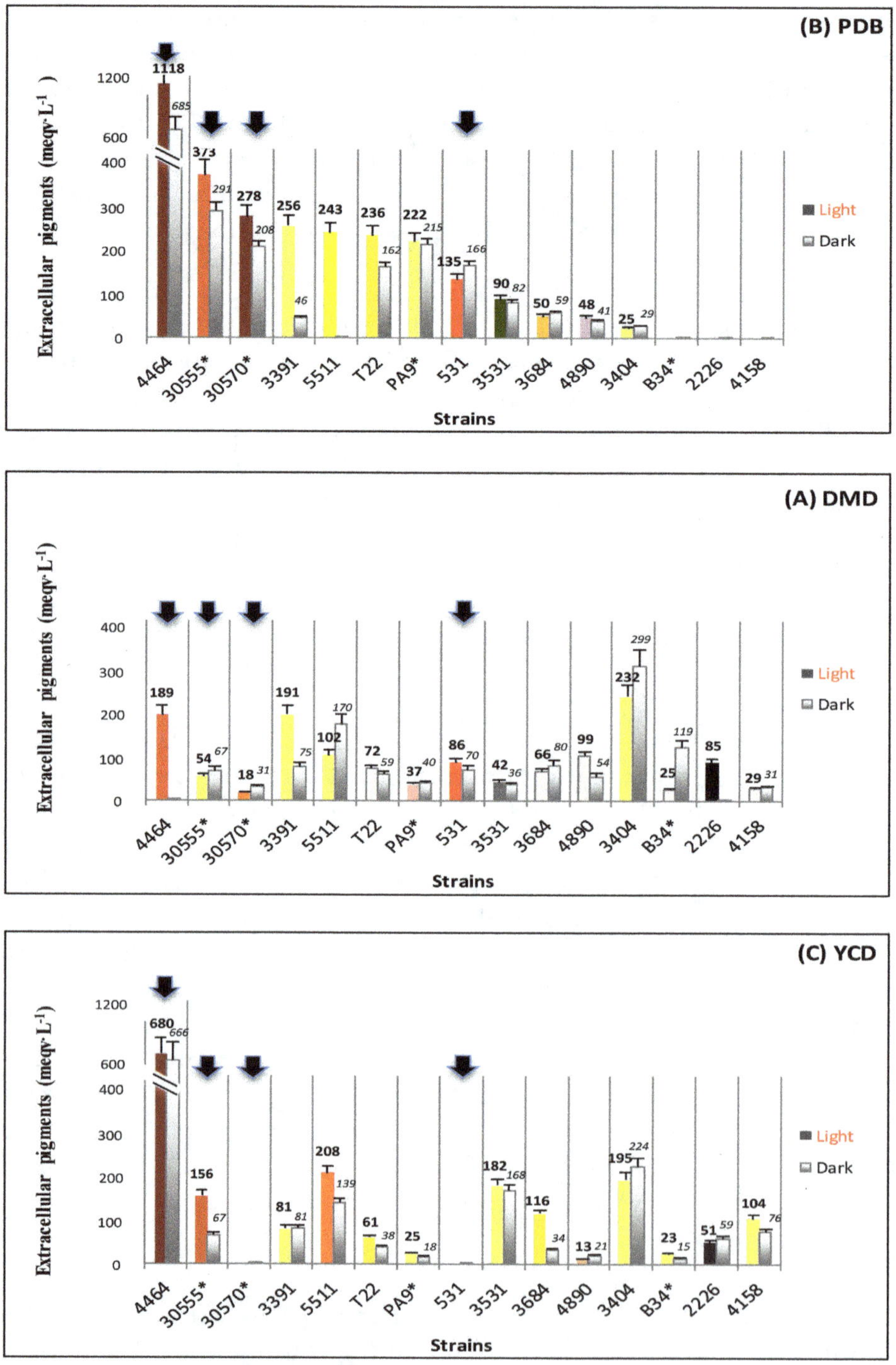

Figure 2. Volumetric production of extracellular pigments (mean in meqv. carmine pigment per liter ± s.d.) by the ascomycetous fungi in submerged culture. (**A**) Volumic production of extracellular pigments obtained in DMD (Defined Minimal Dextrose broth) submerged culture; (**B**) in PDB (Potato Dextrose Broth) submerged culture; (**C**) in YCD (Yeast Casamino Dextrose broth) submerged culture; Culture conditions: in the dark (grey) and under illumination (in color; color in figure for each strain indicates the shade of the fungal culture filtrate after 7 days of fermentation); s.d.: standard deviation; ❶ *Penicillium purpurogenum rubisclerotium* (4464); ❷ *Talaromyces spp.* (marine isolate 30570); ❸ *Fusarium oxysporum* (531); ❹ *Trichoderma atroviride* (marine isolate 30555). For other strain identification (numbers in horizontal axis), see Figure 1. The yield of extracellular pigments was estimated according to a calibration curve of standard carmine solution by measuring the absorbance at 276 nm (λ_{max} of carmine) of the colored culture filtrate (after blank subtraction).

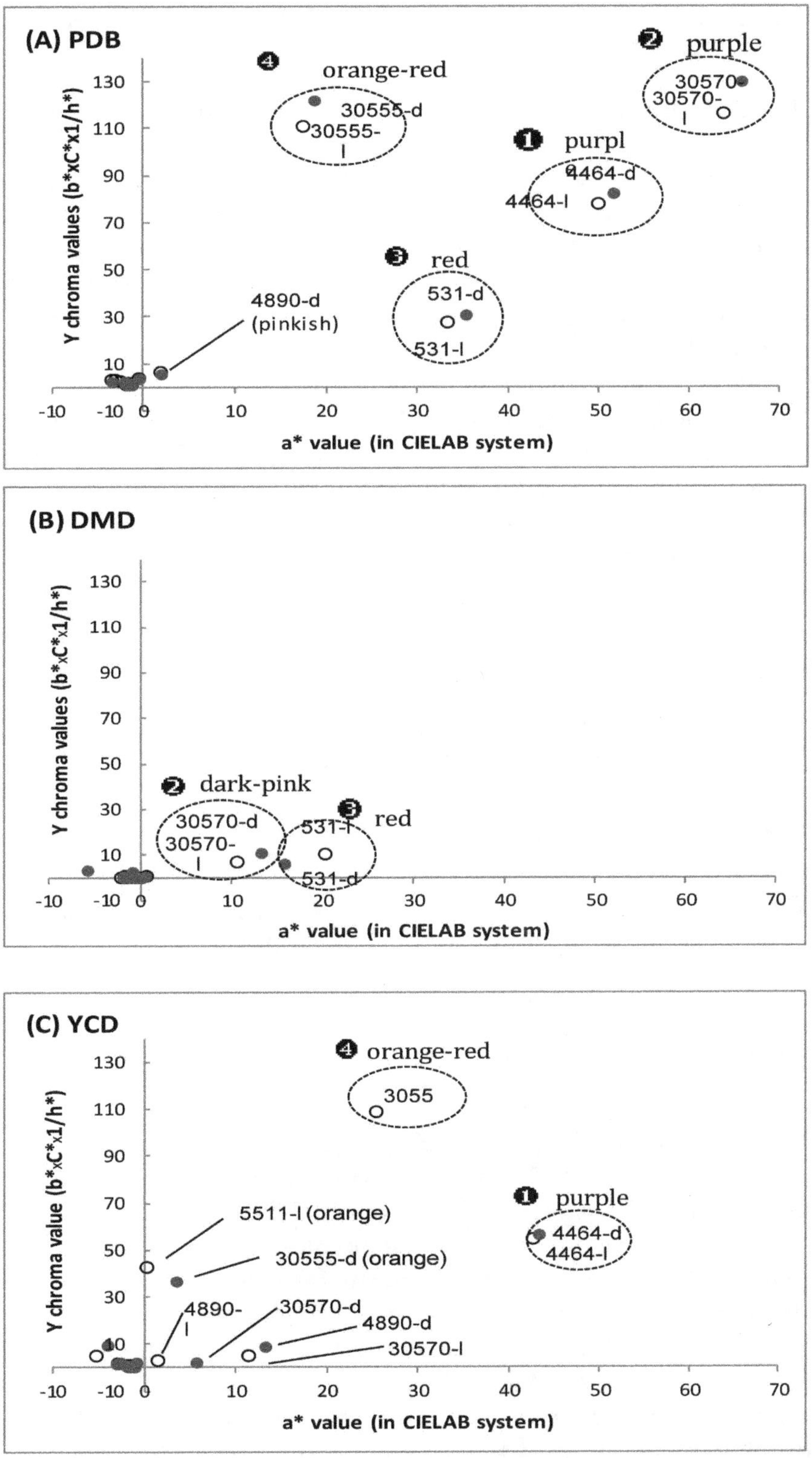

Figure 3. Color coordinates projected in the CIE L*a*b* colorimetric system of the culture filtrates after seven days of cultivation of the ascomycetous fungi. (**A**) Color coordinates of extracellular pigment extracts obtained in PDB (Potato Dextrose Broth) submerged culture; (**B**) in DMD (Defined Minimal Dextrose broth) submerged culture; (**C**) in YCD (Yeast Casamino Dextrose broth) submerged culture; Culture conditions: under illumination (-l), and darkness (-d). Strain identification: ❶ *Penicillium purpurogenum rubisclerotium* (4464); ❷ *Talaromyces spp.* (marine isolate 30570); ❸ *Fusarium oxysporum* (531); ❹ *Trichoderma atroviride* (marine isolate 30555); *Penicillium purpurogenum* (4890).

3.2. New Extraction Procedure and Nature of the Polyketide Red Pigments Produced by *Talaromyces spp.* Marine Isolate and *P. purpurogenum rubisclerotium* Terrestrial Isolate

In order to evaluate the intracellular production and the nature of the red pigments produced by the *Talaromyces* spp. marine isolate and *P. purpurogenum rubisclerotium* terrestrial isolate, a series of six sequential pressurized liquid extractions (PLE) were realized. The sequence of solvents was set to show a decreasing polarity profile, therefore allowing a refined isolation of different pigment molecules, depending on their polarity profiles. The total polyketide secondary metabolite contents extracted from the mycelial biomass, analyzed by measuring the absorbance of all the extracts by spectral analysis at 276 nm and expressed in terms of meqv. of polyketide carmine per g dry cell mass, are shown in Table 1. Thus, the results confirmed that the amount of polyketide pigments produced extra- versus intracellularly varies with different cultivation factors and the choice of fungal strain. The DMD medium favors the production of mycelial biomass, whereas the nutrient composition of the PDB medium favors both the intracellular production of polyketide pigments (up to 69.8 and 116.0 meqv·g^{-1} dry biomass of *Talaromyces* spp. and *P. purpurogenum rubisclerotium*, respectively) and their liberation as extrolites in the liquid broth (up to 278 and 1118 meqv·L^{-1} of culture filtrate of *Talaromyces* spp. and *P. purpurogenum rubisclerotium*, respectively).

Table 1. Biomass and extracellular versus intracellular pigment production by the marine isolate *Talaromyces* spp. and terrestrial isolate *P. purpurogenum rubisclerotium* in submerged cultures.

Fungal Strain	Broth	Biomass (g/L)	Extracellular Pigments (meqv. Carmine/L)	*a*-Value (CIELab) (Red Color)	Intracellular Pigments Extracted (meqv/g Biomass) or (meqv/L Culture)	
30570	in DMD	**6.2–7.1**	18–31	10.8–13.4	23.8–24.7	148–176
	in PDB	5.0–5.5	**208–278**	**63.8–65.8**	**49.4–69.8**	**247–384**
	in YCD	1.6–2.0	-	5.3–5.6	44.3–48.5	71–97
4464	in DMD	**8.4–8.5**	1–189	0.7 – 0.9	57.9–83.2	**487–707**
	in PDB	1.0–1.3	**685–1118**	**49.9–51.5**	**69.2–116.0**	90–116
	in YCD	1.6–1.7	666–680	42.8–43.5	22.5–71.2	36–121

30570: marine isolate of *Talaromyces* spp.; 4464: terrestrial isolate of *Penicillium purpurogenum rubisclerotium*; DMD: Defined Minimal Medium broth; PDB: Potato Dextrose Broth. YCD: Yeast Casamino acids Dextrose Broth; Values corresponding to the main productivities are noticed in bold.

The intracellular extracts of the *Talaromyces* spp. marine isolate and *P. purpurogenum rubisclerotium* terrestrial isolate were analyzed by HPLC-DAD. Concerning the polyketide pigments extracted from the mycelial biomass of *Talaromyces* spp. grown in PDB culture, all the intracellular extracts recovered from the six-stage PLE presented an intense purple-red shade (S4). The chromatograms of the intracellular extracts shown in Figure 4 indicated that the intracellular aqueous extract (Figure 4C) was entirely composed of a major colored metabolite (compound **1**; not tentatively identified, >95% *w*/*w* on total compounds in this intracellular aqueous extract) with the retention time (Rt.) of 1.71 min (red polar compound with λ_{max} 215, 244, 276, 418, 514, 524 nm). Then, the 50% aqueous methanolic solution used as a solvent in the six-stage PLE extracted the largest part of other intracellular pigments from *Talaromyces* spp. Six other major colored metabolites were detected in this reddish liquid sample (Figure 4C): compound **2** with Rt. 26.07 min (3.1% *w*/*w* on total secondary metabolites in this intracellular hydroalcoholic extract), compound **3** with Rt. 29.60 min (4.9%), the main compound **4** with Rt. 30.97 min (54.4%), compound **5** with Rt. 32.66 min (14.8%), compound **6** with Rt. 38.04 min (10.9%), and compound **7** with RT 43.95 min (6.3%). In addition, compound **8** with Rt. 69.78 min, identified as ergosterol by UHPLC-HRMS analysis (see below), was essentially detected in the 100% methanolic extract (solvent polarity index of 5.0) (Figure 4E).

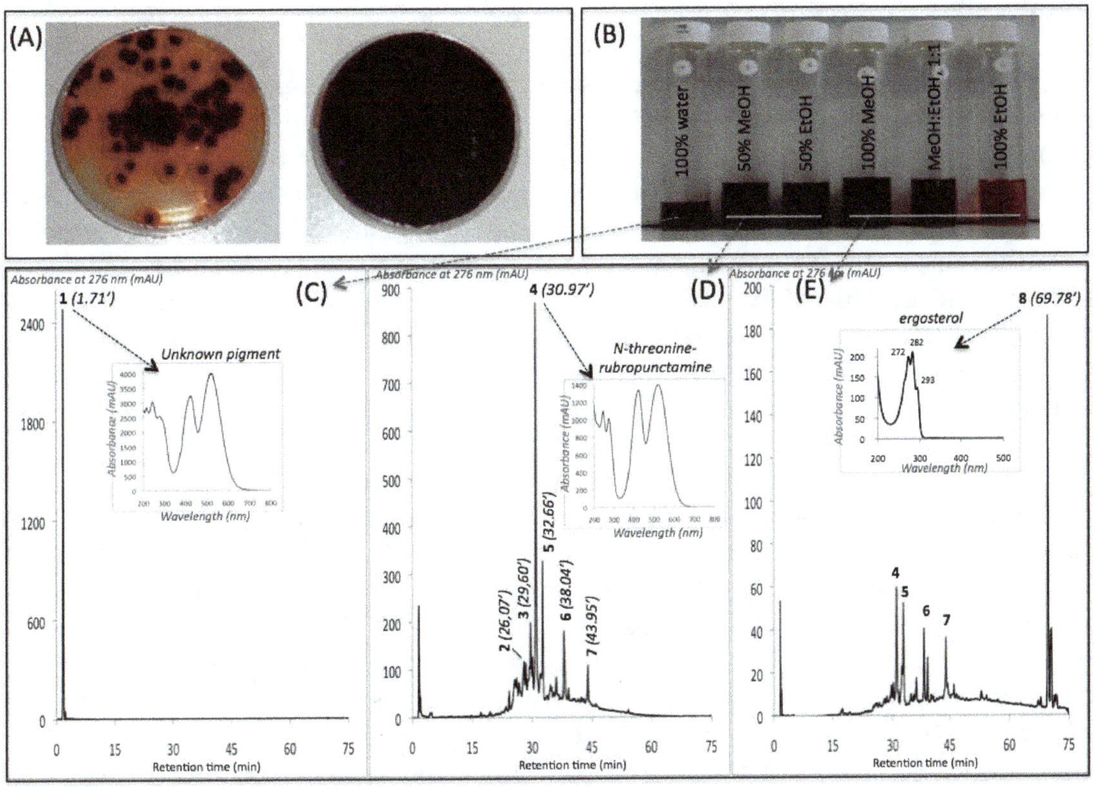

Figure 4. (**A**) Cultures in PDA of the marine isolate *Talaromyces* spp. 305_70; (**B**) liquid extracts obtained after a six-stage pressurized liquid solvent extraction of the mycelium cultivated in PDB submerged culture; (**C–E**) chromatograms* of the overall compounds detected by HPLC-DAD in the extracts. *Captions*: PDA (Potato Dextrose Agar); PDB (Potato Dextrose Broth); HPLC-DAD: high-performance liquid chromatogram combined with photo-diode array detection; Rt.: retention time; * the analysed samples were extracted with different solvents. Table 2 reports on all of the identified or assumed compounds detected in all the different samples and, in particular, the 50% aqueous methanolic extract was the most representative, as shown in Figure 4D. Only compound 1 (Rt. 1.71 min), not tentatively identified, was detected in the intracellular aqueous extract (Figure 4C). The main compound 4 (Rt. 30.97 min), identified as *N*-threoninerubropunctamine, was detected mainly in the 50% aqueous methanolic extract (Figure 4D), and compound 8 (Rt. 69.78 min), identified as ergosterol, was detected mainly in 100% methanolic extract (Figure 4E).

The eluted compounds (from compound **1** to **8**) were analyzed by performing UHPLC-HRMS, their absorption spectra were measured from 200–800 nm, and their mass spectra were determined and compared against the spectral library and literature database for identification. The UV-visible absorption and mass spectra of each identified or assumed compound detected in all the different intracellular extracts of the marine isolate *Talaromyces* spp. are shown in Figure 5. The results suggested that the main compound **4** that presents two absorption maxima in the visible region, at λ_{max} at 424 and 521 nm (red color), was identified as the red polyketide azaphilone pigment *N*-threoninerubropunctamine. Indeed, the molecule showed a similar UV-visible absorption spectrum, as well as an ESI-MS molecular ion in positive mode $[M + H]^+$ at m/z 456 as the red pigment *N*-threoninerubropunctamine previously isolated from some species of *Talaromyces* by other authors [25,26]. Presumably, the compound **3** that presents λ_{max} at 421 and 518 nm in the visible region, and the ESI-MS molecular ion observed at m/z 416 in positive mode $[M + H]^+$ seemed to correspond to the red polyketide azaphilone pigment glycylrubropunctatin, which has previously been isolated from

Monascus cultures [27]. Compound **6** (*N*-glutarylrubropunctamine) and **7** (monascorubramine), which exhibit ESI-MS molecular ions observed at m/z 484 and m/z 381 in positive mode $[M + H]^+$, respectively, were identified in good agreement with the expected mass of the corresponding molecules previously isolated from some species of *Talaromyces* [25,26]. Finally, the results suggested that compound **8** (with Rt. 69.78 min) seemed to correspond to an ergosterol derivate, according to its absorption spectrum, which was similar to that of an ergosterol molecule, and to the ESI-MS molecular ion observed at m/z 393 in positive mode $[M + H]^+$ relatively close to those of the standard molecule (which in addition to the expected ion m/z 397, also yielded the same ion at m/z 393 (major) because ergosterol had undergone desaturation during LC/MS) [28]. This observation is consistent with the findings of previous studies, which have already isolated ergosterol derivates from some fungal species [29,30]. All secondary metabolites detected from intracellular extracts of *Talaromyces* spp. are reported in Table 2, and the chemical structures of all identified or assumed compounds are described in Figure 6.

The three other compounds eluted, i.e., compound **1** (with Rt. 1.71 min; λ_{max} 215, 244, 276, 418, 514, 524 nm; but no signal observed in ESI-MS analysis), compound **2** (with Rt. 26.07 min; λ_{max} 246, 276, 425, 512 nm; ESI-MS molecular ion observed at m/z 488 in positive mode $[M + H]^+$ together with the ion at m/z 550 $[M+Na+CH_3CN]^+$), and compound **5** (with Rt. 32.66 min; λ_{max} 218, 250, 287, 581 424, 546 nm; ESI-MS molecular ion observed at m/z 498 in positive mode $[M + H]^+$ together with the ion at m/z 546), presented UV-visible absorption spectra relatively close to those of the *Monascus*-like azaphilone red pigments, which exhibit two absorption maxima in the visible region near 420 and 520 nm (Figure S5). According to this information, these three not tentatively identified metabolites were classified as water-soluble red azaphilone-derivative pigments. Unfortunately, further experiments with a dereplication purpose using UHPLC-HRMS and comparison with a spectral library were not conclusive enough to fully conclude on the nature of these pigmented compounds. Works are in progress in our laboratory to fully characterize the pigments by NMR analysis.

Table 2. Overall compounds detected by HPLC-DAD and UHPLC-HRMS in intracellular extracts (IE) of the marine isolate of *Talaromyces* spp. and terrestrial isolate of *P. purpurogenum rubisclerotium* cultivated in PDB submerged culture, with reference to the chromatograms shown in Figures 4 and 7.

Compound No.	R.t. (mn)	λmax (nm)	Solvent	Tentative Identification	Mol. Peak (*m/z*)
		IE of *Talaromyces* spp. (305_70)			
1	1.71	215, 244, 276, 418, 514, 524	water	n.i.	n.d.
2	26.07	246, 276, 425, 512	MeOH	n.i.	488 $[M + H]^+$
3	29.60	245, 274, 421, 518	MeOH	Glycylrubropunctatin	416 $[M + H]^+$
4	**30.97**	**245, 274, 424, 521**	**MeOH**	***N*-threoninerubropunctamine**	**456 $[M + H]^+$**
5	32.66	218, 250, 287, 424, 546	MeOH	n.i.	498 $[M + H]^+$
6	38.04	246, 273, 427, 521	MeOH	*N*-glutarylrubropunctamine	484 $[M + H]^+$
7	43.95	245, 272, 424, 519	MeOH	Monascorubramine	381 $[M + H]^+$
8	69.78	272, 282, 293	MeOH	Ergosterol	393 $[M + H]^+$
		IE of *Penicillium purpurogenum rubisclerotium* (4464)			
8	69.78	272, 282, 293	MeOH	Ergosterol	393 $[M + H]^+$
9	1.57	216, 292, 492	MeOH	n.i.	n.d.
10	**22.05**	**262, 322, 430**	**MeOH**	n.i.	n.d.
11	23.50	284, 389	MeOH	n.i.	n.d.

IE: Intracellular extract; PDB: Potato Dextrose Broth; Main compound in bold; n.i. = not tentatively identified.

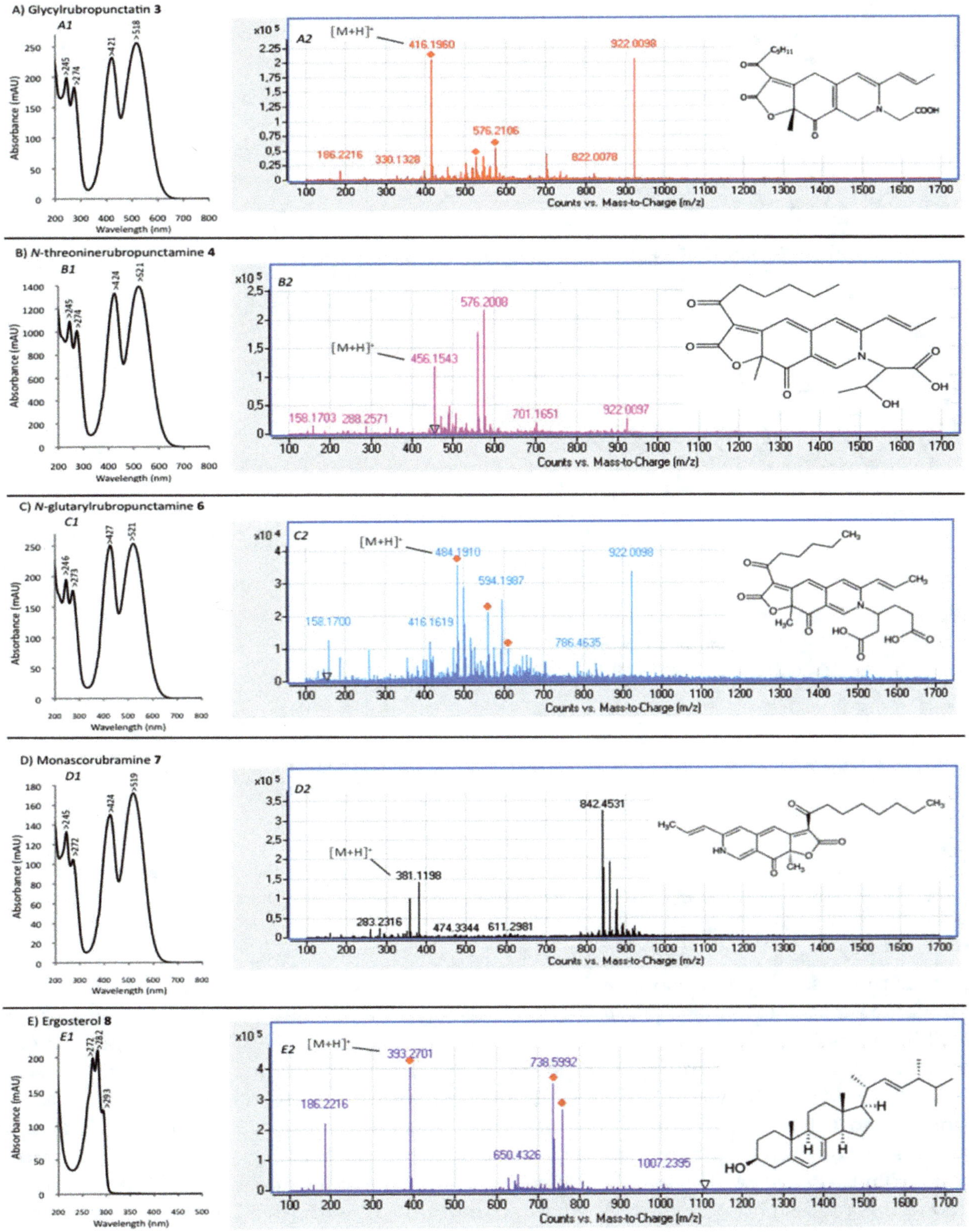

Figure 5. UV-visible absorption (**A1, B1, C1, D1, E1**) and mass spectra (**A2, B2, C2, D2, E2**) of the identified or assumed compounds detected in intracellular extracts of the marine isolate *Talaromyces* spp. with reference to the chromatograms shown in Figure 4: glycylrubropunctatin **3**, *N*-threoninerubropunctamine **4**, *N*-glutarylrubropunctamine **6**, Monascorubramine **7**, and ergosterol **8**. Compounds **1**, **2** and **5** are red azaphilone-derivative pigments not tentatively identified.

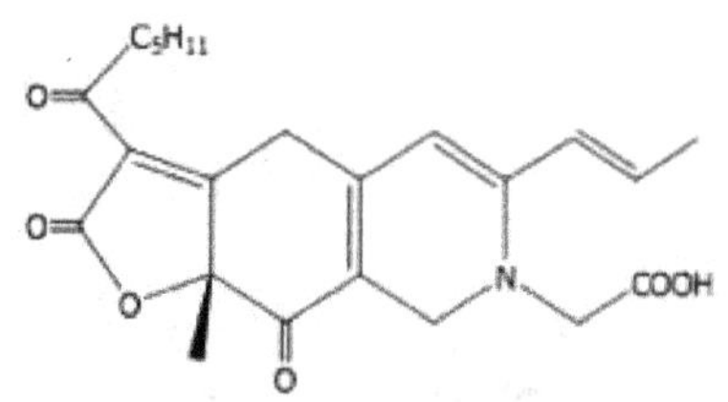

Glycylrubropunctatin **3**
($C_{23}H_{27}NO_6$; Mass: 415)

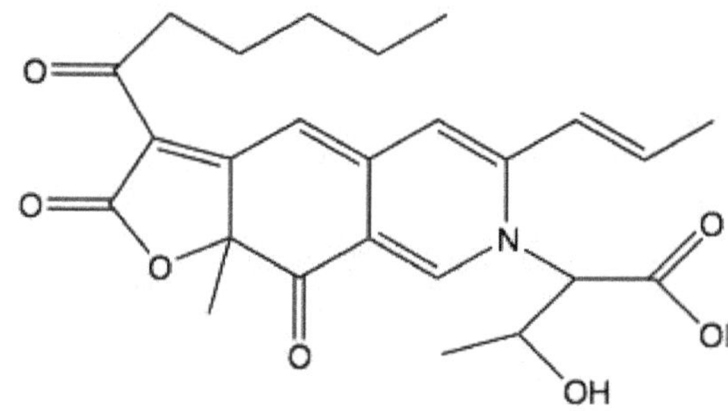

N-threoninerubropunctamine **4**
($C_{25}H_{29}NO_7$; Mass: 455)

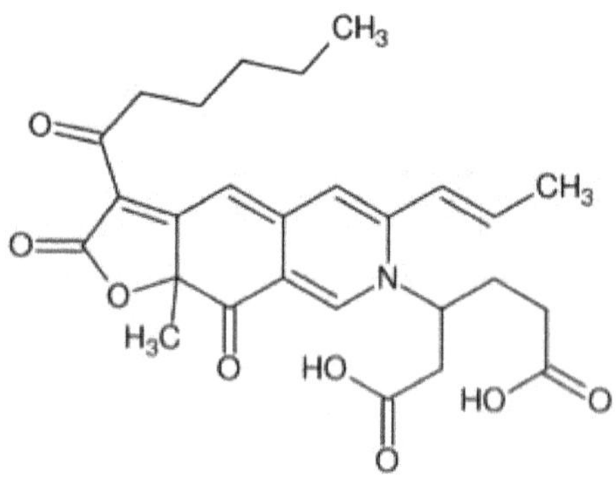

N-glutarylrubropunctamine **6**
($C_{26}H_{29}NO_8$; Mass: 483)

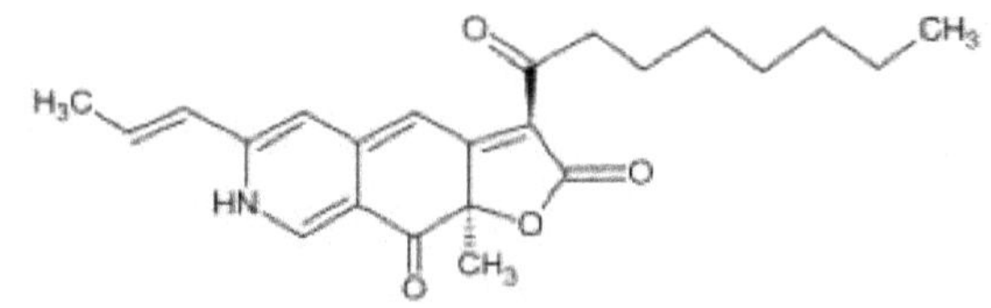

Monascorubramine **7**
($C_{23}H_{27}NO_4$; Mass: 381)

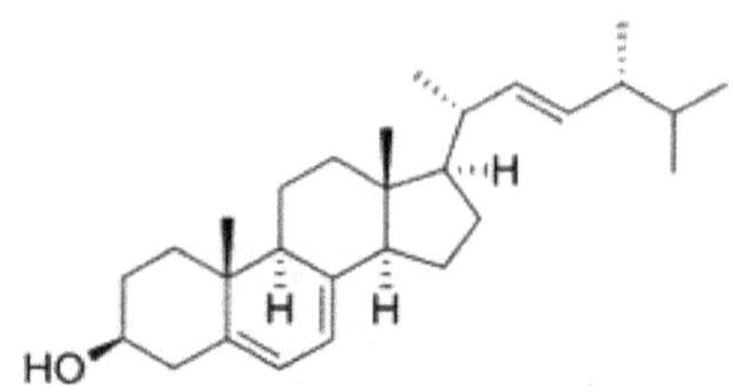

Ergosterol **8**
($C_{28}H_{44}O$; Mass: 396)

Figure 6. Chemical structures of the identified or assumed compounds detected in the present study in intracellular extracts of the marine isolate *Talaromyces* spp. Formula and calculated nominal masses are shown in parentheses.

For the purpose of comparison, Table 2 summarized the compounds eluted in all the different intracellular extracts of the strain *P. purpurogenum rubisclerotium*. Regarding the intracellular pigments extracted from this fungal mycelium cultivated in PDB, the chromatograms shown in Figure 7 indicated that the aqueous extract was entirely composed of a major secondary metabolite with Rt. 1.57 min (compound **9** with the following λ_{max}: 216, 292, 492 nm; not tentatively identified), which is different from compound **1** detected from the aqueous extract of *Talaromyces* spp. Similarly, the 50% aqueous methanolic solution used as a solvent in the six-step PLE extracted the largest part of other intracellular pigments from this *P. purpurogenum rubisclerotium* strain: two other major colored metabolites were detected in this reddish liquid extract: the main compound **10** with Rt. 22.05 min that presents λ_{max} at 262, 322, 430 nm (not tentatively identified); and compound **11** with Rt. 23.50 min exhibited λ_{max} at 284 and 389 nm (not tentatively identified). Further analytical experiments and comparisons with the spectral library were not conclusive enough to fully conclude on the nature of these pigmented compounds from intracellular extracts of *P. purpurogenum rubisclerotium*. The last compound eluted with Rt. 69.78 min was essentially detected in the 100% methanolic extract, and seemed to correspond to the same ergosterol **8** derivate previously isolated from intracellular extracts of *Talaromyces* spp. marine isolate.

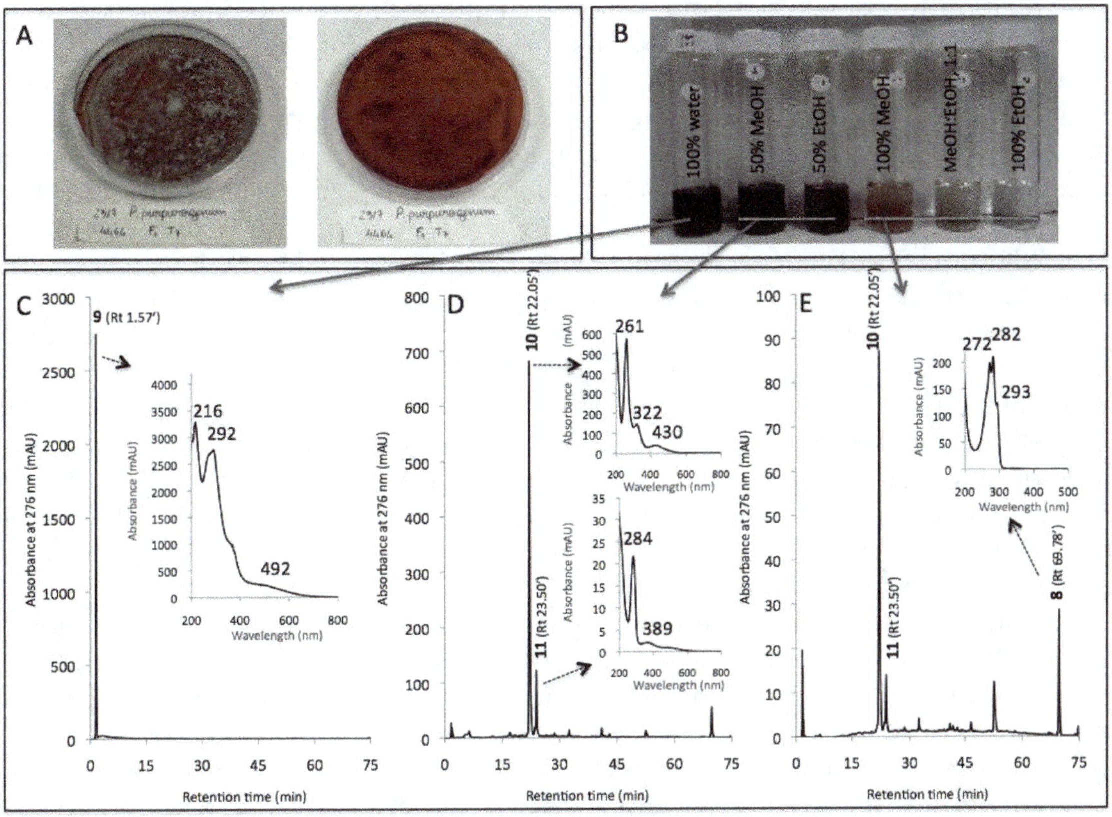

Figure 7. (**A**) Cultures in PDA of the strain *Penicillium purpurogenum rubisclerotium* LCP4464; (**B**) liquid extracts obtained after a six-stage pressurized liquid solvent extraction of the mycelium cultivated in PDB submerged culture; (**C–E**) chromatograms * of the overall compounds detected by HPLC-DAD in the different extracts. *Captions*: PDA (Potato Dextrose Agar); PDB (Potato Dextrose Broth); HPLC-DAD: high-performance liquid chromatography combined with photo-diode array detection; Rt.: retention time; *the analysed samples were extracted with different solvents. Table 2 reports the compounds detected in all the different samples and, in particular, the 50% aqueous methanolic extract was the most representative, as shown in Figure 7D. Only compound **9** (Rt. 1.57 min; not tentatively identified) was detected in the aqueous extract (Figure 7C), the main compound **10** (Rt. 22.05 min; not tentatively identified) was detected mainly in the 50% aqueous methanolic extract (Figure 7D), and ergosterol 8 (Rt. 69.78 min) was detected mainly in 100% MeOH extract (Figure 7E).

4. Discussion

The 11 fungal strains from the terrestrial environment investigated in this study were firstly selected according to the capacity of analogous fungal species cited in the literature to produce pigments in solid and/or submerged cultures. The four local marine isolates were selected according to previous works performed in our laboratory (data not published) based on their extracellular pigment producing ability in submerged cultures. We identified four strains, *P. purpurogenum rubisclerotium*, *F. oxysporum*, and two marine isolates sampled from the lagoon on the west coast of La Reunion Island identified as *Talaromyces* spp. and *Trichoderma atroviride*, as potential pigment producers that produced polyketide red pigments. Regarding the pigment yield, there are generally two natural ways in which the concentration of pigments can be increased: either by improving the fungal growth, or by increasing the intracellular accumulation of pigments. The problem, however, lies with the difficulty in increasing both the biomass and the pigment production, which would be optimal for industrial production. Biomass and pigment yields tend to be negatively correlated. An increase in the biomass yield is connected to the abundance of nutrients in the medium, whereas pigment production tends to be increased under nutrient-poor conditions and/or external stresses, as a protective mechanism [31].

Therefore, the relationship between biomass and pigment production needs to be fully elucidated in order to optimize and control pigment production in submerged fungal culture.

Our results confirmed that same fungal strain did not show the same metabolism of extracellular pigments in submerged cultures (due to the excretion of diffusible pigments in liquid broth) with respect to the medium composition used. For example, the highest water-soluble extrolites production was obtained for the culture filtrate of the strain *P. purpurogenum rubisclerotium* in PDB liquid broth, although its biomass concentration was very low in this medium. As a result, the nutrient composition of the PBD liquid broth allows one to associate the high production and excretion of extracellular water soluble pigments with the low formation of mycelial biomass. Along similar lines, two other strains belonging to the reddish pigment producers that are *F. oxysporum* and *T. atroviride* presented the highest extracellular polyketide pigment productivities in PDB submerged cultures, but the lowest contents of mycelial biomass. These results confirmed the difficulty in increasing both the mycelial biomass and the extracellular pigment production at the same time, which would be optimal for industrial production. Surprisingly, the marine isolate *Talaromyces* spp. presented a different behavior towards the medium nutrient composition. It simultaneously showed a relatively high mycelial biomass content (above 5.5 $g \cdot L^{-1}$ of dry biomass) and a strong content in extracellular polyketide pigments in culture filtrate, using submerged fermentation in PDB.

In general, the secretion of extracellular pigments is favored over intracellular production at an industrial scale, as it implies less downstream extraction processes. Furthermore, there are several submerged fermentation techniques that can be used, such as fed-batch or continuous mode approaches, in order to achieve optimal pigment yields [31]. Our results suggested that an increase in the mycelial biomass yield is mainly connected to the abundance of simple source of carbon and nitrogen in the medium (f.i. in DMD broth), whereas the production and especially secretion of extracellular polyketide pigments seemed to be increased with an abundance of complex sources of carbon (e.g., starch from PDB) and nitrogen (e.g., amino acids and proteins from PDB) combined with some essential cofactors like calcium, magnesium, phosphor, iron, manganese, zinc, and copper (all present in PDB broth). Indeed, it has been previously reported in the literature that the use of a complex nitrogen source or the addition of individual amino acids tends to influence the number, type, and excretion rates of different pigments that are formed in submerged fungal cultures [31]. It has been suggested that the stimulating effect of amino acids on the production or excretion of pigments is caused by an increase in solubility due to the greater hydrosolubility profile of the pigments when bound to an amino acid than the native pigment on its own. It has also been reported that fungal pigments access the aqueous environment by association with proteins or other polar compounds [31].

In this study, the volumetric production of polyketide extrolites secreted by each strain in the culture filtrate was estimated by a spectrophotometer on the basis of the measured absorbance at 276 nm, and expressed as milli-equivalents of polyketide carmine per liter of liquid broth. This estimation is a value proportional to the total polyketide pigment concentration in the culture filtrate. Indeed, most polyketide-derived pigments are characterized by absorption bands in the UV domain (near 240–260 nm) due to the benzene structure and most of them also presented one UV absorption peak near 276 nm, while the maximum absorbencies in the visible region of these polyketide-derived pigments varied from the range of 400 (yellow pigments) to 500 nm (red pigments). Examples include the anthraquinone red pigment carmine which exhibited a λ_{max} at 276 nm, the naphtoquinone red pigment bikaverin (λ_{max} at 253 and 276 nm) isolated from *Fusarium* species, and most *Monascus*-like pigments with a common azaphilone skeleton isolated from *Penicillium* and *Talaromyces* species such as yellow ankaflavin and red monascorubramine (λ_{max} at 276 nm). Furthermore, our data indicate that the positive a*-value, i.e., the red colorimetric coordinate in the CIE L*a*b* color system, and the measured absorbance at 276 nm of the pigmented culture filtrates, tend to be slightly positively correlated for some fungal strains belonging to the reddish pigment producers (data not shown), which justifies the use of this wavelength as the detection value for all polyketide pigments produced by the fungal strains.

For the quantification and characterization of intracellular polyketide pigments, a greener protocol for the pigment extraction, that is a pressurized liquid extraction (PLE) procedure using environmental-friendly solvents, has been investigated. To our knowledge, there is no standard method for polyketide-derived pigment extraction from mycelial biomass. However, based on literature works, the commonly used process involves mixing dried biomass with an organic solvent (like ethyl acetate, chloroform, etc.) or a mixture, followed by the mechanical disruption of the cells, and subsequent centrifugation or filtration (solid-liquid extraction) [6,31]. In general, cell disruption is most necessary for the higher recovery of intracellular fungal pigments. Many different methods for cell disruption have been suggested by authors, including mechanical (sonification) and non-mechanical disruptions (chemical extraction processes) [6,31]. The preferred extraction procedure is based on a quick process that efficiently releases all of the intracellular pigments from the matrix into the solution without altering them, while using environmental-friendly solvents, if possible [6,31]. Moreover, the polarity of the biocompounds to be extracted is the determinant in the selection of the extraction solvent [32]. Therefore, and as the fungal biomass is likely to contain a mixture of pigments of various natures, multi-stage extraction methods using solvents of a different polarity are required, in order to obtain the most exhaustive pigment composition profile for each fungal strain tested.

In our study, lyophilized mycelial biomass was subjected to a six-stage PLE using environmental-friendly solvents including water, methanol, and ethanol, all already allowed and widely used in the EU and in the US for the extraction of natural food colorants. The protocol was illustrated in the intracellular pigment extracts of *P. purpurogenum rubisclerotium* and *Talaromyces* spp. analysed by HPLC-DAD and UHPLC-HRMS. The PLE system offers the advantages of considerably reducing the extraction times and solvent amounts to be used, as well as using more sustainable solvents such as water, ethanol, and methanol with higher extraction efficiencies [32]. It consists of a solid-liquid extraction process carried out at a high temperature (50–100 °C) and elevated pressure (10–150 MPa) in order to maintain the solvent(s) at a liquid state when applied to the sample [18,19]. The high pressures help in forcing the solvent into the cell pores, while the increased temperature enhances the extraction kinetics. Additionally, the use of increased pressure during the extraction protects the compounds of interest from oxygen, and ensures a higher quality of the recovered biochemical. Then, PLE is considered as a promising extraction process particularly appropriate for polar compounds. Moreover, being able to use common and non-toxic solvents is of great interest when dealing with compounds, which are intended to be used in food and cosmetic products, from health and safety, environmental, and economic prospects, as it would result in less downstream purification processes, less waste treatments, and less energy and solvent expenses. Furthermore, solvents such as ethanol and methanol can be produced from carbon neutral homoacetogenic gas fermentation and biogas produced from waste, respectively, strengthening the sustainability of such process. The six-stage PLE procedure gives encouraging results in terms of the efficiency and selectivity of the pigments extracted from the fungal mycelia. It also paves the way for further optimizations of the solvent mixture which can be used to isolate specific pigments. In the meantime, extraction results confirmed that the minimal 'DMD' medium favors the production of mycelial biomass to the detriment of pigment production. On the other hand, the nutrient composition of the PDB medium favors both the intracellular production of pigments (up to 69.8 and 116.0 meqv$\cdot$g^{-1} dry biomass of *Talaromyces* spp. and *P. purpurogenum rubisclerotium*, respectively) and their liberation as extrolites in the liquid broth (up to 277 and 1178 meqv$\cdot$L^{-1} culture filtrate in submerged culture of *Talaromyces* spp. and *P. purpurogenum rubisclerotium*, respectively). Thus, an efficient and sustainable multi-step extraction process has been developed, allowing a sequential extraction of a wide panel of pigmented molecules based on their respective polarity. Such a process could be further applied to other types of sample matrices for biochemical content determination.

The 50% aqueous methanolic solution used as extraction solvent in the six-stage PLE procedure recovered the largest part of intracellular pigments from the two fungal mycelia. Seven major colored metabolites were detected in intracellular pigment extracts of *Talaromyces* spp. marine isolate: all corresponding to *Monascus*-like pigments with a common azaphilone skeleton, including the main water-soluble azaphilone red pigment identified as *N*-threoninerubropunctamine **4** (non-toxic compound), followed by three other assumed *Monascus*-like azaphilone pigments such as *N*-glutarylrubropunctamine **6**, monascorubramine **7**, and glycylrubropunctatin **3**. The generally recognized pathway proposes that the orange *Monascus* and *Monascus*-like azaphilone pigments, including monascorubrin and rubropunctatin (insoluble in water), are formed by the esterification of a beta-ketoacid (from the fatty acid synthase pathway) to the chromophore (derived from the polyketide synthase pathway). Then, the reduction of the orange *Monascus* and *Monascus*-like azaphilone pigments yields the yellow *Monascus*-like azaphilone pigments (monascin, ankaflavin). In contrast, amination of the orange pigments with amino group-containing compounds in the medium (proteins, amino acids and nucleic acids) leads to the water-soluble red *Monascus* and *Monascus*-like azaphilone pigments, including the monascorubramine **7** and rubropunctamine derivates like *N*-threoninerubropunctamine **4** and *N*-glutarylrubropunctamine **6** identified in this study [33].

For the purpose of comparison, three other red colored metabolites (not tentatively identified in this study) were detected in intracellular extracts of *P. purpurogenum rubisclerotium* terrestrial isolate (also known under its *Talaromyces pinophilus* name, according to Frisvad et al. [14]). Ergosterol **8** was also isolated as a minor compound in the intracellular pigment extracts from the two fungal mycelia.

5. Conclusions

We report on this article newly isolated polyketide red pigment producing ascomycetous fungi namely *Penicillium purpurogenum rubisclerotium*, *Fusarium oxysporum*, and two fungal isolates of marine origin identified as *Talaromyces* spp. and *Trichoderma atroviride*, sampled from the Reunion Island marine habitat. A new sustainable pigment extraction method was used, that is a six-stage pressurized liquid extraction protocol, for advanced mycelial pigment extraction. Such methods, which use less and non-toxic solvents, have not been used on fungal matrixes thus far. The more promising strain was the non-mycotoxigenic fungus *Talaromyces* spp. of marine origin. According to its pigment production capacity and the emerging natural red pigments in the global market, this fungal strain represents an interesting example of a microorganism which can produce a variety of interesting bioactive compounds like natural red colorants. In addition, the production of secondary metabolites is important for the chemotaxonomic characterization of this fungal isolate. The main pigment produced by this strain has been characterized as *N*-threoninerubropunctamine, a non-toxic red *Monascus*-like azaphilone pigment. Knowledge of the intracellular and extracellular pigment productions by this fungus in submerged culture is important for the development of a high-performance industrial fermentation process for polyketide red pigment production. Further studies are necessary for future applications in the dye industry.

Acknowledgments: We would like to thank the Conseil Régional de La Réunion, Reunion Island, France, for the financial support of research activities dedicated to microbial pigments. The authors thank Kristian Fog Nielsen of the Department of Systems Biology, Technical University of Denmark, for the UHPLC-HRMS.

Author Contributions: Yanis Caro, Thomas Petit, Mireille Fouillaud, Francesco Vinale and Laurent Dufossé conceived and designed the experiments; Juliana Lebeau and Mekala Venkatachalam performed the experiments. Juliana Lebeau and Yanis Caro analyzed the data and wrote the paper. All authors read and approved the final manuscript.

References

1. Sutthiwong, N.; Caro, Y.; Laurent, P.; Fouillaud, M.; Valla, A.; Dufossé, L. Production of Biocolors. In *Biotechnology in Agriculture and Food Processing: Opportunities and Challenges*, 1st ed.; Panesar, P.S., Marwaha, S.S., Eds.; Francis & Taylor, CRC Press: Boca Raton, FL, USA, 2013; pp. 417–445.

2. Caro, Y.; Anamale, L.; Fouillaud, M.; Laurent, P.; Petit, T.; Dufossé, L. Natural hydroxyanthraquinoid pigments as potent food grade colorants: An overview. *Nat. Prod. Bioprospect.* **2012**, *2*, 174–193. [CrossRef]
3. Dufossé, L. Anthraquinones, the Dr Jekyll and Mr Hyde of the food pigment family. *Food Res. Int.* **2014**, *65*, 132–136. [CrossRef]
4. Durán, N.; Teixera, M.F.; Conti, R.D.; Esposito, E. Ecological-Friendly Pigments From Fungi. *Crit. Rev. Food Sci. Nutr.* **2002**, *42*, 53–66. [CrossRef] [PubMed]
5. Dufosse, L.; Fouillaud, M.; Caro, Y.; Mapari, S.A.; Sutthiwong, N. Filamentous fungi are large-scale producers of pigments and colorants for the food industry. *Curr. Opin. Biotechnol.* **2014**, *26*, 56–61. [CrossRef] [PubMed]
6. Caro, Y.; Venkatachalam, M.; Lebeau, J.; Fouillaud, M.; Dufossé, L. Pigments and Colorants from Filamentous Fungi. In *Fungal Metabolites*, 1st ed.; Mérillon, J.-M., Ramawat, K.G., Eds.; Springer International Publishing: Berlin, Germany, 2015; pp. 499–568.
7. Fouillaud, M.; Venkatachalam, M.; Girard-Valenciennes, E.; Caro, Y.; Dufossé, L. Anthraquinones and Derivatives from Marine-Derived Fungi: Structural Diversity and Selected Biological Activities. *Mar. Drugs* **2016**, *14*, 64. [CrossRef] [PubMed]
8. Yilmaz, N.; Houbraken, J.; Hoekstra, E.S.; Frisvad, J.C.; Visagie, C.M.; Samson, R.A. Delimitation and characterisation of *Talaromyces purpurogenus* and related species. *Persoonia* **2012**, *29*, 39–54. [CrossRef] [PubMed]
9. Santos-Ebinuma, V.C.; Teixeira, M.F.; Pessoa, A., Jr. Submerged culture conditions for the production of alternative natural colorants by a new isolated *Penicillium purpurogenum* DPUA 1275. *J. Microbiol. Biotechnol.* **2013**, *23*, 802–810. [CrossRef] [PubMed]
10. Mapari, S.A.; Meyer, A.S.; Thrane, U.; Frisvad, J.C. Identification of potentially safe promising fungal cell factories for the production of polyketide natural food colorants using chemotaxonomic rationale. *Microb. Cell. Fact.* **2009**, *8*, 24. [CrossRef] [PubMed]
11. Mapari, S.A.; Thrane, U.; Meyer, A.S. Fungal polyketide azaphilone pigments as future natural food colorants? *Trends Biotechnol.* **2010**, *28*, 300–307. [CrossRef] [PubMed]
12. Mapari, S.A.S.; Meyer, A.S.; Frisvad, J.C.; Thrane, U. Production of *Monascus*-like azaphilone pigments. European patent EP2262862 B1, 28 March 2012.
13. Arai, T.; Koganei, K.; Umemura, S.; Kojima, R.; Kato, J.; Kasumi, T.; Ogihara, J. Importance of the ammonia assimilation by *Penicillium purpurogenum* in amino derivative *Monascus* pigment, PP-V, production. *AMB Express* **2013**, *3*, 19. [CrossRef] [PubMed]
14. Frisvad, J.C.; Yilmaz, N.; Thrane, U.; Rasmussen, K.B.; Houbraken, J.; Samson, R.A. Talaromyces atroroseus, a new species efficiently producing industrially relevant red pigments. *PLoS ONE* **2013**, *8*, e84102. [CrossRef] [PubMed]
15. Frandsen, R.J.; Nielsen, N.J.; Maolanon, N.; Sorensen, J.C.; Olsson, S.; Nielsen, J.; Giese, H. The biosynthetic pathway for aurofusarin in *Fusarium graminearum* reveals a close link between the naphthoquinones and naphthopyrones. *Mol. Microbiol.* **2006**, *61*, 1069–1080. [CrossRef] [PubMed]
16. Limon, M.C.; Rodriguez-Ortiz, R.; Avalos, J. Bikaverin production and applications. *Appl. Microbiol. Biotechnol.* **2010**, *87*, 21–29. [CrossRef] [PubMed]
17. Gessler, N.N.; Egorova, A.S.; Belozerskaya, T.A. Fungal anthraquinones. *Appl. Biochem. Microbiol.* **2013**, *49*, 85–99. [CrossRef]
18. Vazquez-Roig, P.; Picó, Y. Pressurized liquid extraction of organic contaminants in environmental and food samples. *Trend. Anal. Chem.* **2015**, *71*, 55–64. [CrossRef]
19. Carabias-Martínez, R.; Rodríguez-Gonzalo, E.; Revilla-Ruiz, P.; Hernández-Méndez, J. Pressurized liquid extraction in the analysis of food and biological samples. *J. Chromatogr. A* **2005**, *1089*, 1–17. [CrossRef] [PubMed]
20. Velmurugan, P.; Lee, Y.H.; Venil, C.K.; Lakshmanaperumalsamy, P.; Chae, J.-C.; Oh, B.-T. Effect of light on growth, intracellular and extracellular pigment production by five pigment-producing filamentous fungi in synthetic medium. *J. Biosci. Bioeng.* **2010**, *109*, 346–350. [CrossRef] [PubMed]
21. Guyomarc'h, F.; Binet, A.; Dufossé, L. Production of carotenoids by *Brevibacterium linens*: variation among strains, kinetic aspects and HPLC profiles. *J. Ind. Microbiol. Biotechnol.* **2000**, *24*, 64–70. [CrossRef]
22. Dufossé, L.; Galaup, P.; Carlet, E.; Flamin, C.; Valla, A. Spectrocolorimetry in the CIE L*a*b* color space as useful tool for monitoring the ripening process and the quality of PDO red-smear soft cheeses. *Food Res. Int.* **2005**, *38*, 919–924. [CrossRef]

23. Sutthiwong, N.; Caro, Y.; Milhau, C.; Valla, A.; Fouillaud, M.; Dufossé, L. *Arthrobacter arilaitensis* strains isolated from ripened cheeses: Characterization of their pigmentation using spectrocolorimetry. *Food Res. Int.* **2014**, *65*, 184–192. [CrossRef]
24. Klitgaard, A.; Iversen, A.; Andersen, M.R.; Larsen, T.O.; Frisvad, J.C.; Nielsen, K.F. Aggressive dereplication using UHPLC-DAD-QTOF: Screening extracts for up to 3000 fungal secondary metabolites. *Anal. Bioanal. Chem.* **2014**, *406*, 1933–1943. [CrossRef] [PubMed]
25. Jung, H.; Kim, C.; Kim, K.; Shin, C.S. Color Characteristics of *Monascus* Pigments Derived by Fermentation with Various Amino Acids. *J. Agric. Food Chem.* **2003**, *51*, 1302–1306. [CrossRef] [PubMed]
26. Mapari, S.A.S.; Hansen, M.E.; Meyer, A.S.; Thrane, U. Computerized Screening for Novel Producers of *Monascus*-like Food Pigments in *Penicillium* Species. *J. Agric. Food Chem.* **2008**, *56*, 9981–9989. [CrossRef] [PubMed]
27. Yuliana, A.; Singgih, M.; Julianti, E.; Blanc, P.J. Derivates of azaphilone *Monascus* pigments. *Biocatal. Agric. Biotechnol.* **2017**, *9*, 183–194. [CrossRef]
28. Slominski, A.; Wortsman, J.; Plonka, P.M.; Schallreuter, K.U.; Paus, R.; Tobin, D.J. Hair follicle pigmentation. *J. Investig. Dermatol.* **2005**, *124*, 13–21. [CrossRef] [PubMed]
29. Torres, S.; Cjas, D.; Palfner, G.; Astuya, A.; Aballay, A.; Pérez, C.; Hernandez, V.; Becerra, J. Steroidal composition and cytotoxic activity from fruiting body of Cortinarius xiphidipus. *Nat. Prod. Res.* **2017**, *31*, 473–476. [CrossRef] [PubMed]
30. Dame, Z.T.; Silima, B.; Gryzenhout, M.; van Ree, T. Bioactive compounds from the endophytic fungus *Fusarium proliferatum*. *Nat. Prod. Res.* **2016**, *30*, 1301–1304. [CrossRef] [PubMed]
31. Gmoser, R.; Ferreira, J.A.; Lennartsson, P.R.; Taherzadeh, M.J. Filamentous ascomycetes fungi as a source of natural pigments. *Fungal Biol. Biotechnol.* **2017**, *4*, 4. [CrossRef]
32. Azmir, J.; Zaidul, I.S.M.; Rahman, M.M.; Sharif, K.M.; Mohamed, A.; Sahena, F.; Jahurul, M.H.A.; Ghafoor, K.; Norulaini, N.A.N.; Omar, A.K.M. Techniques for extraction of bioactive compounds from plant materials: A review. *J. Food Eng.* **2013**, *117*, 426–436. [CrossRef]
33. Chen, W.; Chen, R.; Liu, Q.; He, Y.; He, K.; Ding, X.; Kang, L.; Guo, X.; Xie, N.; Zhou, Y.; et al. Orange, red, yellow: Biosynthesis of azaphilone pigments in *Monascus* fungi. *Chem. Sci.* **2017**. [CrossRef]

12

News from the Sea: A New Genus and Seven New Species in the Pleosporalean Families Roussoellaceae and Thyridariaceae

Anna Poli, Elena Bovio, Lucrezia Ranieri, Giovanna Cristina Varese * and Valeria Prigione

Mycotheca Universitatis Taurinensis, Department of Life Sciences and Systems Biology, University of Torino, Viale Mattioli 25, 10125 Torino, Italy; anna.poli@unito.it (A.P.); elena.bovio@inrae.fr (E.B.); lucrezia.ranieri@edu.unito.it (L.R.); valeria.prigione@unito.it (V.P.)

* Correspondence: cristina.varese@unito.it.

Abstract: Nineteen fungal strains associated with the seagrass *Posidonia oceanica*, with the green alga *Flabellia petiolata*, and the brown alga *Padina pavonica* were collected in the Mediterranean Sea. These strains were previously identified at the family level and hypothesised to be undescribed species. Strains were examined by deep multi-loci phylogenetic and morphological analyses. Maximum-likelihood and Bayesian phylogenies proved that *Parathyridariella* gen. nov. is a distinct genus in the family Thyriadriaceae. Analyses based on five genetic markers revealed seven new species: *Neoroussoella lignicola* sp. nov., *Roussoella margidorensis* sp. nov., *R. mediterranea* sp. nov., and *R. padinae* sp. nov. within the family Roussellaceae, and *Parathyridaria flabelliae* sp. nov., *P. tyrrhenica* sp. nov., and *Parathyridariella dematiacea* gen. nov. et sp. nov. within the family Thyridariaceae.

Keywords: marine fungi; new taxa; phylogeny; lignicolous fungi

---

1. Introduction

Marine fungi are a relevant and active component of the microbial communities that inhabit the oceans [1]. Fungi in the marine environment live as mutualists, parasites, pathogens and saprobes, and are pivotal to marine food webs because of the recycling of recalcitrant substrata [2]; besides, these widely dispersed organisms are a source of novel bioactive compounds [3].

Marine fungi have been recovered worldwide from a broad range of biotic and abiotic substrata, such as driftwood algae, sponges, corals, sediments, etc. [4,5]. Following the definition of Pang et al [6] that considered "a marine fungus" to be any fungus retrieved repeatedly from marine environment and that reproduces in the marine environment, Jones et al. [7] listed 1680 fungal species belonging to 693 genera, 223 families, 87 orders, 21 classes and six phyla. However, considering that the total number of marine fungi has been estimated to exceed 10,000 taxa [8], fungal diversity remains largely undescribed. With more than 900 species [9], the Ascomycota are the dominant fungal phylum in the sea; the most represented lineages include the order Pleosporales (class Dothideomycetes) with 36 families, 95 genera and 194 species described to date (www.marinefungi.org).

In recent surveys aimed to uncover the underwater fungal diversity, 19 unidentified Roussoellaceae were isolated from several substrates, as follows: 12 from the brown alga *Padina pavonica* (L.) Thivy [10], 4 from the green alga *Flabellia petiolata* (Turra) Nizamuddin [11], 2 from the seagrass *Posidonia oceanica* (L.) [12] Delile, and 1 from the Atlantic sponge *Dysidea fragilis* (Montagu) [13]. The Roussoellaceae is a well-resolved family in the Pleosporales [14]. Others [15] have treated the family Roussoellaceae as a synonym of Thyridariaceae, based on phylogenetic affinities. However, following the discovery of new genera in this group, delineated by high resolution multi-locus phylogenetic analyses, the Roussoellaceae and Thyridariaceae are now recognized as two distinct but closely related families [16–20].

Many new species of Roussoellaceae and Thyridariaceae have recently been described on terrestrial plants including bamboo, palms and mangroves [14,17,20,21]. This paper provides a more precise phylogenetic placement of the 19 strains isolated from marine substrata together with morphological insights of those strains that represent new species within these two families.

2. Materials and Methods

2.1. Fungal Isolates

The fungal isolates analyzed in this paper were retrieved in the Mediterranean Sea from *P. oceanica* (2), collected in Riva Trigoso bay and Elba island, *P. pavonica* (12), and *F. petiolata* (3) from the coastal waters of Elba island [10–12]. A single isolate was previously retrieved in association with *D. fragilis* in the Atlantic Ocean [13] (Table 1).

Table 1. Dataset used for phylogenetic analysis. Genbank sequences including newly generated nrITS, nrLSU, nrSSU, TEF1-α and RPB2 amplicons relative to the novel species of Roussoellaceae and Thyridariaceae, to *Parathyridaria robiniae* MUT 2452 and MUT 4893 and to *Parathyridaria ramulicola* MUT 4397.

Species	Strain Code	Source	nrITS	nrSSU	nrLSU	TEF-1α	RPB2
Roussoellaceae							
Arthopyrenia salicis **Massal**	CBS 368.94	*Salix bark*	KF443410	AY538333	AY538339	KF443404	KF443397
Neoroussoella bambusae **Liu and Hyde**	MFLUCC 11-0124	Dead branch of *Bambusa*	KJ474827	–	KJ474839	KJ474848	KJ474856
N. alishanense **Karunarathna, Kuo, Phookamsak and Hyde**	AKTW 03 FU31016	*Pennisetum purpureum*	MK503816	MK503828	MK503822	MK336181	MN037756
	AKTW 11 FU31018	*Pennisetum purpureum*	MK503818	MK503830	MK503824	MK336182	MN037757
N. entadae **Jones and Hyde**	MFLUCC 18-0243	*Leucaena sp.*	MK347786	MK347893	MK348004	MK360065	MK434866
N. heveae **Senwanna, Phookamsak and Hyde**	MFLUCC 17-1983	Twig of *Hevea brasiliensis*	MH590693	–	MH590689	–	–
N. leucaenae **Jones and Hyde**	MFLUCC 18-1544	Decaying pod of *Leucaena*	MK347767	MK347874	MK347984	MK360067	MK434876
	MFLUCC 17-0927	*Pterocarpus* sp.	MK347733	MK347841	MK347950	MK360066	MK434896
Neoroussoella lignicola**sp. nov.**	**MUT 4904**	*P. pavonica*	KT699129	**MN556307***	**MN556319***	**MN605894***	**MN605914***
	MUT 5008	*P. oceanica* leaves	**MN556317***	**MN556308***	**MN556320***	**MN605895***	**MN605915***
	MUT 5373	*P. pavonica*	KU314953	KU314954	**MN556321***	**MN605896***	**MN605916***
Pararoussoella juglandicola **Crous and Schumach**	CBS 145037		MK442607	–	MK442543	MK442699	MK442671
P. mukdahanensis **(Phookamsak, Dai and Hyde) Crous**	MFLUCC 11-0201	Bamboo	KU940129	KU872121	KU863118	–	–
P. rosarum **Jones and Hyde**	MFLUCC 17-0796	*Rosa* sp.	MG8289391	MG829154	MG829048	MG829224	–
Pseudoneoconiothyrium rosae **(Phukhams., Camporesi and Hyde) Phukhams., Camporesi and Hyde**	MFLUCC 15-0052	Dead aerial spines of *Rosa canina*	MG828922	MG829138	MG829032	–	–

Table 1. *Cont.*

Species	Strain Code	Source	nrITS	nrSSU	nrLSU	TEF-1α	RPB2
Roussoella chiangraina Phookamsak, Liu and Hyde	MFLUCC 10-0556	Dead branch of bamboo	KJ474828	–	KJ474840	KJ474849	KJ474857
R. doimaesalongensis Thambug. and Hyde	MFLUCC 14–0584	Dead branch of bamboo	KY026584	–	KY000659	KY651249	KY678394
R. elaeicola Konta. and Hyde	MFLUCC 15-0276a	Dead petiole of *Elaeis guineensis*	MH742329	–	MH742326	–	–
	MFLUCC 15-0276b	Dead petiole of *Elaeis guineensis*	MH742330	–	MH742327	–	–
R. euonymi Crous and Akulov	CBS 143426	Fallen branches of *Euonymus europaeus*	MH107915	–	MH107961	–	MH108007
R. hysterioides (Ces.) Höhn.	CBS 546.94	*Phyllostachys*	KF443405	AY642528	KF443381	KF443399	KF443392
R. intermedia Ju, Rogers and Huhndorf	CBS 170.96	Bamboo	KF443407	KF443390	KF443382	KF443398	KF443394
*R. japanensis*Kaz. Tanaka, Liu and Hyde	MAFF 239636	Twigs of *Sasa veitchii*	KJ474829	AB524480	AB524621	AB539114	AB539101
R. kunmingensis Jiang, Phookamsak and Hyde	KUMCC 18-0128	Dead bamboo	MH453491	–	MH453487	MH453480	MH453484
R. mangrovei Phukhams. and Hyde	MFLUCC 16-0424	Dead branches of *Rhizophora*	MH025951	–	MH023318	MH028246	MH028250
Roussoella margidorensis sp. nov.	**MUT 5329**	*P. pavonica*	KU314944	**MN556309***	**MN556322***	**MN605897***	**MN605917***
Roussoella mediterranea sp. nov.	**MUT 5306**	*P. pavonica*	KU255054	**MN556310***	**MN556323***	**MN605898***	**MN605918***
	MUT 5369	*P. pavonica*	KU314947	KU314948	**MN556324***	**MN605899***	**MN605919***
R. mexicana Crous and Yáñez-Mor.	CPC 25355	Leaf spots of Coffea arabica	KT950848	–	KT950862	–	–
R. neopustulans Dai, Liu and Hyde	MFLUCC 11-0609	Bamboo	KJ474833	–	KJ474841	KJ474850	
	MFLUCC 12-0003	Bamboo	KU940130	KU872122	KU863119	–	–
R. nitidula Sacc. and Paol.	MFLUCC 11-0182	Bamboo	KJ474835	–	KJ474843	KJ474852	KJ474859
	MFLUCC 11-0634	Bamboo	KJ474834	–	KJ474842	KJ474851	KJ474858
Roussoella padinae sp. nov.	**MUT 5341**	*P. pavonica*	KU158153	KU158176	**MN556325***	**MN605900***	**MN605920***
	MUT 5365	*P. pavonica*	KU158170	KU158179	**MN556326***	**MN605901***	**MN605921***
	MUT 5503	*P. pavonica*	KU158170	**MN556312***	**MN556327***	**MN605902***	**MN605922***
R. pseudohysterioides Dai and Hyde	MFLUCC 13-0852	Bamboo	KU940131	KU872123	KU863120	KU940198	–
R. pustulans (Ellis and Everh.) Ju, Rogers and Huhndorf	KT 1709	Culms of *Sasa kurilensis*	KJ474830	AB524482	AB524623	AB539116	AB539103
R. scabrispora (Höhn.) Aptroot	MFLUCC 11-0624	Bamboo	KJ474836	–	KJ474844	KJ474853	KJ474860
	RSC	Bamboo	KX650566	–	KX650566	KX650537	–
R. siamensis Phookamsak, Liu and Hyde	MFLUCC 11-0149	Bamboo	KJ474837	KU872125	KJ474845	KJ474854	KJ474861
R. thailandica Dai, Liu and Hyde	MFLUCC 11-0621	Bamboo	KJ474838	–	KJ474846	–	
R. tuberculata Dai and Hyde	MFLUCC 13-0854	Bamboo	KU940132	KU872124	KU863121	KU940199	
R. verrucispora Kaz. Tanaka, Liu and Hyde	CBS 125434	*Sasa kurilensis*	KJ474832	AB52448	AB524622	AB539115	–

Table 1. *Cont.*

Species	Strain Code	Source	nrITS	nrSSU	nrLSU	TEF-1α	RPB2
R. yunnanensis **Jiang, Phookamsak and Hyde**	KUMCC 18-0115	Dead bamboo	MH453492	–	MH453488	MH453481	–
Roussoellopsis macrospora **(Hino and Katum.) Hinoand Katum**	MFLUCC 12-0005	Bamboo	KJ739604	KJ739608	KJ474847	KJ474855	KJ474862
Ro. tosaensis **(Hino and Katum.) Hino and Katum**	KT 1659	Culms of bamboo	–	AB524484	AB524625	AB539117	AB539104
Thyridariaceae							
Cycasicola goaensis **Jones and Hyde**	MFLUCC 17-0754	*Cycas* sp.	MG828885	MG829112	MG829001	MG829198	–
C. leucaenae **Jones and Hyde**	MFLUCC 17-0914	Leucaena leucocephala	MK34772	MK347833	MK347942	MK360046	MK434900
Liua muriformis **Phookamsak, Jiang and Hyde**	KUMCC 18-0177	Dead hanging branches of *Lonicera maackii*	MK433599	MK433595	MK433598	MK426798	MK426799
Parathyridaria percutanea **(Ahmed, Stevens, van de Sande and de Hoog) Jaklitsch and Voglmayr**	CBS 868.95	Human	KF322118	KF366451	KF366449	KF407987	KF366452
	CBS 128203	Human	KF322117	KF366450	KF366448	KF407988	KF366453
P. ramulicola **Jaklitsch, Fourn and Voglmayr**	CBS 141479	Twigs of *Ribes rubrum*	NR_147657	KX650514	KX650565	KX650536	KX650584
	MUT 4397	*P. oceanica*	KC339235	**MN556311***	KF636775	**MN605913***	**MN605933***
P. robiniae**Mapook, Camporesi and Hyde**	MFLUCC 14-1119	Dead branch of *Robinia pseudoacacia*	KY511142	–	KY511141	KY549682	–
	MUT 2452	*Dysidea fragilis*	MG813183	**MN556312***	MG816491	**MN605903***	**MN605923***
	MUT 4893	*P. pavonica*	KM355998	KM355993	**MN556328***	**MN605904***	**MN605924***
Parathyridaria flabelliae **sp. nov.**	**MUT 4859**	*F. petiolata*	KR014355	KT587315	KP671716	**MN605909***	**MN605929***
	MUT 4886	*F. petiolata*	KR014358	KT587317	KP671720	**MN605910***	**MN605930***
Parathyridaria tyrrhenica **sp. nov.**	**MUT 4966**	*F. petiolata,*	KR014366	KT587309	KP671740	**MN605911***	**MN605931***
	MUT 5371	*P. pavonica*	KU314951	KU314952	**MN556329***	**MN605912***	**MN605932***
Parathyridariella dematiacea **sp. nov.**	**MUT 4419**	*P. oceanica* rhizomes	KC339245	**MN556313***	KF636786	**MN605905***	**MN605925***
	MUT 4884	*F. petiolata*	**MN556317***	KT587329	KP671726	**MN605906***	**MN605926***
	MUT 5310	*P. pavonica*	KU255057	**MN556314***	**MN556330***	**MN605907***	**MN605927***
	MUT 5381	*P. pavonica*	KU314959	KU314960	**MN556331***	**MN605908***	**MN605928***
Thyridaria acaciae **(Crous and Wingf.) Jaklitsch and Voglmayr**	CBS 138873	Leaves of *Acacia tortilis*	KP004469	–	KP004497	–	–
T. broussonetiae **(Sacc.) Traverso**	TB	*Hippocrepis emerus*	KX650567	–	KX650567	KX650538	KX650585
	TB1	*Amorpha fruticosa*	KX650568	KX650515	KX650568	KX650539	KX650586
Thyridariella mahakoshae **Devadatha, Sarma, Wanas., Hyde and Jones**	NFCCI 4215	Decaying wood *Avicennia marina*	MG020435	MG020441	MG020438	MG023140	MG020446
Th. mangrovei **Devadatha, Sarma, Hyde, Wanas. and Jones**	NFCCI 4213	Decaying wood *Avicennia marina*	MG020434	MG020440	MG020437	MG020443	MG020445
	NFCCI 4214	Decaying wood *Avicennia marina*	MG020436	MG020442	MG020439	MG020444	MG020447

Table 1. *Cont.*

Species	Strain Code	Source	nrITS	nrSSU	nrLSU	TEF-1α	RPB2
Occultibambusaceae							
***Occultibambusa bambusae* Dai and Hyde**	MFLUCC 11-0394	Bamboo	KU940124	–	KU863113	KU940194	KU940171
	MFLUCC 13-0855	Bamboo	KU940123	KU872116	KU863112	KU940193	KU940170
Ohleriaceae							
***Ohleria modesta* Fuckel**	MGC		KX650562	–	KX650562	KX650533	KX650582
	OM	Branches of *Chamaecytisus proliferus*	KX650563	KX650513	KX650563	KX650534	KX650583
Torulaceae							
***Dendryphion europaeum* Crous and Schumacher**	CPC 22943	*Heracleum sphondylium*	KJ869146	–	KJ869203	–	–
***Torula herbarum* (Pers.) Link**	CBS 111855	n.a.	KF443409	KF443391	KF443386	KF443403	KF443396
	CBS 595.96		KF443408	KF443387	KF443385	KF443402	KF443395
***Torula hollandica* Crous**	CBS 220.69	Delphinium dead stem	KF443406	–	MH877717	–	–

* = newly generated sequences; n.a. = not available.

The strains investigated were originally isolated on Corn Meal Agar medium supplemented with sea salts (CMASS; 3.5% w/v sea salt mix, Sigma-Aldrich, Saint Louis, USA, in ddH_2O) and are preserved at the *Mycotheca Universitatis Taurinensis* (MUT), Italy.

2.2. Morphological Analysis

All isolates were pre-grown on Malt Extract Agar-sea water (MEASW; 20 g malt extract, 20 g glucose, 2 g peptone, 20 g agar in 1 L of sea water) for one month at 24 °C prior to inoculation in triplicate onto new Petri dishes (9 cm Ø) containing (i) MEASW, (ii) Oatmeal Agar-sea water (OASW; 30 g oatmeal, 20 g agar in 1 L of sea water), or iii) Potato Dextrose Agar-sea water (PDASW; 4 g potato extract, 20 g dextrose, 20 g agar in 1 L of sea water). Petri dishes were incubated at 15 and/or 24 °C. The colony growth was monitored periodically for 28 days. Macroscopic and microscopic traits, were assessed for strains grown on MEASW at the end of the incubation period.

In an attempt to induce sporulation, sterile pieces of *Quercus ruber* cork and *Pinus pinaster* wood (species autochthonous to the Mediterranean area) were placed on 3 week old fungal colonies grown on MEASW ([22], modified). Petri dishes were further incubated for 4 weeks at 24 °C. Subsequently, cork and wood pieces were transferred to 50 mL tubes containing 20 mL of sterile sea water. Samples were incubated at 24 °C for one month. In parallel, the strains were also plated on Syntetic Nutrient Agar-sea water (SNASW; 1 g KH_2PO_4, 1 g KNO_3, 0.5 g $MgSO_4 \bullet 7H_2O$, 0.5 g KCl, 0.2 g glucose, 0.2 g sucrose, 20 g agar in 1 L of sea water) supplemented with sterile pine needles. Petri dishes were incubated at 24 °C for one month.

Morphological structures were observed, and images captured using an optical microscope (Leica DM4500B, Leica microsystems GmbH, Wetzlar) equipped with a camera (Leica DFC320, Leica microsystems GmbH, Wetzlar). Macro- and microscopic features were compared with the available description of Roussoellaceae and Thyridariaceae [14,15,17,18,20].

2.3. DNA Extraction, PCR Amplification, and Data Assembling

Genomic DNA was extracted from about 100 mg of fresh mycelium grown on MEASW plates. Mycelium was disrupted by the mean of a MM400 tissue lyzer (Retsch GmbH, Haan, Germany) and DNA extracted using a NucleoSpin kit (Macherey Nagel GmbH, Duren, DE, USA) following the manufacturer's instructions. The quality and quantity of DNA were measured spectrophotometrically (Infinite 200 PRO NanoQuant; TECAN, Männedorf); DNA was stored at −20 °C.

The partial sequences of five genetic markers were amplified by PCR. Primer pairs ITS1/ITS4 [23], LR0R/LR7 [24], NS1/NS4 [23] were used to amplify the internal transcribed spacers, including the 5.8S rDNA gene (nrITS), 28S large ribosomal subunit (nrLSU) and 18S small ribosomal subunit (nrSSU). The translation elongation factor (TEF1α) and RNA polymerase II subunit (RPB2) were amplified by using primer pairs EF1-1018F/EF1-1620R [25] and fRPB2-5F/fPB2-7R [26].

Amplifications were run in a T100 Thermal Cycler (Bio-Rad, Hercules, CA, USA) programmed as described in Table 2. Reaction mixtures consisted of 20–40 ng DNA template, 10× PCR Buffer (15 mM $MgCl_2$, 500 mM KCl, 100 mM Tris-HCl, pH 8.3), 200 μM each dNTP, 1 μM each primer, 2.5 U Taq DNA Polymerase (Qiagen, Chatsworth, CA, USA), in 50 μL final volume. For problematic cases, additional $MgCl_2$ and/or 2.5% DMSO facilitated the reaction.

Table 2. Primers and PCR conditions used to amplify specific gene marker.

	Forward and Reverse Primers	Thermocycler Conditions	References
ITS	ITS1- ITS4	95 °C: 5 min, (95 °C: 40 s, 55 °C: 50 s, 72 °C: 50 sec) × 35 cycles; 72 °C: 8 min; 4 °C: ∞	[23]
LSU	LR0R-LR7	95 °C: 5 min, (95 °C: 1 min, 50 °C: 1 min, 72 °C: 2 min) × 35 cycles; 72 °C: 10 min; 4 °C: ∞	[24]
SSU	NS1-NS4	95 °C: 5 min, (95 °C: 1 min, 50 °C: 1 min, 72 °C: 2 min) × 35 cycles; 72 °C: 10 min; 4 °C: ∞	[23]
TEF-1α	1018F/1620R	95 °C: 5 min, (95 °C: 1 min, 50 °C: 1 min; 72 °C: 2 min) × 40 cycles, 72 °C: 10 min; 4 °C: ∞	[25]
RPB2	fRPB2-5F/fPB2-7cR	94 °C: 3 min, (94 °C: 30 s; 55 °C: 30 s; 72 °C: 1 min) × 40 cycles, 72 °C: 10 min; 4 °C: ∞	[26]

Amplicons, together with a GelPilot 1 kb plus DNA Ladder, were visualized on a 1.5% agarose gel stained with 5 mL 100 mL^{-1} ethidium bromide; PCR products were purified and sequenced at the Macrogen Europe Laboratory (Madrid, Span). The resulting Applied Biosystem (ABI) chromatograms were inspected, trimmed and assembled to obtain consensus sequences using Sequencer 5.0 (GeneCodes Corporation, Ann Arbor, Michigan, USA http://www.genecodes.com). Newly generated sequences were deposited in GenBank (Table 1).

2.4. Sequence Alignment and Phylogenetic Analysis

A dataset consisting of nrSSU, nrITS, nrLSU, TEF1α and RPB2 was assembled on the basis of BLASTn results and of recent phylogenetic studies focused on Roussoellaceae and Thyridariaceae [18,20]. Reference sequences were retrieved from GenBank (Table 1).

Sequences were aligned using MUSCLE (default conditions for gap openings and gap extension penalties), implemented in MEGA v. 7.0 (Molecular Evolutionary Genetics Analysis), visually inspected and trimmed by TrimAl v. 1.2 (http://trimal.cgenomics.org) to delimit and discard ambiguously aligned regions. Since no incongruence was observed among single-loci phylogenetic trees, alignments were concatenated into a single data matrix with SequenceMatrix [27]. The best evolutionary model under the Akaike Information Criterion (AIC) was determined with jModelTest 2 [28].

Phylogenetic inference was estimated using Maximum Likehood (ML) and Bayesian Inference (BI) criteria. The ML analysis was generated using RAxML v. 8.1.2 [29] under GTR + I + G evolutionary model and 1000 bootstrap replicates. Support values from bootstrapping runs (MLB) were mapped on the globally best tree using the "-f a" option of RAxML and "-x 12345" as a random seed to invoke the novel rapid bootstrapping algorithm. BI was performed with MrBayes 3.2.2 [30] with the same substitution model (GTR + I + G). The alignment was run for 10 million generations with two independent runs each containing four Markov Chains Monte Carlo (MCMC) and sampling every 100 iterations. The first 25% of generated trees were discarded as "burn-in". A consensus tree was generated using the "sumt" function of MrBayes and Bayesian posterior probabilities (BPP) were calculated. Consensus trees were visualized in FigTree v. 1.4.2 (http://tree.bio.ed.ac.uk/software/figtree).

Two strains of *Occutibambusa bambusae* (Occultibambusaceae) were used to root the tree. Due to topological similarity of the two resulting trees, only ML analysis with MLB and BPP values was reported (Figure 1).

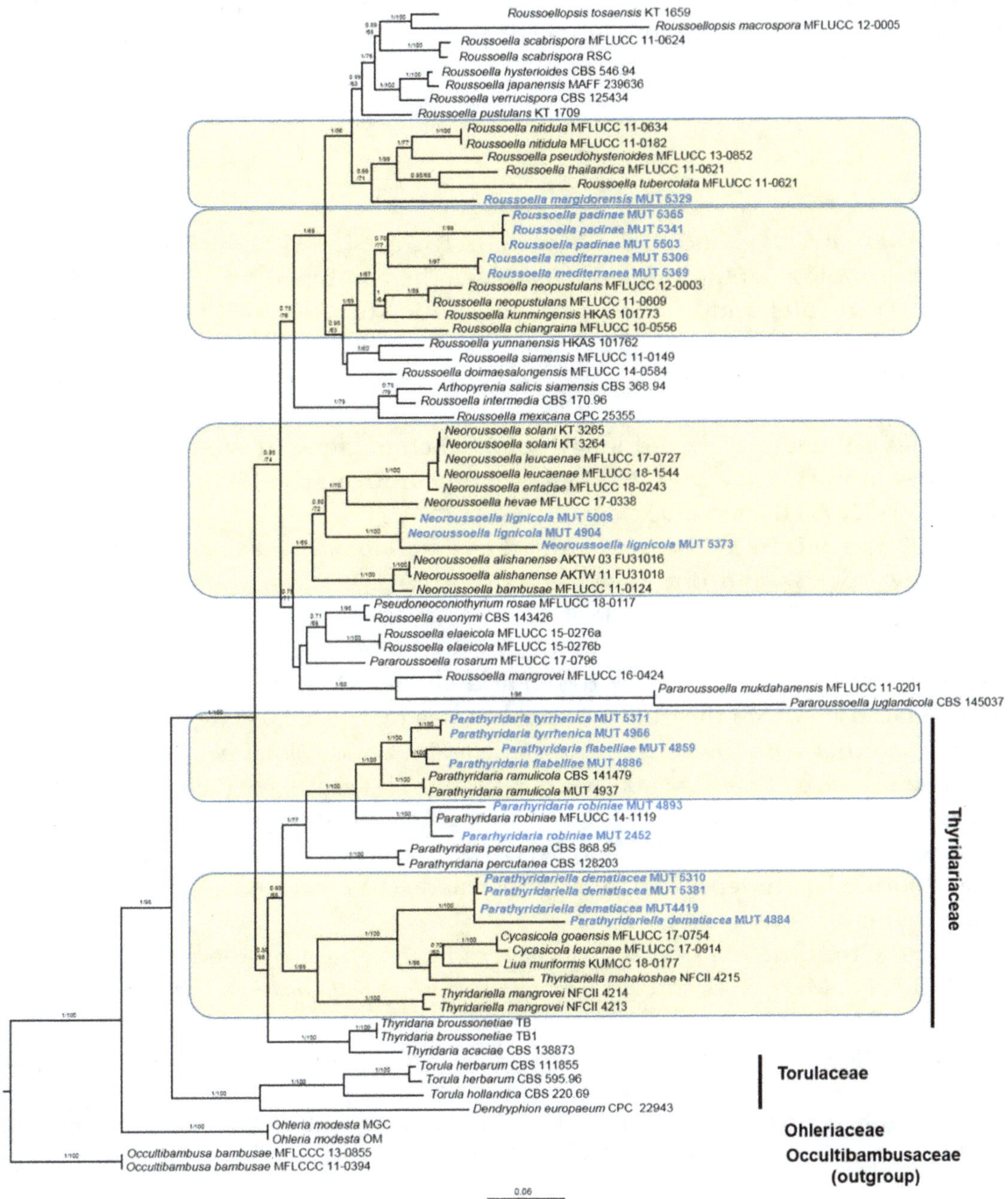

Figure 1. Phylogram generated from RAxML analysis based on a combined dataset of nrITS, nrSSU, nrLSU, TEF1α and RPB2 partial sequences. The tree is rooted to Occultibambusa bambusae. Branch numbers indicate BYPP/MLB values; Bar = expected changes per site (0.06).

DNA diagnostic characters were visually identified by the presence of heterozygous bases. For each locus, aligned sequences of the individual clusters containing new species, were inspected. Nucleotide diversities of the novel species were annotated when occurred (Tables S1–S18).

Sequence alignments and phylogenetic trees were deposited in TreeBASE (http://www.treebase.org, submission number S24773).

Following phylogenetic tree inspection, isolates that clustered in the same group and that derived from the same substrate were subjected to PCR-fingerprinting by using the micro- and mini-satellite primers $(GTG)_5$ and M13 [31,32] to exclude duplicates from further analysis. DNA fingerprints were visualized with 1.5% agarose gel stained with 5 mL 100 mL^{-1} ethidium bromide while a GelPilot 1 kb plusDNA Ladder was used as a reference. Images were acquired with a Gel Doc1000 System (Bio-Rad, Hercules, CA, USA) and fingerprints analyzed using Bionumerics v 7.6 (http: //www.applied-maths.com).

3. Results

3.1. Phylogenetic Inference

Preliminary analyses carried out individually with nrITS, nrSSU, nrLSU, TEF1α and RPB2 denoted no incongruence in the topology of the single-locus trees. The combined five-markers dataset—built on the basis of BLASTn results and of recent phylogenetic studies [18,20]—consisted of 81 taxa (including MUT isolates) that represented 16 genera and 56 species (Table 1). A total of 63 sequences (2 nrITS, 8 nrSSU, 13 nrLSU, 20 TEF1α and 20 RPB2) were newly generated while 261 were retrieved from GenBank.

The combined dataset had an aligned length of 3390 characters, of which 1683 were constant, 657 were parsimony-uninformative and 1050 parsimony informative (TL = 218, CI = 0.422018, RI = 0.825243, RC = 0.348267, HI = 0.877952).

Strains MUT 4893 and MUT 2452 were identified as *Parathyridaria robiniae*, the rest of the strains represented seven new species and one new genus (Figure 1). *Parathyridaria tyrrhenica* sp. nov. (MUT 5371 and MUT 4966) formed a sister clade to *Parathyridaria flabelliae* sp. nov. (MUT 4859 and MUT 4886) with high statistical support (BYPP = 1.00; MLB = 100%); these two novel species are closely related to *P. ramulicola* (BYPP = 1.00; MLB = 100%) and clustered with other *Parathyridaria* species in the Thyridariaeae family. Within this family, four isolates (MUT 5310, MUT 5381, MUT 4419 and MUT 4884) clustered together with the genera *Thyridariella*, *Liua* and *Cycasicola*, and formed a strongly supported monophyletic lineage (BYPP = 1.00; MLB = 100%). Therefore, we have introduced the novel genus *Parathyridariella*, typified by the new species *Parathyridariella dematiacea* sp. nov.

The three strains, MUT 4904, MUT 5373 and MUT 5008, represented a novel species *Neoroussoella lignicola* sp. nov. and formed an independent and robust clade (BYPP = 1.00; MLB = 100%), within the *Neoroussoella* group in the Roussoellaceae.

Two sister clades within the *Roussoella* group were represented by the new species *Roussoella padinae* sp. nov. (MUT 5503, MUT 5341 and MUT 5365) and *Roussoella mediterranea* sp. nov. (MUT 5306 and MUT 5369). Finally, MUT 5329 *Roussoella margidorensis* sp. nov. clustered together with *R. nitidula*, *R. pseudohysterioides*, *R. thailandica* and *R. tubercolata* (BYPP = 0.99; MLB = 71%) but was phylogenetically distant from these species.

Nucleotide divergence between each novel species and members of the same clusters were annotated for each locus, when occurred (Tables S1–S18).

3.2. Taxonomy

Parathyridariella gen. nov. V. Prigione, A. Poli, E. Bovio and G.C. Varese
MYCOBANK: MB 832836
Type species. Parathyridariella dematiacea sp. nov.
Etymology. In reference to the phylogenetic proximity to the genus *Thyridariella*.
Phylogenetic placement. Thyridariaceae, Sordariomycetes, Ascomycota. The genus *Parathyridariella* gen. nov. clusters together with genera *Cycasicola*, *Liua* and *Thyridariella* (Figure 1).

Parathyridariella dematiacea sp. nov. V. Prigione, A. Poli, E. Bovio and G.C. Varese
MYCOBANK: MB 832837
Figure 2

Type. Italy, Tuscany, Mediterranean Sea, Elba Island (LI), Ghiaie ISL, 14–15 m depth, 42°49′04″N, 10°19′20″E, form the green alga *Flabellia petiolata*, 20 March 2010, R. Mussat-Sartor and N. Nurra, MUT 4884 holotype, living culture permanently preserved in metabolically inactive state by deep-freezing at Mycotheca Universitatis Taurinensis (MUT).

Additional material examined. Italy, Ligury, Mediterranean Sea, Riva Trigoso, Punta Manara (GE), 5–21 m depth, 44°15′08.62″N 9°24′17.64″E, from the seagrass *Posidonia oceanica*, March 2008, MUT 4419.

Etymology. In reference to the color of the colony on culture media.

Description. Growing actively on *Pinus pinaster* and *Quercus ruber* cork. Showing a floccose growth mainly on *Pinus pinaster*. *Hyphae* 2.8–4.8 µm wide, septate, hyaline to lightly pigmented. *Chlamydospores* numerous, mostly in chain, intercalary or solitary, globose to subglobose, from brownish to dark brown, 7–10 × 6–8 µm diameter.

Sexual morph not observed. Asexual morph with differentiated conidiogenesis not observed.

Colony description. Colonies on MEASW attaining 28–34 mm diam after 28 days at 24 °C, mycelium from dark grey/black to dark green, dense with radial grooves and concentric rings, submerged edges; reverse dark green. Brown exudate present above the concentric rings. Growth on OASW reaching 40–54 mm diam at 24 °C and 21–29 mm diam at 15 °C; colonies on PDA attaining 36–49 mm diam and 15.5–22.5 mm diam at 24 °C and 15 °C, respectively.

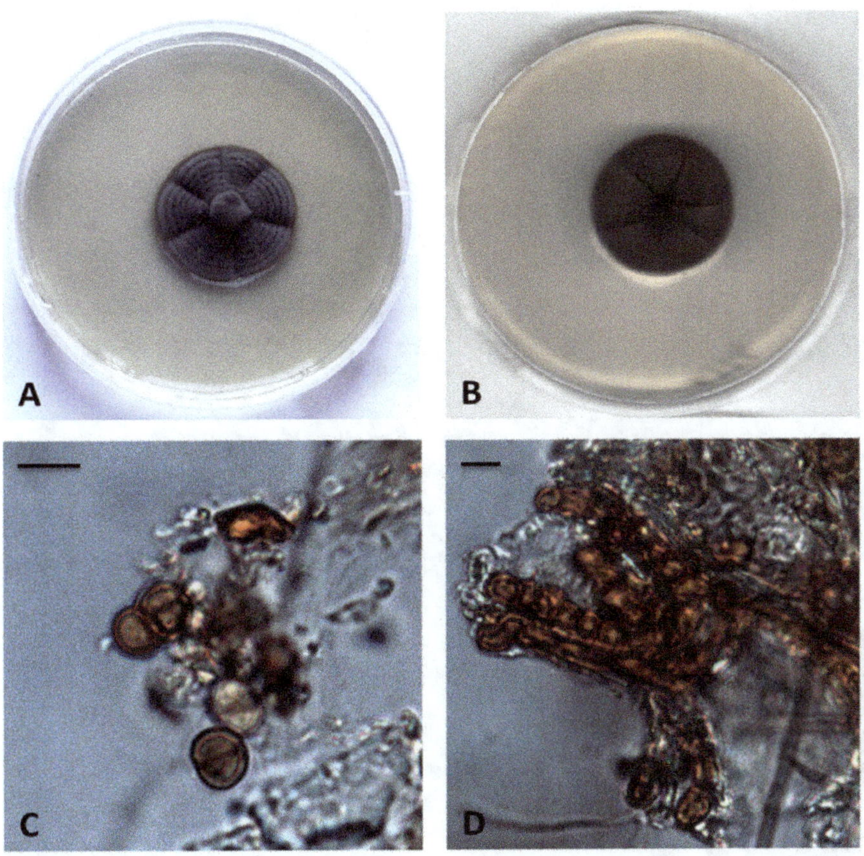

Figure 2. *Parathyridariella dematiacea* sp. nov. 28-days-old colony at 21 °C on MEASW (**A**) and reverse (**B**); solitary (**C**) and in chain (**D**) chlamydospores. Scale bars: 10 µm (C, D).

Parathyridaria tyrrhenica sp. nov. A. Poli, V. Prigione, E. Bovio and G.C. Varese
MYCOBANK: MB 832838
Figure 3

Type. Italy, Tuscany, Mediterranean Sea, Elba Island (LI), Ghiaie ISL, 14–15 m depth, 42°49′04″N, 10°19′20″E, from the brown alga *Padina pavonica*, March 2010, R. Mussat-Sartor and N. Nurra, MUT 5371 holotype, living culture permanently preserved in metabolically inactive state by deep-freezing at MUT.

Additional material examined. Italy, Tuscany, Mediterranean Sea, Elba Island (LI), Ghiaie ISL, 14–15 m depth, 42°49′04″N, 10°19′20″E, from the green alga *Flabellia petiolata*, March 2010, R. Mussat-Sartor and N. Nurra, MUT 4966.

Etymology. In reference to Tyrrhenian Sea.

Description. Growing actively on *Pinus pinaster* wood and *Quercus ruber* cork. *Hyphae* 5 μm diameter, septate, hyaline to brownish, sometimes wavy or swollen, forming hyphal strands.

Sexual morph not observed. Asexual morph with differentiated conidiogenesis: not observed.

Colony description. Colonies growing on MEASW, reaching 10 mm diam after 28 days, at 21 °C, mycelium funiculose, yellowish, lightly ochre at the edges; reverse light yellow, lighter at the edges. Growth on OASW reaching 48–50 mm diam at 24 °C and 26–29 mm diam at 15 °C; colonies on PDA attaining 31–46 mm diam and 16–19 mm diam at 24 °C and 15 °C, respectively.

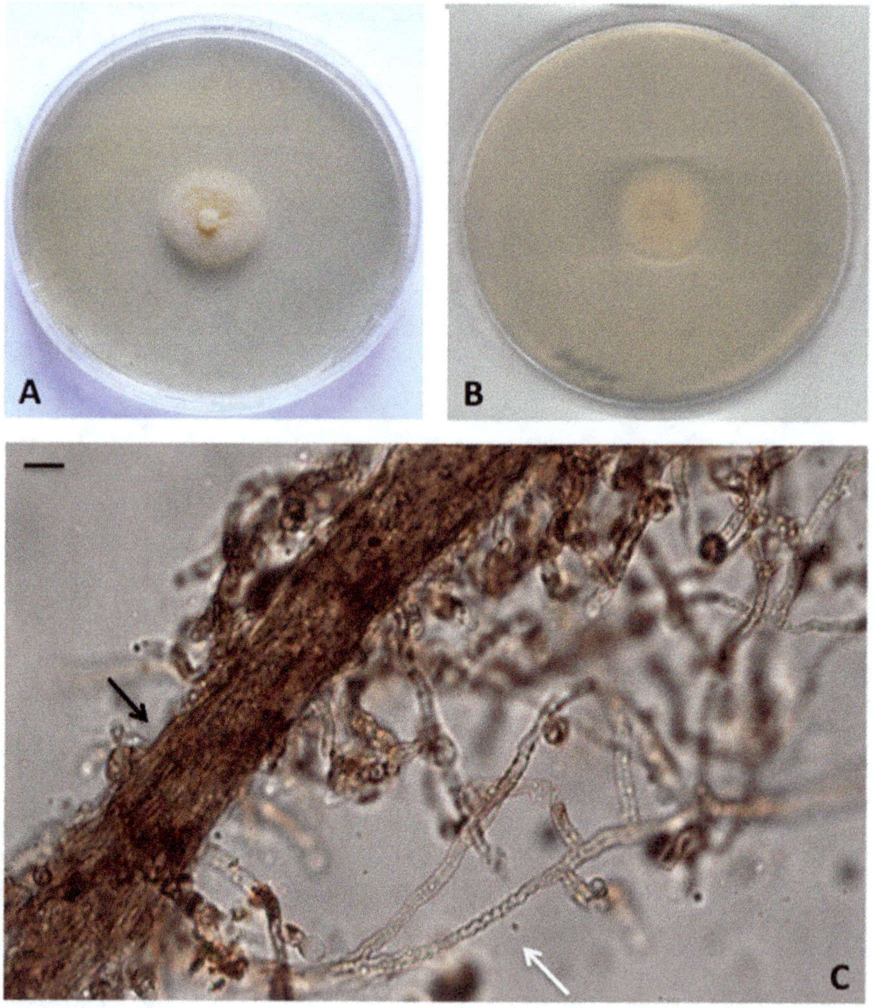

Figure 3. *Parathyridaria tyrrhenica* sp. nov. 28-days-old colony at 21 °C on MEASW (**A**) and reverse (**B**); mycelium (**C**), black and white arrows indicate hyphal strands and wavy hyphae, respectively. Scale bar: 10 μm.

Parathyridaria flabelliae sp. nov. E. Bovio, A. Poli, V. Prigione and G.C. Varese
MYCOBANK: MB 832839

Figure 4

Type. Italy, Tuscany, Mediterranean Sea, Elba Island (LI), Ghiaie ISL, 14–15 m depth, 42°49′04″N, 10°19′20″E, from the green alga *Flabellia petiolata*, March 2010, R. Mussat-Sartor and N. Nurra, MUT 4859 holotype, living culture permanently preserved in metabolically inactive state by deep-freezing at MUT.

Additional material examined. Italy, Tuscany, Mediterranean Sea, Elba Island (LI), Ghiaie ISL, 14–15 m depth, 42°49′04″N, 10°19′20″E, from the green alga *Flabellia petiolata*, March 2010, R. Mussat-Sartor and N. Nurra, MUT 4886.

Etymology. In reference to the original substratum, the green alga *Flabellia petiolata*.

Description. Growing actively on *Pinus pinaster* and on *Quercus ruber* cork. *Hyphae* 2.6–5 μm wide, septate and hyaline. *Chlamydospores* numerous, globose or subglobose, from light to dark brown, unicellular (4 × 5 μm diameter) and multicellular (up to four-celled; 8 × 12 μm diameter).

Sexual morph not observed. Asexual morph with differentiated conidiogenesis not observed.

Colony description. Colonies growing on MEASW, reaching 37–44 mm diam after 28 days at 21 °C, funiculose, whitish with submerged edges; reverse brown in the middle, lighter at edges. Growth on OASW reaching 60 mm diam at 24 °C and 33–35 mm diam at 15 °C; colonies on PDA attaining 53–64 mm diam and 23–24 mm diam at 24 °C and 15 °C, respectively.

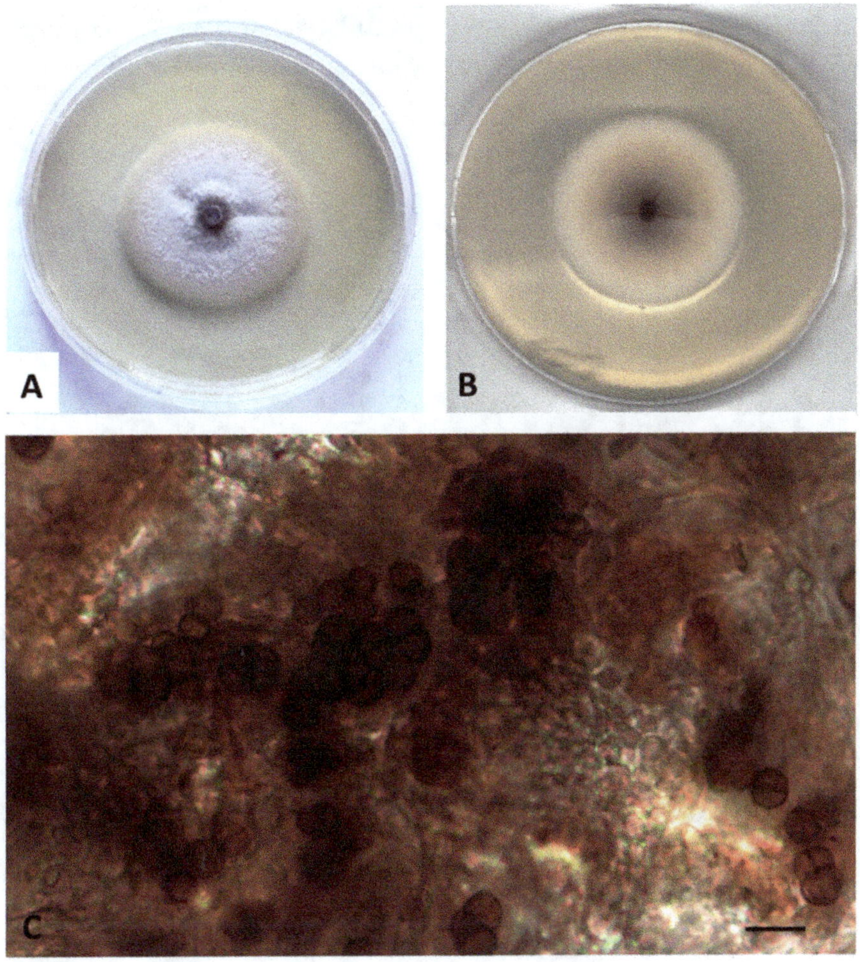

Figure 4. *Parathyridaria flabelliae* sp. nov. 28-days-old colony at 21 °C on MEASW (**A**) and reverse (**B**); unicellular and multicellular chlamydospores (**C**). Scale bar: 10 μm.

Neoroussoella lignicola sp. nov. A. Poli, E. Bovio, V. Prigione and G.C. Varese
MYCOBANK: MB 832840

Figure 5

Type. Italy, Tuscany, Mediterranean Sea, Elba Island (LI), Margidore ISL, 14–15 m depth, UTM WGS84 42°45′29″N, 10°18′24″E, from the brown alga *Padina pavonica*, March 2010, R. Mussat-Sartor and N. Nurra, MUT 5373 holotype, living culture permanently preserved in metabolically inactive state by deep-freezing at MUT.

Additional material examined. Italy, Tuscany, Mediterranean Sea, Elba Island (LI), Margidore ISL, 14–15 m depth, UTM WGS84 42°45′29″N, 10°18′24″E, from the brown alga *Padina pavonica*, March 2010, R. Mussat-Sartor and N. Nurra, MUT 4904.

Italy, Tuscany, Mediterranean Sea, Elba Island (LI), Margidore ISL, 14–15 m depth, UTM WGS84 42°45′29″N, 10°18′24″E, from the seagrass *Posidonia oceanica*, March 2010, R. Mussat-Sartor and N. Nurra, MUT 5008.

Etymology. In reference to the lignicolous behavior.

Description. Growing efficiently on *Pinus pinaster* wood. *Hyphae* 2–4.4 μm wide, septate, hyaline, assuming toruloid aspect when growing into wood vessels and forming chains of two-celled chlamydospores which, at maturity, protrude from the vessels. *Chlamydospores* 7.4 × 5.2 μm, from light to dark brown, globose or subglobose.

Sexual morph not observed. Asexual morph with differentiated conidiogenesis not observed.

Colony description. Colonies growing on MEASW, reaching 28–29 mm diam after 28 days at 21° C, from grey to dark green, floccose with irregular edges, reverse dark grey. Clear exudate often present. Growth on OASW reaching 27–40 mm diam at 24 °C and 14.5–26 mm diam at 15 °C; colonies on PDA attaining 38–45 mm diam and 19–29 mm diam at 24 °C and 15 °C, respectively.

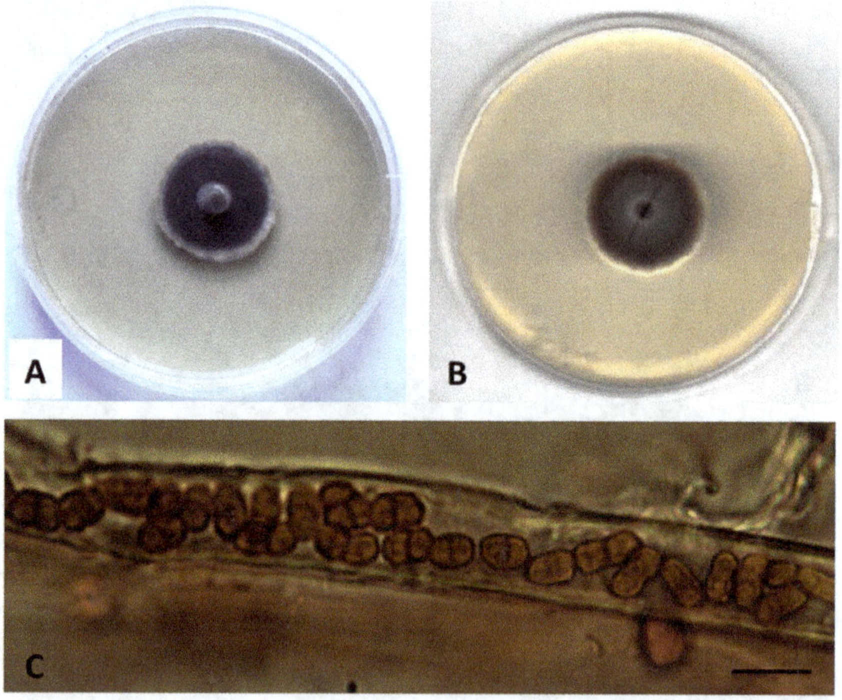

Figure 5. *Neoroussoella lignicola* sp. nov. 28-days-old colony at 21 °C on MEASW (**A**) and reverse (**B**); two-celled chlamydospores inside wood vessels (**C**). Scale bar: 10 μm.

Roussoella margidorensis sp. nov. E. Bovio, V. Prigione, A. Poli and G.C. Varese
MYCOBANK: MB 832841
Figure 6

Type. Italy, Tuscany, Mediterranean Sea, Elba Island (LI), Margidore ISL, 14–15 m depth, UTM WGS84 42°45′29″N, 10°18′24″E, from the brown alga *Padina pavonica*, March 2010, R. Mussat-Sartor and N. Nurra, MUT 5329 holotype, living culture permanently preserved in metabolically inactive state by deep-freezing at MUT.

Etymology. In reference to the area of origin, Margidore.

Description. Growing actively on *Pinus pinaster* wood. *Hyphae* approx. 2 µm wide, septate, brownish.

Sexual morph not observed. Asexual morph and differentiated conidiogenesis not observed.

Colony description. Colonies growing on MEASW, attaining 33–34 mm diam after 28 days at 21 °C; whitish, lighter to the edge, umbonate in the middle, reverse ochre. Caramel diffusible pigment produced. Growth on OASW reaching 45 mm diam at 24 °C and 27 mm diam at 15 °C; colonies on PDA attaining 45 mm diam and 23 mm diam at 24 °C and 15 °C, respectively.

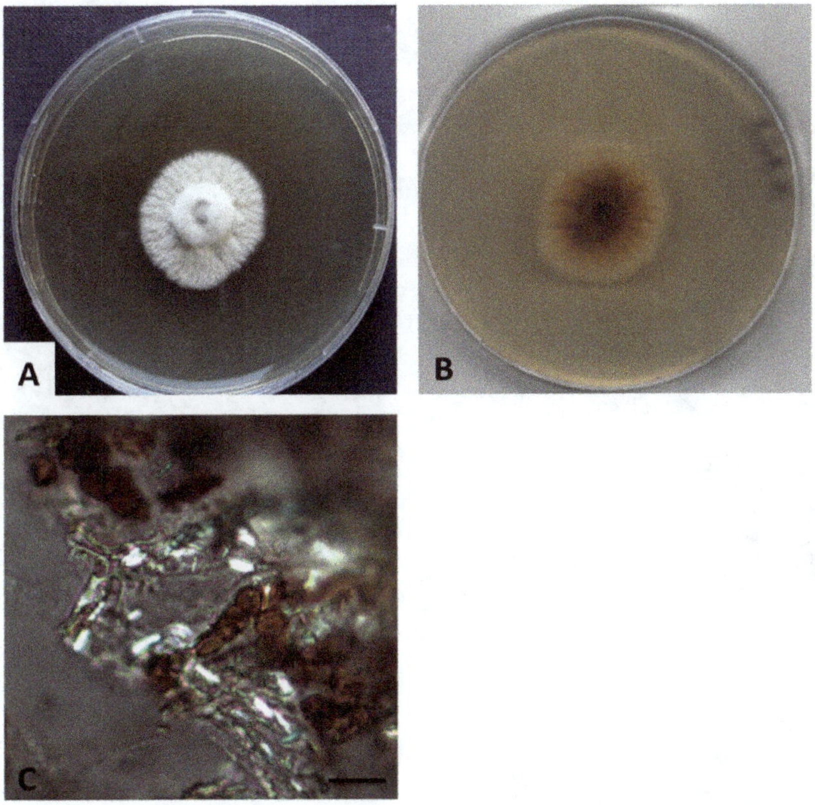

Figure 6. *Roussoella margidorensis* sp. nov. 28-days-old colony at 21 °C on MEASW (**A**) and reverse (**B**); chlamydospores (**C**). Scale bar: 10 µm.

Roussoella mediterranea sp. nov. A. Poli, E. Bovio, V. Prigione, and G.C. Varese
MYCOBANK: MB 832842
Figure 7

Type. Italy, Tuscany, Mediterranean Sea, Elba Island (LI), Margidore ISL, 14–15 m depth, UTM WGS84 42°45′29″N, 10°18′24″E, from the brown alga *Padina pavonica*, March 2010, R. Mussat-Sartor and N. Nurra, MUT 5369 holotype, living culture permanently preserved in metabolically inactive state by deep-freezing at MUT.

Additional material examined. Italy, Tuscany, Mediterranean Sea, Elba Island (LI), Margidore ISL, 14–15 m depth, UTM WGS84 42°45′29″N, 10°18′24″E, from the brown alga *Padina pavonica*, March

2010, R. Mussat-Sartor and N. Nurra, MUT 5306 (identical to MUT 5306 on the basis of micro- and minisatellite analyses)

Etymology. In reference to the geographical origin, Mediterranean Sea.

Description in culture. Growing actively on *Pinus pinaster* wood and poorly colonizing *Quercus ruber* cork. *Hyphae* 2.4 μm wide, septate, dematiaceous. *Chlamydospores* 4.5 × 5.7 μm, from unicellular to 4-celled; branched chains of light to dark brown chlamydospores often present.

Sexual morph not observed. Asexual morph with differentiated conidiogenesis not observed.

Colony description. Colonies growing on MEASW, reaching 55 mm diam after 28 days at 21 °C, light grey, floccose, with umbonate area in the middle, reverse brown with lighter edges. Dark exudate present. Growth on OASW reaching 67–72 mm diam at 24 °C and 33–38 mm diam at 15 °C; colonies on PDA attaining 69–76 mm diam and 32.5–39 mm diam at 24 °C and 15 °C, respectively.

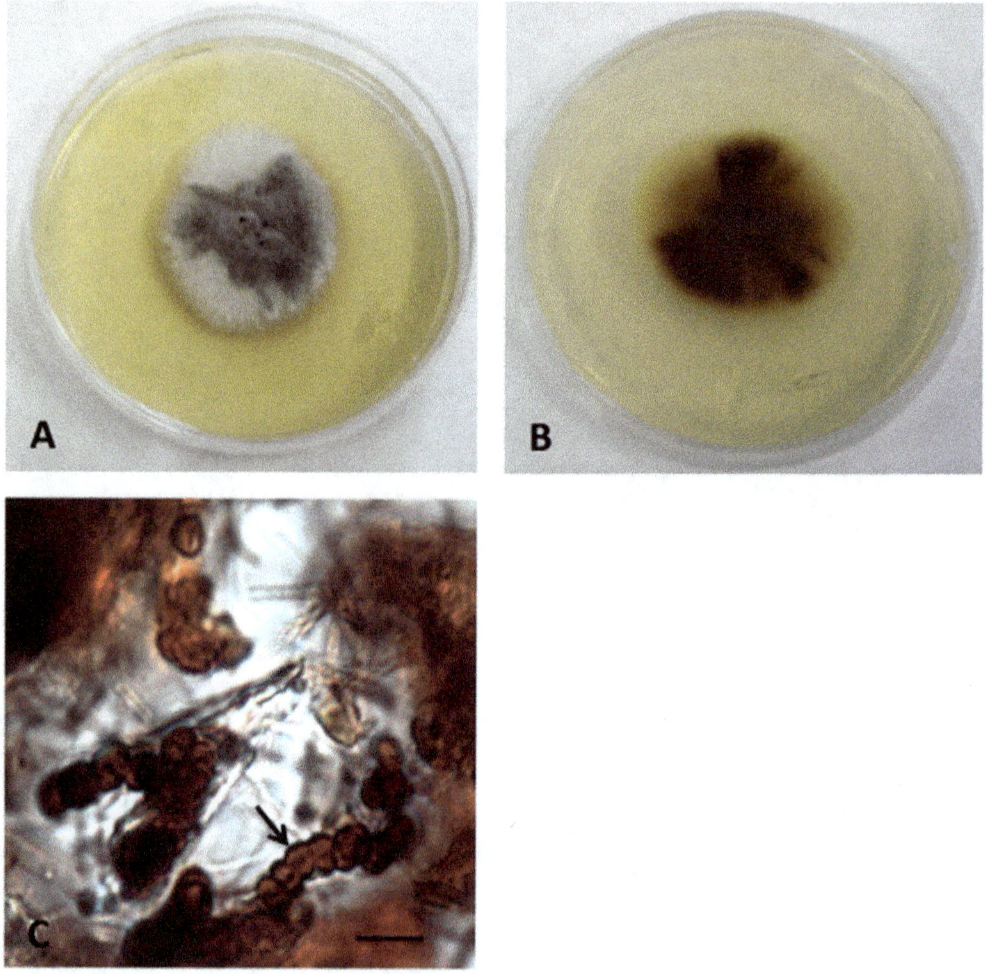

Figure 7. *Roussoella mediterranea* sp. nov. 28-days-old colony at 21 °C on MEASW (**A**) and reverse (**B**); unicellular and multicellular chlamydosporesn indicated by a black arrow (**C**). Scale bar: 10 μm.

Roussoella padinae sp. nov. V. Prigione, E. Bovio, A. Poli and G.C. Varese
MYCOBANK: MB 832843
Figure 8

Type. Italy, Tuscany, Mediterranean Sea, Elba Island (LI), Margidore ISL, 14–15 m depth, UTM WGS84 42°45′29″N, 10°18′24″E, from the brown alga *Padina pavonica*, March 2010, R. Mussat-Sartor and N. Nurra, MUT 5503 holotype, living culture permanently preserved in metabolically inactive state by deep-freezing at MUT.

Additional material examined. Italy, Tuscany, Mediterranean Sea, Elba Island (LI), Margidore ISL, 14–15 m depth, UTM WGS84 42°45′29″N, 10°18′24″E, from the brown alga *Padina pavonica*, March 2010, R. Mussat-Sartor and N. Nurra, MUT 5341 and MUT 5345 (identical to MUT 5503 on the basis of micro- and minisatellite analyses)

Etymology. In reference to the original substratum, *Padina pavonica*.

Description in culture. Growing efficiently on *Quercus ruber* cork and poorly colonizing *Pinus pinaster* wood. *Hyphae* 3 µm wide, septate, brownish, assuming toruloid aspect when growing into wood vessels and forming chains of two-celled chlamydospores which, at maturity, protrude from the vessels. *Chlamydospores* 5–7 × 4 µm, from light to dark brown, subglobose, ellipsoidal or cylindrical.

Sexual morph not observed. Asexual morph n with differentiated conidiogenesis not observed.

Colony description. Colonies growing on MEASW, reaching 53 mm diam after 28 days at 21 °C, from grey to dark green, floccose in the middle, with radial grooves, fimbriate edges, reverse brown. Growth on OASW reaching 57.5–65 mm diam at 24 °C and 30–35 mm diam at 15 °C; colonies on PDA attaining 60–69 mm diam and 30–34 mm diam at 24 °C and 15 °C, respectively.

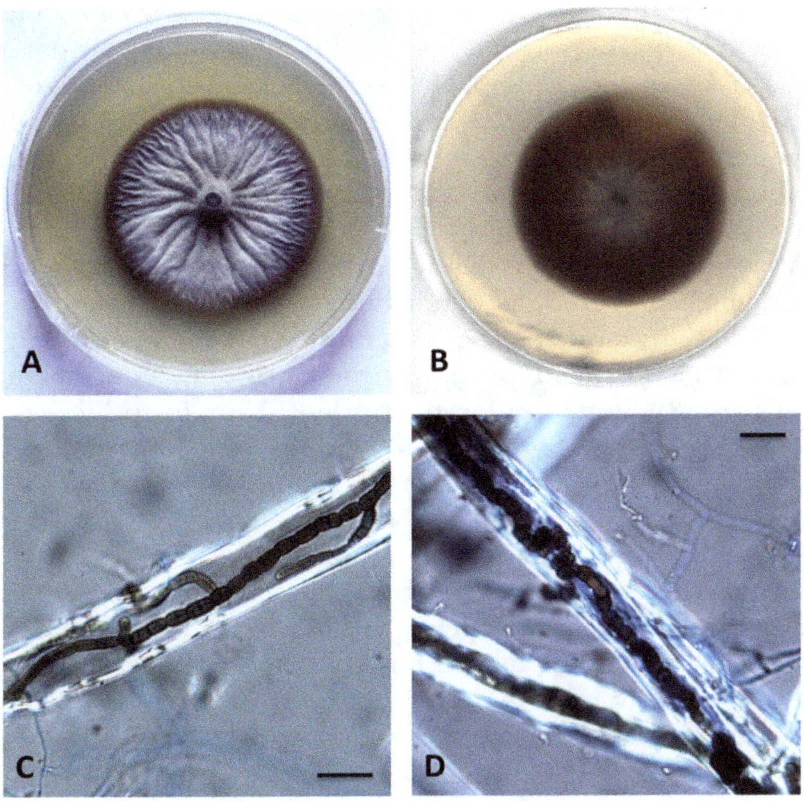

Figure 8. *Roussoella padinae* sp. nov. 28-days-old colony at 21 °C on MEASW (**A**) and reverse (**B**); toruloid hyphae (**C**) and two-celled chlamydospores (**D**) inside wood vessels. Scale bars: 10 µm.

4. Discussion

The description of these new taxa was particularly challenging because neither asexual nor sexual reproductive structures developed in axenic conditions. Therefore, we were unable to describe the range of anatomical variations and diagnostic features among these newly recognized phylogenetic lineages. Indeed, strictly vegetative growth without sporulation is a common feature of many marine fungal strains [10,11,33]. Possibly, these organisms rely on hyphal fragmentation for their dispersal, or alternatively, the differentiation of reproductive structures may be obligatorily dependent on the peculiar environmental conditions under which they live (e.g., wet-dry cycles, high salinity, low temperature, high pressure, etc.). During the study of these fungi, we tried to mimic the saline environment by using different culture media supplemented with natural sea water or sea salts. Although these culture methods were applied to induce sporulation, we observed that only media supplemented with sea water supported a measurable growth of vegetative mycelium (data not shown). The method introduced by Panebianco et al. [22] to induce sporulation by placing wood and cork specimens on the colony surface with their subsequent transfer into sea water, was only partially successful: out of seven species, three (*P. dematiacea, P. flabelliae, R. mediterranea*) developed chlamydospores in the mycelium above the wood surface, two (*N. lignicola, R. padinae*) gave rise to resting spores inside wood vessels. Most of the strains preferred to colonize *P. pinaster* wood rather than *Q. ruber* cork. These structures were interpreted as "chlamydospores" instead of "conidia" for the following reasons: (i) They were characterized by a very thick cell wall, a typical feature of resting spores; (ii) conidiogenous cells were never observed. Additional efforts to force the development of reproductive structures by using SNASW and pine needles, were also unsuccessful.

Both *R. padinae* and *N. lignicola* displayed a similar lignicolous behavior, growing and producing chlamydospores inside wooden vessels, although of different size and shape. The ability to form hyphae and to grow inside the wood vessels has been reported for a number of dark septate endophyte fungi in terrestrial environment [34] and, recently, for *Posidoniomyces atricolor* Vohník and Réblová, a marine endophyte that lives in association with the roots of *P. oceanica* [35]. By definition, endophytes live inside living plant tissues. To induce sporulation, sterilized specimens of dead wood were employed, therefore *R. padinae* and *N. lignicola* were inferred to be "lignicolous fungi" rather than "endophytes". The observation of this growth characteristic in two different genera, may find its reason in an evolutionary adaptation to marine life in association with lignocellulosic matrices. Therefore, we can hypothesize their ecological role as saprobes involved in degrading organic matter.

Notwithstanding the lack of exhaustive descriptions of morphological features, the strongly supported phylogenetic and molecular analysis, conducted with five different genetic markers (nrSSU, nrITS, nrLSU, TEF1α and RPB2) undoubtedly pointed out the differences among these species and their belonging to new taxa. This is also supported by the DNA diagnostic characters identified in the individual loci (Tables S1–S18). In particular, the present study introduces four new species of Roussoellaceae and three new species of Thyridariaceae. Indeed, only MUT 2452 and MUT 4893 were ascribable to the previously described *P. robiniae* (Figure 1). In the case of MUT 4884, the holotype of *P. dematiacea*, a novel genus was proposed since it formed a defined cluster with MUT 5310 and MUT 4419, well separated by the genera *Cycasicola*, *Liua* and *Thyridariella*.

Most of the Roussoellaceae and Thyridariaceae described to date are associated with terrestrial plants, especially bamboo and palm species [15,16]. In fact, only two species, *R. mangrovei* and *R. nitidula* have been retrieved from the marine environment (www.marinefungi.org). However, considering the present study, we can infer that these families are well represented in the sea, thus improving our knowledge on the largely unexplored fungal marine biodiversity.

Supplementary Materials:
Table S1: The eight variable sites detected in the nrITS region among *P. dematiacea* and its neighbor species, Table S2: The single variable site detected in the nrLSU region among *P. dematiacea* and its neighbor species, Table S3: The five variable sites detected in the nrSSU region among *P. dematiacea* and its neighbor species, Table S4: The six variable sites detected in the TEF1α partial gene among *P. dematiacea* and its neighbor species, Table S5: The six variable sites detected in the nrITS region among *P. tyrrhenica, P. flabelliae* and their neighbor species, Table S6: The eight variable sites detected in the nrLSU region among *P. tyrrhenica, P. flabelliae* and their neighbor species, Table S7: The eight variable sites detected in the TEF1α partial gene among *P. tyrrhenica, P. flabelliae* and their neighbor species, Table S8: The 33 variable sites detected in the RPB2 partial gene among *P. tyrrhenica, P. flabelliae* and their neighbor species, Table S9: The two variable sites detected in nrITS region among *R. mediterranea, R. padinae* and the neighbor species, Table S10: The single variable site detected in nrLSU region among *R. mediterranea, R. padinae,* and the neighbor species, Table S11: The six sites detected in the TEF1α partial gene among *R. mediterranea, R. padinae* and the neighbor species, Table S12: The six sites detected in the RPB2 partial gene among *R. mediterranea, R. padinae* and the neighbor species, Table S13: The eight variable sites detected in the nrITS region among *N. lignicola* and its neighbor species, Table S14: The three variable sites detected in the nrLSU region among *N. lignicola* and its neighbor species, Table S15: The eight variable sites detected in the nrSSU region among *N. lignicola* and its neighbor species, Table S16: The ten sites detected in the TEF1α partial gene among *N. lignicola* and its

neighbor species, Table S17: The three variable sites detected in the nrITS region among *R. margidoriensis* and its neighbor species, Table S18: The 29 variable sites detected in the TEF1α partial gene among *R. margidoriensis* and its neighbor species

Author Contributions: Conceptualization, A.P., E.B., V.P., G.C.V.; methodology, A.P., E.B., V.P., G.C.V.; software, A.P.; validation, A.P., E.B., V.P., L.R., G.C.V.; formal analysis, A.P., E.B., V.P., L.R.; investigation, A.P., E.B., V.P., L.R., G.C.V.; resources, V.P., G.C.V.; data curation, A.P., V.P.; writing—original draft preparation, A.P., V.P.; writing—review and editing, A.P., E.B., V.P., G.C.V.; visualization, A.P., V.P.; supervision, V.P., G.C.V.; project administration, A.P., E.B., V.P., G.C.V.; funding acquisition, G.C.V. All authors have read and agreed to the published version of the manuscript.

Acknowledgments: The authors are grateful to Pelagosphera s.c.r.l. for harvesting algal and seagrass samples.

References

1. Richards, T.A.; Jones, M.D.; Leonard, G.; Bass, D. Marine fungi: Their ecology and molecular diversity. *Ann. Rev. Mar. Sci.* **2012**, *4*, 495–522. [CrossRef] [PubMed]
2. Amend, A.; Burgaud, G.; Cunliffe, M.; Edgcomb, V.P.; Ettinger, C.L.; Gutierrez, M.H.; Heitman, J.; Hom, E.F.Y.; Ianiri, G.; Jones, A.C.; et al. Fungi in the marine environment: Open questions and unsolved problems. *Mbio* **2019**, *10*, 15. [CrossRef] [PubMed]
3. Carroll, A.R.; Copp, B.R.; Davis, R.A.; Keyzers, R.A.; Prinsep, M.R. Marine natural products. *Nat. Prod. Rep.* **2019**, *36*, 122–173. [CrossRef] [PubMed]
4. Jones, E.G.; Pang, K.-L. *Marine fungi and fungal-like organisms*; Jones, E.G., Pang, K.-L., Eds.; Walter de Gruyter: Berlin, Germany, 2012.
5. Raghukumar, S. *Fungi in Coastal and Oceanic Marine Ecosystems: Marine Fungi*; Springer: Cham, Switzerland, 2017.
6. Pang, K.L.; Overy, D.P.; Jones, E.B.G.; Calado, M.D.; Burgaud, G.; Walker, A.K.; Johnson, J.A.; Kerr, R.G.; Cha, H.J.; Bills, G.F. 'Marine fungi' and 'marine-derived fungi' in natural product chemistry research: Toward a new consensual definition. *Fungal Biol. Rev.* **2016**, *30*, 163–175. [CrossRef]
7. Jones, E.B.G.; Pang, K.-L.; Abdel-Wahab, M.A.; Scholz, B.; Hyde, K.D.; Boekhout, T.; Ebel, R.; Rateb, M.E.; Henderson, L.; Sakayaroj, J.; et al. An online resource for marine fungi. *Fungal Divers.* **2019**, *96*, 347–433. [CrossRef]
8. Jones, E.B.G. Fifty years of marine mycology. *Fungal Divers.* **2011**, *50*, 73–112. [CrossRef]
9. Jones, E.B.G.; Suetrong, S.; Sakayaroj, J.; Bahkali, A.H.; Abdel-Wahab, M.A.; Boekhout, T.; Pang, K.L. Classification of marine Ascomycota, Basidiomycota, Blastocladiomycota and Chytridiomycota. *Fungal Divers.* **2015**, *73*, 1–72. [CrossRef]
10. Garzoli, L.; Poli, A.; Prigione, V.; Gnavi, G.; Varese, G.C. Peacock's tail with a fungal cocktail: First assessment of the mycobiota associated with the brown alga *Padina pavonica*. *Fungal Ecol.* **2018**, *35*, 87–97. [CrossRef]
11. Gnavi, G.; Garzoli, L.; Polil, A.; Prigione, V.; Burgaud, G.; Varese, G.C. The culturable mycobiota of *Flabellia petiolata*: First survey of marine fungi associated to a Mediterranean green alga. *PLoS ONE* **2017**, *12*. [CrossRef]
12. Panno, L.; Bruno, M.; Voyron, S.; Anastasi, A.; Gnavi, G.; Miserere, L.; Varese, G.C. Diversity, ecological role and potential biotechnological applications of marine fungi associated to the seagrass *Posidonia oceanica*. *New Biotechnol.* **2013**, *30*, 685–694. [CrossRef]
13. Bovio, E.; Garzoli, L.; Poli, A.; Prigione, V.; Firsova, D.; McCormack, G.; Varese, G. The culturable mycobiota associated with three Atlantic sponges, including two new species: *Thelebolus balaustiformis* and *T. spongiae*. *Fungal Syst. Evol.* **2018**, *1*, 141–167. [CrossRef]
14. Liu, J.-K.; Phookamsak, R.; Dai, D.-Q.; Tanaka, K.; Jones, E.; Xu, J.-C.; Chukeatirote, E.; Hyde, K.D. Roussoellaceae, a new pleosporalean family to accommodate the genera *Neoroussoella* gen. nov., *Roussoella* and *Roussoellopsis*. *Phytotaxa* **2014**, *181*, 1–33. [CrossRef]
15. Jaklitsch, W.M.; Voglmayr, H. Hidden diversity in *Thyridaria* and a new circumscription of the Thyridariaceae. *Stud. Mycol.* **2016**, *85*, 35–64. [CrossRef]
16. Tibpromma, S.; Hyde, K.D.; Jeewon, R.; Maharachchikumbura, S.S.N.; Liu, J.K.; Bhat, D.J.; Jones, E.B.G.; McKenzie, E.H.C.; Camporesi, E.; Bulgakov, T.S.; et al. Fungal diversity notes 491–602: Taxonomic and phylogenetic contributions to fungal taxa. *Fungal Divers.* **2017**, *83*, 1–261. [CrossRef]

17. Devadatha, B.; Sarma, V.V.; Jeewon, R.; Wanasinghe, D.N.; Hyde, K.D.; Jones, E.B.G. *Thyridariella*, a novel marine fungal genus from India: Morphological characterization and phylogeny inferred from multigene DNA sequence analyses. *Mycol. Prog.* **2018**, *17*, 791–804. [CrossRef]
18. Jayasiri, S.; Hyde, K.; Jones, E.; McKenzie, E.; Jeewon, R.; Phillips, A.; Bhat, D.; Wanasinghe, D.; Liu, J.; Lu, Y.; et al. Diversity, morphology and molecular phylogeny of Dothideomycetes on decaying wild seed pods and fruits. *Mycosphere* **2019**, *10*, 1–186. [CrossRef]
19. Phookamsak, R.; Hyde, K.D.; Jeewon, R.; Bhat, D.J.; Jones, E.B.G.; Maharachchikumbura, S.S.N.; Raspe, O.; Karunarathna, S.C.; Wanasinghe, D.N.; Hongsanan, S.; et al. Fungal diversity notes 929–1035: Taxonomic and phylogenetic contributions on genera and species of fungi. *Fungal Divers.* **2019**, *95*, 1–273. [CrossRef]
20. Jiang, H.B.; Hyde, K.D.; Jayawardena, R.S.; Doilom, M.; Xu, J.C.; Phookamsak, R. Taxonomic and phylogenetic characterizations reveal two new species and two new records of *Roussoella* (Roussoellaceae, Pleosporales) from Yunnan, China. *Mycol. Prog.* **2019**, *18*, 577–591. [CrossRef]
21. Dai, D.Q.; Phookamsak, R.; Wijayawardene, N.N.; Li, W.J.; Bhat, D.J.; Xu, J.C.; Taylor, J.E.; Hyde, K.D.; Chukeatirote, E. Bambusicolous fungi. *Fungal Divers.* **2017**, *82*, 1–105. [CrossRef]
22. Panebianco, C.; Tam, W.Y.; Jones, E.B.G. The effect of pre-inoculation of balsa wood by selected marine fungi and their effect on subsequent colonisation in the sea. *Fungal Divers.* **2002**, *10*, 77–88.
23. White, T.J.; Bruns, T.; Lee, S.; Taylor, J. Amplification and direct sequencing of fungal ribosomal RNA genes for phylogenetics. In *PCR protocols: A guide to methods and applications*; Academic Press: New York, NY, USA, 1990; Volume 18, pp. 315–322.
24. Vilgalys, R.; Hester, M. Rapid genetic identification and mapping of enzymatically amplified ribosomal DNA from several *Cryptococcus* species. *J. Bacteriol.* **1990**, *172*, 4238–4246. [CrossRef]
25. Stielow, J.; Lévesque, C.; Seifert, K.; Meyer, W.; Iriny, L.; Smits, D.; Renfurm, R.; Verkley, G.; Groenewald, M.; Chaduli, D. One fungus, which genes? Development and assessment of universal primers for potential secondary fungal DNA barcodes. *Persoonia* **2015**, *35*, 242–263. [CrossRef]
26. Liu, Y.J.; Whelen, S.; Hall, B.D. Phylogenetic relationships among ascomycetes: Evidence from an RNA polymerse II subunit. *Mol. Biol. Evol.* **1999**, *16*, 1799–1808. [CrossRef]
27. Vaidya, G.; Lohman, D.J.; Meier, R. SequenceMatrix: Concatenation software for the fast assembly of multi-gene datasets with character set and codon information. *Cladistics* **2011**, *27*, 171–180. [CrossRef]
28. Darriba, D.; Taboada, G.L.; Doallo, R.; Posada, D. jModelTest 2: More models, new heuristics and parallel computing. *Nat. Methods* **2012**, *9*, 772. [CrossRef]
29. Stamatakis, A. RAxML version 8: A tool for phylogenetic analysis and post-analysis of large phylogenies. *Bioinformatics* **2014**, *30*, 1312–1313. [CrossRef]
30. Ronquist, F.; Teslenko, M.; van der Mark, P.; Ayres, D.L.; Darling, A.; Hohna, S.; Larget, B.; Liu, L.; Suchard, M.A.; Huelsenbeck, J.P. MrBayes 3.2: Efficient Bayesian Phylogenetic Inference and Model Choice Across a Large Model Space. *Syst. Biol.* **2012**, *61*, 539–542. [CrossRef]
31. Lieckfeldt, E.; Meyer, W.; Borner, T. Rapid identification and differentiation of yeasts by DNA and PCR fingerprinting. *J. Basic Microbiol.* **1993**, *33*, 413–426. [CrossRef]
32. Poli, A.; Lazzari, A.; Prigione, V.; Voyron, S.; Spadaro, D.; Varese, G.C. Influence of plant genotype on the cultivable fungi associated to tomato rhizosphere and roots in different soils. *Fungal Biol.* **2016**, *120*, 862–872. [CrossRef]
33. Poli, A.; Vizzini, A.; Prigione, V.; Varese, G.C. Basidiomycota isolated from the Mediterranean Sea-Phylogeny and putative ecological roles. *Fungal Ecol.* **2018**, *36*, 51–62. [CrossRef]
34. Oses, R.; Valenzuela, S.; Freer, J.; Sanfuentes, E.; Rodriguez, J. Fungal endophytes in xylem of healthy Chilean trees and their possible role in early wood decay. *Fungal Divers.* **2008**, *33*, 77–86.
35. Vohnik, M.; Borovce, O.; Kolarikova, Z.; Sudova, R.; Reblova, M. Extensive sampling and high-throughput sequencing reveal *Posidoniomyces atricolor* gen. et sp. nov. (Aigialaceae, Pleosporales) as the dominant root mycobiont of the dominant Mediterranean seagrass Posidoniaoceanica. *Mycokeys* **2019**, *55*, 59–86. [CrossRef] [PubMed]

13

Combinatorial Biosynthesis of Novel Multi-Hydroxy Carotenoids in the Red Yeast *Xanthophyllomyces dendrorhous*

Hendrik Pollmann [1], Jürgen Breitenbach [1], Hendrik Wolff [2], Helge B. Bode [2,3] and Gerhard Sandmann [1,*]

1 Biosynthesis Group, Molecular Biosciences, Fachbereich Biowissenschaften, Goethe Universität Frankfurt, Frankfurt am Main 60438, Germany; pollmann@bio.uni-frankfurt.de (H.P.); Breitenbach@em.uni-frankfurt.de (J.B.)
2 Merck Stiftungsprofessur für Molekulare Biotechnologie, Fachbereich Biowissenschaften, Goethe Universität Frankfurt, Frankfurt am Main 60438, Germany; wolff@bio.uni-frankfurt.de (H.W.); h.bode@bio.uni-frankfurt.de (H.B.B.)
3 Buchmann Institute for Molecular Life Sciences (BMLS), Goethe Universität Frankfurt, Max-von-Laue-Strasse 15, Frankfurt am Main 60438, Germany
* Correspondence: sandmann@bio.uni-frankfurt.de.

Academic Editors: Laurent Dufossé, Yanis Caro and Mireille Fouillaud

Abstract: The red yeast *Xanthophyllomyces dendrorhous* is an established platform for the synthesis of carotenoids. It was used for the generation of novel multi oxygenated carotenoid structures. This was achieved by a combinatorial approach starting with the selection of a β-carotene accumulating mutant, stepwise pathway engineering by integration of three microbial genes into the genome and finally the chemical reduction of the resulting 4,4′-diketo-nostoxanthin (2,3,2′,3′-tetrahydroxy-4,4′-diketo-β-carotene) and 4-keto-nostoxanthin (2,3,2′,3′-tetrahydroxy-4-monoketo-β-carotene). Both keto carotenoids and the resulting 4,4′-dihydroxy-nostoxanthin (2,3,4,2′,3′,4′-hexahydroxy-β-carotene) and 4-hydroxy-nostoxanthin (2,3,4,2′3′-pentahydroxy-β-carotene) were separated by high-performance liquid chromatography (HPLC) and analyzed by mass spectrometry. Their molecular masses and fragmentation patterns allowed the unequivocal identification of all four carotenoids.

Keywords: carotenoid biosynthesis; 4,4′-dihydroxy-nostoxanthin; 4,4′-diketo-nostoxanthin; genetic engineering; HPLC separation; MS-MS spectra; *Xanthophyllomyces dendrorhous*

---

1. Introduction

Xanthophyllomyces dendrorhous (with *Phaffia rhodozyma* as its anamorphic state) is a basidiomycetous red yeast that accumulates astaxanthin [1], which is a unique feature among fungi [2]. This carotenoid is formed via the mevalonate pathway starting with a condensation of two molecules of geranylgeranyl pyrophosphate, a 4-step desaturation, cyclization and a final 4-ketolation plus 3-hydroxylation [3,4]. In contrast to other organisms, only three genes are involved in the whole pathway from phytoene to astaxanthin, which facilitates genetic modification of carotenoid biosynthesis in *X. dendrorhous.* This yeast has the potential to be engineered as a cell-factory for the production of industrially valuable carotenoids [5]. Tools and techniques for genetic manipulations of *X. dendrorhous* are available [6], including several integrative transformation plasmids based on four different selection markers [7,8]. It is also advantageous for the development of high-yield carotenoid producers since a carotenoid pathway that can be manipulated is already established, a carotenoid storage system exists and a very active acetyl-CoA metabolism can be utilized. The published genomic sequence of *X. dendrorhous* CBS6938 [9] is also very helpful for genetic engineering of the carotenoid pathway.

The potential of *X. dendrorhous* for the production of economically interesting carotenoids like astaxanthin for feed and zeaxanthin as a nutraceutical for eye care has been demonstrated. The highest production of astaxanthin was reached by combining classical mutagenesis with genetic pathway engineering [7]. Starting from a high-yield astaxanthin mutant, genes of three limiting enzymes were over-expressed, enhancing metabolite flow toward carotenoid biosynthesis and into the astaxanthin pathway [7].

Zeaxanthin is another carotenoid of interest as a nutraceutical, which is important for protection of our vision. For the engineering of a zeaxanthin producing strain, a mutant with inactive astaxanthin synthase accumulating β-carotene [10,11] was used to extend the carotenoid pathway to zeaxanthin by expression of a bacterial β-carotene hydroxylase gene and engineering of an enhanced metabolite flow into the carotenoid pathway [12].

Hydroxylated carotenoids such as the dihydroxy compound zeaxanthin [13], tetrahydroxy-β-carotene derivative nostoxanthin [14] and various other hydroxylated acyclic carotenoids [15] have been generated in *Escherichia coli* by combination of carotenogenic genes from different organisms. It has been shown that not only zeaxanthin but also several other hydroxyl derivatives have superior antioxidative activity. In our engineering approach with *X. dendrorhous*, we utilized β-carotene mutants to integrate hydroxylase and ketolase genes from bacteria and algae and finally chemically reduced the resulting tetrahydroxy-monoketo and tetrahydroxy-diketo products to a pentahydroxy and a hexahydroxy β-carotene derivative, respectively.

2. Materials and Methods

2.1. Strains and Cultivation

The β-carotene accumulation mutant of *X. dendrorhous* strain PR1-104 was generated by ethyl-methane sulphonate mutagenesis and selection for carotenoid content [10]. This mutant and transformants were grown as shaking cultures (50 mL in 500 mL baffled Erlenmeyer flasks, 180 rpm) over 7 days in YM medium (0.3% yeast extract, 0.3% malt extract, 0.5% peptone, 1.0% glucose) with white light illumination. For selection of transformants on agar plates, geneticin (G418 sulfate, 100 μg/mL), hygromycin (60 μg/mL) or nourseothricin (30 μg/mL) were added. *Escherichia coli* strains DH5α and JM110 used for genetic manipulations and generation of carotenoid standards were grown in LB medium containing ampicillin (100 μg/mL) and chloramphenicol (34 μg/mL) according to the plasmids involved.

2.2. Plasmid Construction and Transformantion of X. dendrorhous, Combinatorial Biosynthesis of Carotenoid Standards in E. coli

The integrative plasmids pPR2TNo-crtZo [12] and pPRcDNA1bkt830 [11] were previously described. Details on the origin of genes and the primers for the amplification of the *X. dendrorhous* transformation plasmids are shown in Table 1. For the construction of plasmid pPR2TNH-crtG, the *crtG* gene from plasmid pUC-Bre-O11 [16] was amplified, which also generated an *Eco*RI and *Xho*I restriction site (Table 1), and was cloned into the *Xcm*I site of the *E. coli* plasmid pMon38201 by its a-overhang [17]. From there, the *crtG* gene was cut out with *Eco*RI and *Xho*I and ligated into pUC8ΔEcoRI-HNNH [7] for fusion with the promoter and terminator of the glyceraldehyde phosphate dehydrogenase gene from *X. dendrorhous*. Then, the whole cassette was cut out by restriction with *Hin*dIII and ligated into the *Hin*dIII site of pPR2TNH [7], yielding pPR2TNH-crtG. Transformation of *X. dendrorhous* was performed by electroporation as described in Visser et al. [6] with 10 μg plasmid DNA. After phenol-chloroform purification, plasmid DNA was linearized by digestion with *Sfi*I.

Carotenoid standards were generated in *E. coli* by combinatorial transformation of plasmids pACCAR25ΔcrtX [18] plus pUC-Bre-O11 [16] for nostoxanthin, caloxanthin and zeaxanthin, pACCAR25ΔcrtX [18] plus pCRBKT [19] for canthaxanthin, echinenone and β-carotene and pACCAR25ΔcrtX plus pPEU30crtO [20] for the synthesis of 4-keto-zeaxanthin.

Table 1. Genes and oligonucleotides for PCR amplification of the *Xanthophyllomyces dendrorhous* transformation plasmids.

Plasmid	Primers
pPRcDNA1bkt830	β-Carotene 4-ketolase gene *bkt* from *Haematococcus pluvialis* (D45881) Primers: 5′-ATATGAATTCATGCACGTCGCATCGGCACT-3′ (forward) 5′-TATACTCGAGTCATGCCAAGGCAGGCACCAG-3′ (reverse)
pPR2TNH-crtG	β-Carotene 2-hydroxylase gene *crtG* from *Brevundimonas* SD212 [16] (AB181388) Primers: 5′-GGAATTCATGTTGAGGGATCTGCTCATC-3′ (forward) 5′-GCTCGAGTCACCGAAGAGGCGCTGAG-3′ (reverse)
pPR2TNo-crtZo	β-Carotene 3-hydroxylase gene *crtZ* from *Brevundimonas* SD212 [21]), codon optimized (KX063854) Sequences: 5′-GAATTCATGGCTTGGCTCACCTGGA-3′ (start) 5′-ATCTTCTTCTTCTGGAGCT TGAGTCGAC-3′ (end)

2.3. Carotenoid Extraction, Purification and Chemical Derivatization

For carotenoid extraction, 20 mg of freeze-dried *X. dendrorhous* cells were mixed with 500 μL glass beads (0.25–0.5 mm), 675 μL methanol and 75 μL of a 60% KOH solution for saponification and broken in a cell mill (Retsch MM 400) for 8 min at a frequency of 30 per second and heated for 20 min at 60 °C. Samples containing keto carotenoid were extracted without addition of KOH. After partitioning of the carotenoid extract into 50% diethyl ether in petrol (bp 40–60 °C), the upper phase was collected and dried in a stream of nitrogen. Carotenoids were quantified from three independently grown cultures.

For the purification of multi-hydroxy keto carotenoids, the extract of PR1-104-ZGbkt was fractionated by TLC on activated silica plates developed with toluene/ethyl acetate/methanol (65:35:5, by volume). The bands with an R_f value of 0.20 and 0.25 were collected and extracted with acetone.

The reduction of carotenoid ketones dissolved in ethanol to the corresponding alcohol was performed according to Eugster [22] with $NaBH_4$. The reaction mixture was transferred to 65% diethyl ether in petrol (bp 40–60 °C) and the upper phase collected and dried in stream of nitrogen. The same procedure was applied to canthaxanthin, yielding 4,4′-diketo-β-carotene together with 4-HO-4′-keto-β-carotene and to echinenone yielding 4-HO-β-carotene.

2.4. HPLC and HR-ESI-MS Analysis

HPLC analysis was performed on three different HPLC systems. For separation of the non-ketolated multi-hydroxy carotenoids, system I with a 15 × 0.4 cm Nucleosil 100 C18, 3 μm column and acetonitrile (ACN)/methanol/2-propanol (85:10:5, by volume) plus 3% H_2O [23] was used as mobile phase with a flow rate of 0.8 mL/min at 10 °C. System II was employed for the separation of the hydroxy-keto carotenoids and their reduction products on a 25 cm C30 RP, 3 μm column (YMC, Wilmington, NC, USA) with a mobile phase of 3% methyl tertiary-butyl ether in methanol for 48 min followed by an increase to 20% with a flow rate of 0.8 mL/min at 20 °C. This system was also used for the quantification of the keto derivatives. HR-ESI-MS analysis was carried out in system III on a 2.1 mm × 50 mm ACQUITY UPLC BEH C18, 1.7 μm column. A binary gradient was applied with ACN (+0.1% formic acid) and H_2O (+0.1% formic acid) for 12 min in the following steps: 0–2 min 5% ACN , 2–2.5 min 40% ACN, 2.5–4 min 40% ACN, 4–14 min 40%–95% at a flow rate of 0.4 mL/min at 40 °C. This HPLC system was coupled to an Impact II QTOF (Bruker) mass spectrometer using Na-formiate as an internal calibration standard [24]. Carotenoids were detected in a positive ion mode with scanning range from 100–1200 m/z. Optical spectra were recorded online with a photodiode array detector 994 (Waters, Milford, CT, USA). Carotenoid standards for identification were generated in *E. coli* by the combination of different *crt* genes as previously described [25].

Although fungi are in general unable to synthesize zeaxanthin or other hydroxy-carotenoids, formation of the 3,3′-dihydroxy-β-carotene can be engineered into *X. dendrorhous* [12]. This potential was extended for the synthesis of other derivatives with up to six hydroxyl groups. As outlined in Figure 1, this was achieved by a strategy involving the use of a β-carotene accumulating mutant of *X. dendrorhous*, its consecutive transformation with three microbial transgenes and finally the reduction of the 4- and 4′-keto groups. Figure 2 shows the HPLC analysis to identify some of the β-carotene-derived oxo compounds from the transgenic lines. Transformation with a bacterial 3-hydroxylase gene changed the PR1-104 from a β-carotene accumulator (trace A) to a transformant synthesizing zeaxanthin together with small amounts of the intermediate β-cryptoxanthin (trace B). In a second transformation step, with a bacterial 2-hydroxylase gene, the line PR1-104-ZG was obtained in which β-carotene was converted to nostoxanthin (trace C) as indicated by the nostoxanthin standard in trace D. Oxygenation of nostoxanthin was increased by a third transformation step with an algal 4-ketolase gene. In the resulting transformant PR1-104-ZGbkt, several carotenoids were detected. For base-line HPLC separation of these very polar oxo carotenoids, system I had to be changed to system II (Figure 2, trace E). In addition to the expected nostoxanthin-derived keto carotenoids, other carotenoids such as β-carotene (peak 7) and its keto derivatives echinenone (peak 6) and canthaxanthin (peak 5) as well as 4-keto-zeaxanthin (peak 4) and nostoxanthin (peak 3) were generated and could be identified with standard carotenoids (Figure 2, traces F and G). Compounds **1** and **2** of trace E are highly polar diketo and mono keto derivatives, respectively, as indicated by their optical absorbance spectra exhibiting a typical shape for keto carotenoids and an absorbance maximum of 475 or 468 nm (Figure 3).

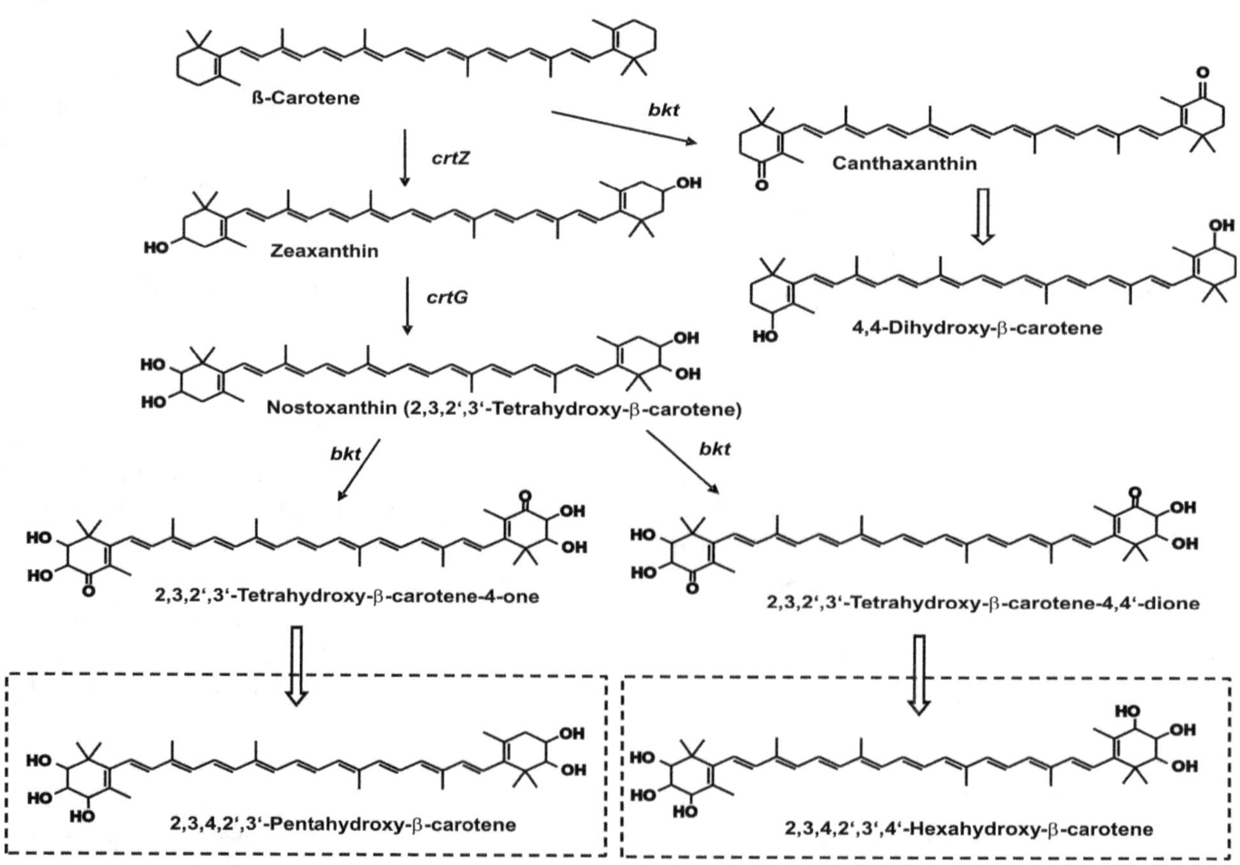

Figure 1. Pathway construction by genetic engineering of *Xanthophyllomyces dendrorhous* for the synthesis of multi-oxygenated β-carotene derivatives. Open arrows indicate chemical reduction. Novel carotenoid structures are boxed.

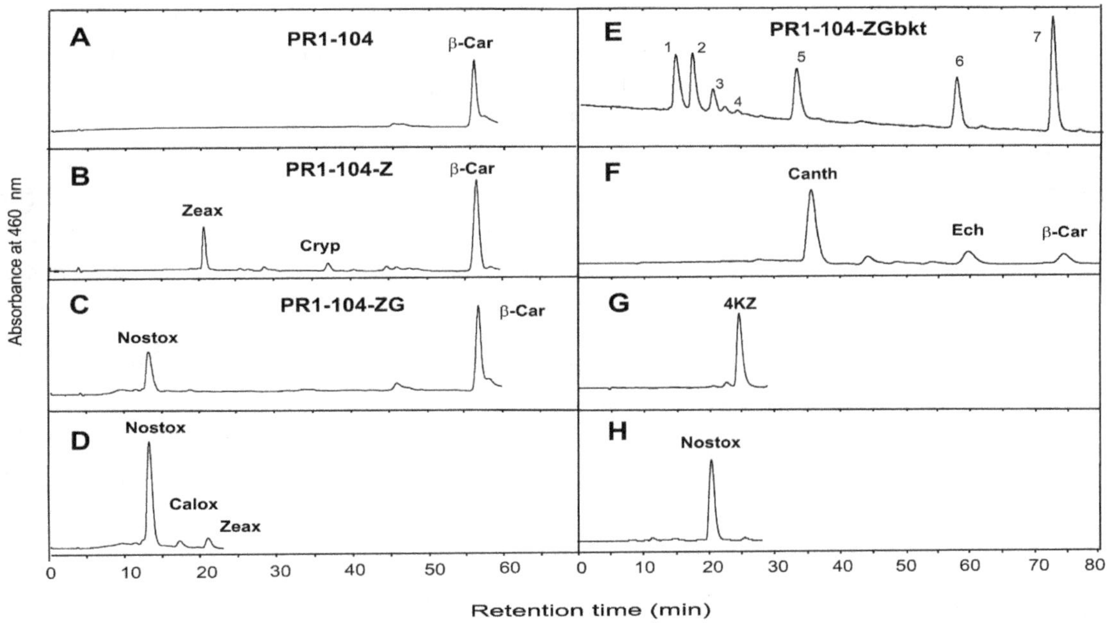

Figure 2. HPLC separation of hydroxy and keto carotenoids from *Xanthophyllomyces dendrorhous* lines obtained by consecutive transformation with different *trans* genes. Traces **A–D** separated in HPLC system I, **E–H** in system II. Standards are shown in traces **D**, **F–H**. Abbreviations (and assignment of corresponding peaks): β-Car β-carotene (7), Zeax zeaxanthin, Cryp β-cryptoxanthin, Nostox nostoxanthin (3), Calox caloxanthin, Canth canthaxanthin (5), Ech echinenone (6), 4KZ 4-keto-zeaxanthin (4).

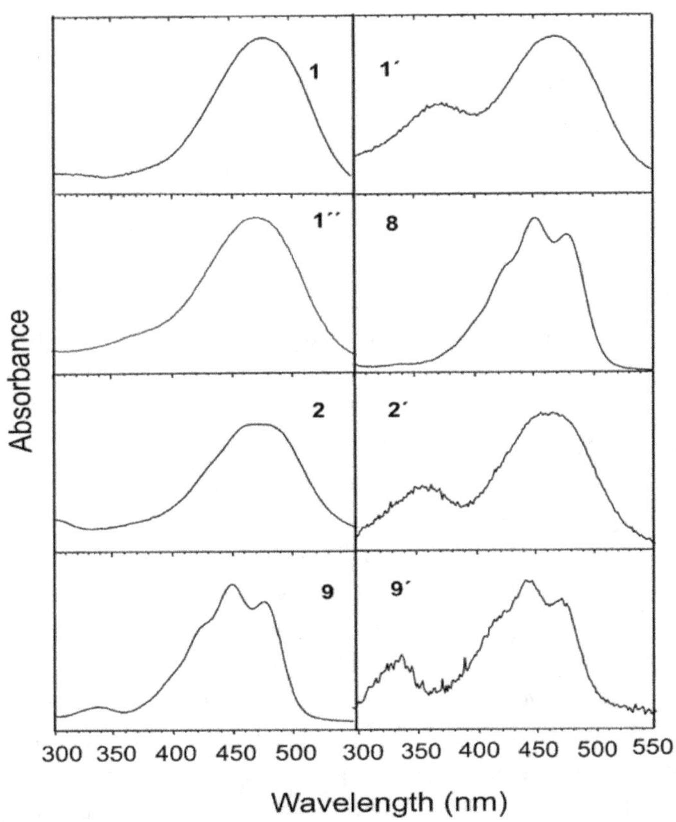

Figure 3. Absorbance spectra of hydroxy-keto carotenoids before (compounds **1** and **2** with *cis* isomers **1′**, **1″** and **2′**) and after reduction (compounds **8** and **9** with *cis* isomer **9′**).

The carotenoid extract from transformant PR1-104-ZGbkt was reduced and separated by HPLC (Figure 4A). New peaks 8 to 12 emerged, which were not present in the non-reduced sample. Peaks 10 to 12 could be identified by chromatography of standards as 4,4′-dihydroxy-β-carotene (trace C), 4-hydroxy-4′-keto-β-carotene (trace B) and 4-hydroxy-β-carotene (trace D). Peaks 8 and 9 exhibited similar spectra with absorbance maxima and shoulder values at 424, 450 and 477 nm (Figure 3). Both compounds were obtained when isolated individual compounds from the non-reduced fraction were reduced. Isolated compound **1** changed into compound **8** and **2** into **9** (Figure 4, right part). The absorbance spectra of the all-*trans* isomers 1, 2, 8 and 9 together with individual *cis* carotenoids indicated with same primed number are shown in Figure 3.

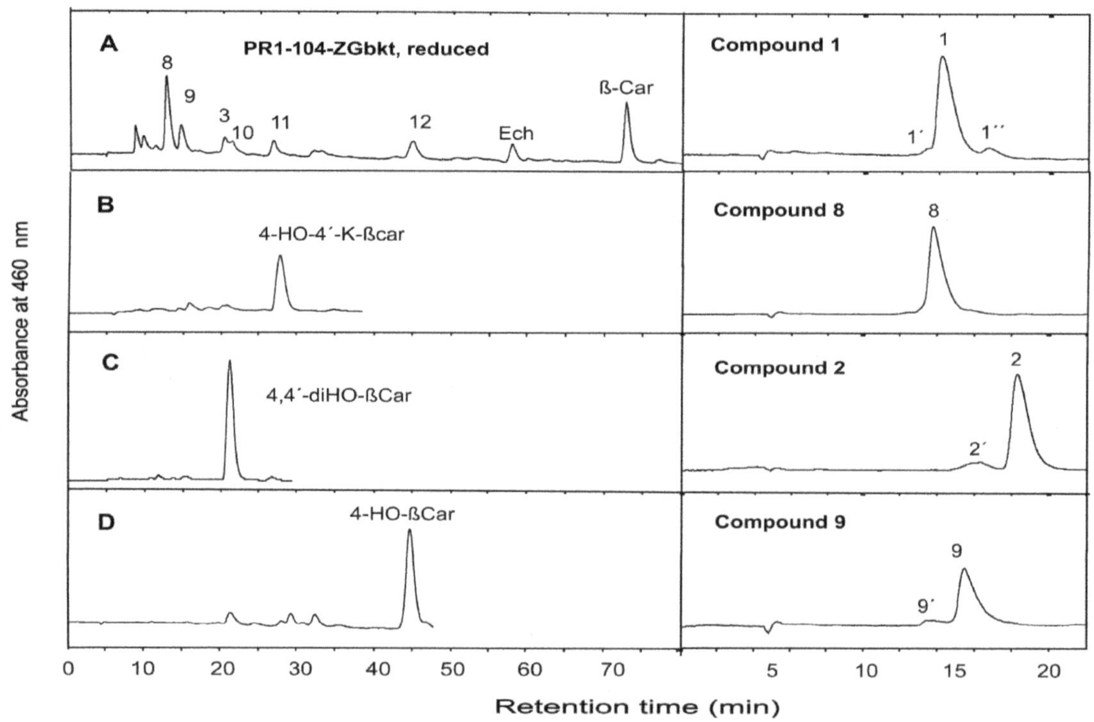

Figure 4. HPLC separation of hydroxy carotenoids after reduction (**A–D**) and isolated peaks 1 and 2 from Figure 2 and their reduction products in system II. Standards are shown in traces **B–D**. Abbreviations (and assignment of corresponding peaks): β-Car, β-carotene; Ech, echinenone; 4-HO-4′-K-βcar, 4-hydroxy-4′-keto-β-carotene (11); 4,4′-diHO-βCar, 4,4′-dihydroxy-β-carotene (10); 4-HO-β-Car, 4-hydroxy-β-carotene (12).

Final identification of compounds **1**, **2**, **8** and **9** was performed using high resolution mass spectrometry (Figure 5). Compound **1** showed a molecular mass of 628.3745 Da (Table 2) and fragments of M-17, M-17-17 and M-92 (Figure 5A). This molecular mass identifies compound **1** as 4,4′-diketo-nostoxanthin. Its reduction product, compound **8**, with a molecular mass of 632.4077 Da (Table 2) is regarded as 4,4′-dihydroxy-nostoxanthin. Instead of the M-17 and M-17-17 fragments of compound **1**, the fragments M-18 and M-18-18 were present in its mass spectrum. (Figure 5B). Therefore, compound **8** is regarded as 4,4′-dihydroxy-nostoxanthin. Compound **2** exhibits a fragmentation pattern in which the M-17-17 fragment of compound **1** is replaced by M-17-18 (Figure 5C). It has a molecular mass of 614.3947 Da (Table 2). This identifies compound **2** as 4-keto-nostoxanthin. Upon its reduction, the molecular mass increased to 616.4102 Da (Table 2). The M-92 fragment is retained but instead of M-17, an M-18 fragment and instead of M-17-18, an M-18-18 fragment appears (Figure 5D). These features of compound **9** correspond to 4-hydroxy-nostoxanthin. Common to the keto carotenoid **1** and **2** is a prominent peak at 147.12, which is much less pronounced in the reduced compounds **8** and **9**. The concentrations of these keto carotenoids in X. *dendrorhous* were determined as 37.4 ± 1.1 (μg/g dw) for 4,4′-diketo-nostoxanthin and 48.8 ± 1.4 (μg/g dw) for 4-keto-nostoxanthin. Their chemical reduction was almost complete (Figure 4A), which implies similar concentrations for 4,4′-dihydroxy-nostoxanthin and 4-hydroxy-nostoxanthin, respectively.

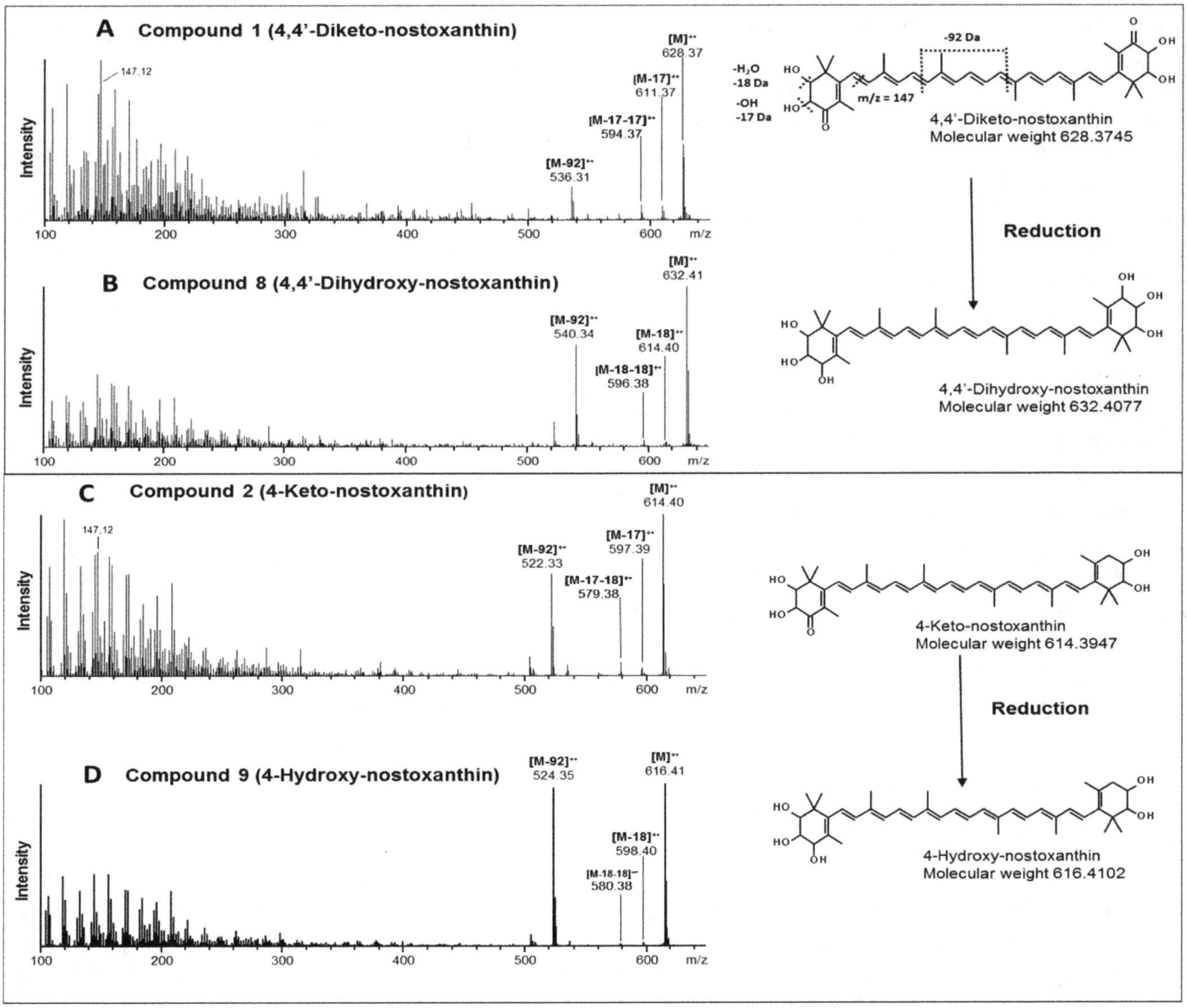

Figure 5. (A–D) MS-MS analysis of de-novo generated hydroxyl and keto-hydroxy β-carotene derivatives before and after reduction. The structures are indicated including the fragmentation pattern exemplified for 4-keto-nostoxanthin.

Table 2. Overview of identified multi-oxygenated nostoxanthins using UPLC-ESI-HRMS.

Compound	Trivial Name	Sum Formula $[M]^+$	Calculated m/z	Detected m/z	Mass Error [Δppm]
1	4,4′-Diketo-nostoxanthin	$C_{40}H_{52}O_6$	628.3758	628.3745	1.3
2	4-Keto-nostoxanthin	$C_{40}H_{54}O_5$	614.3966	614.3947	1.9
8	4,4′-Dihydroxy-nostoxanthin	$C_{40}H_{56}O_6$	632.4071	632.4077	0.6
9	4-Hydroxy-nostoxanthin	$C_{40}H_{56}O_5$	616.4122	616.4102	2.0

4. Discussion

Production of carotenoids by genetically engineered yeasts proved to be a promising alternative to chemical synthesis or extraction from plants [26]. The red yeast *X. dendrorhous* is the most versatile host with the highest carotenoid yield among fungi [5]. It is possible to construct and implement pathways to different carotenoid structures into this yeast. As an example, the synthesis to multi-oxygenated carotenoids was chosen in this publication (Figure 1) to demonstrate the potential of *X. dendrorhous* as a production platform for complex carotenoid structures. This was possible by extension of the pathway from accumulating β-carotene. The step-by-step transformation resulted in intermediary lines accumulating zeaxanthin or nostoxanthin as major carotenoids (Figure 2). Nostoxanthin is a carotenoid found in cyanobacteria [27] and was accumulated in recombinant *Escherichia coli* [14]. A final *X. dendrorhous* line PR1-104-ZGbkt transformed with three microbial genes produced 4-keto-nostoxanthin and 4,4′-diketo-nostoxanthin (Figure 2). Both carotenoids are extremely rare in nature and have been identified before only from two bacteria *Brevundimonas* SD212 and *Rhizobium lupine* [28,29]. By reduction of both keto carotenoids isolated from our line PR1-104-ZGbkt, the novel carotenoids 4-hydroxy-nostoxanthin and 4,4′-dihydroxy-nostoxanthin were obtained (Figure 4). Some of the isolated oxo carotenoids separated into several geometrical isomers on the C30 column (Figure 3). In each case, the all-*trans* isomer dominated. The isomers **1′** and **2′** showed a *cis* peak at 370 nm and **9′** at 330 nm. According to their height in relation to the dominating absorbance maximum, their position in front of the all-*trans* isomer on a C30 column and in comparison to astaxanthin for the keto derivatives [6] and to zeaxanthin for 4-hydroxy-nostoxanthin [30], these isomers are most likely 13-*cis*. In contrast, **1″** without a *cis* peak may be a 9-*cis* 4,4′-diketo-nostoxanthin isomer.

The keto carotenoids and their reduction products were identified by high resolution mass spectrometry (Figure 5). For all of them, the correct molecular masses and the typical prominent fragments were obtained (Table 2). As indicated in the right part of Figure 5A, fragment M-92 originates from an in chain elimination of a toluene unit and is an indication of the central polyene chain [31]. All analyzed carotenoids show the elimination of hydroxyl groups either as water (M-18) or as neutral loss of 17 Da. In addition, an intensified peak was found at 147.12 Da in the spectra of the mono and diketo derivaties (Figure 5A,C). This is typical for a dehydrated 4-keto-β-ionone ring with cleavage of the C7,8 bond [31].

Our combined approach of mutant selection, genetic engineering and chemical modification is set as a general example of how novel carotenoids can be generated in *X. dendrorhous*. For the production of 4-keto-nostoxanthin and 4,4′-diketo-nostoxanthin, it is a proof of concept, which also indicates how to improve their yields in future studies. Judging from the relative low conversion of β -carotene to zeaxanthin (Figure 2B) and complete conversion of zeaxanthin to nostoxanthin (Figure 2C), the 3-hydroxylation step is regarded limiting in the formation of ketolated nostoxanthin. Formation of ketolated β-carotene derivatives echinenone and canthaxanthin demonstrate that the ketolation step is not limited (Figure 2E). It has previously been shown that conversion rates of transgenic reactions in *X. dendrorhous* are dependent on the number of *trans* gene copies integrated into the genome [32]. Therefore, either transformation with a plasmid carrying two copies of the 3-hydroxylase gene *crtZ* as demonstrated by Pollmann et al. [12] or repeated transformation with the *crtZ* gene is a promising way to improve intermediate conversion to the end product. In addition, overall carotenoid synthesis can be enhanced up to 3-fold by improvement of precursor supply in combination with increased flux into the carotenoid pathway [12] and in combination with a high carotenoids producing *X. dendrorhous* mutant, a total increase of carotenoid formation of up to 90-fold can be achieved [7].

Acknowledgments: This work was supported as the ERA-IB project PROCAR through the German Federal Ministry of Education and Research (BMBF) (FKZ 031A569) and in part by COST Action CA15136. The high resolution ESI-MS system was partially funded by the DFG.

Author Contributions: Hendrik Pollmann and Gerhard Sandmann designed the experiments and carried out the HPLC analysis, Jürgen Breitenbach constructed the transformation plasmids, Hendrik Pollmann carried out transformation and cultivation of *X. dendrorhous* including sample preparation, Hendrik Wolff and Helge B. Bode performed high resolution MS-MS measurement and data analysis. All authors commented on the completed manuscript.

References

1. Andrewes, A.G.; Starr, M.P. (3R,3′R)-Astaxanthin from the yeast *Phaffia rhodozyma*. *Phytochemistry* **1976**, *15*, 1009–1011. [CrossRef]
2. Sandmann, G.; Misawa, N. Fungal carotenoids. In *The Mycota X; industrial application*; Osiewacz, H.D., Ed.; Springer: Berlin, Germany, 2002; pp. 247–262.
3. Rodriguez-Saiz, M.; de la Fuente, J.L.; Barredo, J.L. *Xanthophyllomyces dendrorhous* for the industrial production of astaxanthin. *Appl. Microbiol. Biotechnol.* **2010**, *88*, 645–658. [CrossRef] [PubMed]
4. Schmidt, I.; Schewe, H.; Gassel, S.; Jin, C.; Buckingham, J.; Hümbelin, M.; Sandmann, G.; Schrader, J. Biotechnological production of astaxanthin with *Phaffia rhodozyma/Xanthophyllomyces dendrorhous*. *Appl. Microbiol. Biotechnol.* **2010**, *8*, 555–571. [CrossRef] [PubMed]
5. Sandmann, G. Carotenoids of biotechnological importance. In *Biotechnology of Isoprenoids, Advances in Biochemical Engineering/Biotechnology*; Schrader, J., Bohlmann, J., Eds.; Springer: Berlin/Heidelberg, Germany, 2014; Volume 148, pp. 449–467.
6. Visser, H.; Sandmann, G.; Verdoes, J.C. Xanthophylls in fungi: metabolic engineering of the astaxanthin biosynthetic pathway in *Xantophyllomyces. dendrorhous*. In *Methods in Biotechnology: Microbial Processes and Products*; Humana: Totowa, NJ, USA, 2005.
7. Gassel, S.; Breitenbach, J.; Sandmann, G. Genetic engineering of the complete carotenoid pathway towards enhanced astaxanthin formation in *Xanthophyllomyces dendrorhous* starting from a high-yield mutant. *Appl. Microbiol. Biotechnol.* **2014**, *98*, 345–350. [CrossRef] [PubMed]
8. Hara, K.Y.; Morita, T.; Mochizuki, M.; Yamamoto, K.; Ogino, C.; Araki, M.; Kondo, A. Development of a multi-gene expression system in *Xanthophyllomyces dendrorhous*. *Microb. Cell Factories* **2014**, *13*, 175. [CrossRef] [PubMed]
9. Sharma, R.; Gassel, S.; Steiger, S.; Xia, X.; Bauer, R.; Sandmann, G.; Thines, M. The genome of the basal agaricomycete *Xanthophyllomyces dendrorhous* provides insights into the organization of its acetyl-CoA derived pathways and the evolution of agaricomycotina. *BMC Genom.* **2015**, *16*, 233. [CrossRef] [PubMed]
10. Girard, P.; Falconnier, B.; Bricout, J.; Vladescu, B. β-Carotene producing mutants of *Phaffia rhodozyma*. *Appl. Microbiol. Biotechnol.* **1994**, *41*, 183–191. [CrossRef]
11. Ojima, K.; Breitenbach, J.; Visser, H.; Setoguchi, Y.; Tabata, K.; Hoshino, T.; van den Berg, J.; Sandmann, G. Cloning of the astaxanthin synthase gene from *Xanthophyllomyces dendrorhous* (*Phaffia rhodozyma*) and its assignment as a β-carotene 3-hydroxylase/4-ketolase. *Mol. Genet. Genom.* **2006**, *275*, 148–158. [CrossRef] [PubMed]
12. Pollmann, H.; Breitenbach, J.; Sandmann, G. Engineering of the carotenoid pathway in *Xanthophyllomyces dendrorhous* leading to the synthesis of zeaxanthin. *Appl. Microbiol. Biotechnol.* **2016**, *101*, 103–111. [CrossRef] [PubMed]
13. Albrecht, M.; Misawa, N.; Sandmann, G. Metabolic engineering of the terpenoid biosynthetic pathway of *Escherichia coli* for production of the carotenoids β-carotene and zeaxanthin. *Biotechnol. Lett.* **1999**, *21*, 791–795. [CrossRef]
14. Osawa, A.; Harada, H.; Choi, S.-K.; Misawa, N.; Shindo, K. Production of caloxanthin-β-D-glucoside, zeaxanthin 3,3-β-D-diglucoside, and nostoxanthin in a recombinant *Escherichia coli* expressing system harbouring seven carotenoid biosynthesis genes, including *crtX* and *crtG*. *Phytochemistry* **2011**, *72*, 711–716. [CrossRef] [PubMed]
15. Albrecht, M.; Takaichi, S.; Steiger, S.; Wang, Z.Y.; Sandmann, G. Novel hydroxycarotenoids with improved antioxidative properties produced by gene combination in *Escherichia coli*. *Nat. Biotechnol.* **2000**, *18*, 843–846. [CrossRef] [PubMed]
16. Nishida, Y.; Adachi, K.; Kasai, H.; Shizuri, Y.; Shindo, K.; Sawabe, A.; Komemushi, S.; Miki, W.; Misawa, N. Elucidation of a carotenoid biosynthesis gene cluster encoding a novel enzyme, 2,2′-β-hydroxylase, from *Brevundimonas.* sp. strain SD212 and combinatorial biosynthesis of new or rare xanthophylls. *Appl. Environ. Microbiol.* **2005**, *71*, 4286–4296. [CrossRef] [PubMed]
17. Borovkov, A.Y.; Rivkin, M.I. XcmI-Containing vector for direct cloning of PCR products. *BioTechniques* **1997**, *22*, 812–814.

18. Misawa, N.; Satomi, Y.; Kondo, K.; Yokoyama, A.; Kajiwara, S.; Saito, T.; Ohtani, T.; Miki, W. Structure and functional analysis of a marine bacterial carotenoid biosynthesis gene cluster and astaxanthin biosynthetic pathway proposed at the gene level. *J. Bacteriol.* **1995**, *177*, 6575–6584. [CrossRef] [PubMed]
19. Zhong, Y.J.; Huang, J.C.; Liu, J.; Li, Y.; Jiang, Y.; Xu, Z.F.; Sandmann, G.; Chen, F. Functional characterization of various algal carotenoid ketolases reveals that ketolating zeaxanthin efficiently is essential for high production of astaxanthin in transgenic *Arabidopsis*. *J. Exp. Bot.* **2011**, *62*, 3659–3669. [CrossRef] [PubMed]
20. Breitenbach, J.; Gerjets, T.; Sandmann, G. Catalytic properties and reaction mechanism of the CrtO carotenoid ketolase from the cyanobacterium *Synechocystis* sp. PCC 6803. *Arch. Biochem. Biophys.* **2013**, *529*, 86–91. [CrossRef] [PubMed]
21. Choi, S.K.; Matsuda, S.; Hoshino, T.; Peng, X.; Misawa, N. Characterization of bacterial β-carotene 3,3'-hydroxylases, CrtZ, and P450 in astaxanthin biosynthetic pathway and adonirubin production by gene combination in *Escherichia coli*. *Appl. Microbiol. Biotechnol.* **2006**, *72*, 1238–1246. [CrossRef] [PubMed]
22. Eugster, C.H. Chemical derivatization: Microsomale tests for the presence of common functional groups in carotenoids. In *Carotenoid Volume 1A Isolation and Analysis*; Britton, G., Liaaen-Jensen, S., Pfander, H., Eds.; Birkhäuser Verlag: Basel, Switzerland, 1995; pp. 71–80.
23. Steiger, S.; Perez-Fons, L.; Cutting, S.M.; Fraser, P.D.; Sandmann, G. Annotation and functional assignment of the genes for the C30 carotenoid pathways from the genomes of two bacteria *Bacillus indicus* and *Bacillus firmus*. *Microbiology* **2015**, *161*, 194–202. [CrossRef] [PubMed]
24. Schimming, O.; Fleischhacker, F.; Nollmann, F.I.; Bode, H.B. Yeast homologous recombination cloning leading to the novel peptides ambactin and xenolindicin. *ChemBioChem* **2014**, *15*, 1290–1294. [CrossRef] [PubMed]
25. Sandmann, G. Combinatorial biosynthesis of carotenoids in a heterologous host: A powerfull approach for the biosynthesis of novel structures. *ChemBioChem* **2002**, *3*, 629–635. [CrossRef]
26. Mata-Gómez, L.C.; Montañez, J.C.; Méndez-Zavala, A.; Aguilar, C.N. Biotechnological production of carotenoids by yeasts: an overview. *Microb. Cell Factories* **2014**, *13*, 12. [CrossRef] [PubMed]
27. Buchecker, R.; Liaaen-Jensen, S.; Borch, G.; Siegelman, H.W. Carotenoids of *Anacystis nidulans*, structures of caloxanthin and nostoxanthin. *Phytochemistry* **1976**, *15*, 1015–1018. [CrossRef]
28. Yokoyama, A.; Miki, W.; Zumida, H.; Shizuri, Y. New trihydroxy-keto-carotenoids Isolated from an astaxanthin-producing marine bacterium. *Biosci. Biotech. Biochem.* **1996**, *60*, 200–203. [CrossRef] [PubMed]
29. Beyer, P.; Kleinig, H.; Englert, G.; Meister, W.; Noaek, K. Carotenoids of Rhizobia. IV. Isolation and structure elucidation of the carotenoids of a mutant of *Rhizobium lupine*. *Helvetica Chim. Acta* **1979**, *260*, 2551–2557. [CrossRef]
30. Milanowska, J.; Gruszecki, W.I. Heat-induced and light-induced isomerization of the xanthophyll pigment zeaxanthin. *J. Photochem. Photobiol.* **2005**, *80*, 178–186. [CrossRef] [PubMed]
31. Rivera, S.M.; Christou, P.; Canela-Garayoa, R. Identification of carotenoids using mass spectrometry. *Mass Spectromet. Rev.* **2014**, *33*, 353–372. [CrossRef] [PubMed]
32. Ledetzky, N.; Osawa, A.; Iki, K.; Pollmann, H.; Gassel, S.; Breitenbach, J.; Shindo, K.; Sandmann, G. Multiple transformation with the *crtYB* gene of the limiting enzyme increased carotenoid synthesis and generated novel derivatives in *Xanthophyllomyces dendrorhous*. *Arch. Biochem. Biophys.* **2014**, *545*, 141–147. [CrossRef] [PubMed]

14

Genetic Structure and Phylogeography of *Tuber magnatum* Populations

Beatrice Belfiori [1], Valentina D'Angelo [2], Claudia Riccioni [1], Marco Leonardi [2], Francesco Paolocci [1], Giovanni Pacioni [2] and Andrea Rubini [1,*]

1 National Research Council, Institute of Biosciences and Bioresources–Perugia Division, Via Madonna Alta n. 130, 06128 Perugia, Italy; beatrice.belfiori@ibbr.cnr.it (B.B); claudia.riccioni@ibbr.cnr.it (C.R.); francesco.paolocci@ibbr.cnr.it (F.P.)

2 Department of Life, Health & Environmental Sciences, University of L'Aquila, 67010 Coppito-L'Aquila, Italy; valentina.dangelo@accordphoenix.it (V.D.); marco.leonardi@univaq.it (M.L.); giovanni.pacioni@univaq.it (G.P.)

* Correspondence: andrea.rubini@ibbr.cnr.it

Abstract: The ectomycorrhizal fungus *Tuber magnatum* produces the white truffle appreciated worldwide for its unique aroma. With respect to other *Tuber* spp. of economic interest, *T. magnatum* presents a narrower geographical range. This species has, in fact, long been considered endemic to Italy. However, over the last few decades several reports have documented the presence of white truffles in different Mediterranean countries and in particular in various areas of south-east Europe. In this study, samples from several Pannonian and Balkan countries such as Hungary, Serbia, Romania, Bulgaria and Greece have been collected and genotyped with microsatellite markers and the data merged with those available for Italian populations. Our objectives were to test whether Italian and south-east European populations are differentiated and to evaluate the genetic diversity of *T. magnatum* all over its distributional range. We show the genetic structure of *T. magnatum* populations with the differentiation of four main groups: northern Italy, central-northern Italy, southern Italy and the Balkan/Pannonian region. The present study allowed us to refine the evolutionary history of *T. magnatum* and track the possible post-glacial expansion route of this species. The assessment of *T. magnatum*'s genetic structure is not only of scientific relevance, but it is also important for the conservation and market traceability of this prestigious fungus.

Keywords: microsatellite; SSR; white truffle; genetic diversity

---

1. Introduction

Species of the genus *Tuber* (Ascomycota, Pezizales, Tuberaceae) establish symbiosis with the roots of several tree and shrub species by forming structures for nutrients exchange, known as ectomycorrhiza [1]. In virtue of this mutualistic relationship these fungi produce hypogeous fruiting bodies (ascomata) known as truffles, that produce their spores sequestered within the surrounding tissues [2]. *Tuber* spp. thus rely on mycophagists for spore dispersal. Truffles produced by several *Tuber* spp. hold distinctive aromatic properties, which make them appreciated and marketed worldwide as food delicacies. Among edible *Tuber* spp. the black truffles harvested in Europe (*T. melanosporum* Vittad. and *T. brumale* Vittad.), the black summer truffle *T. aestivum* Vittad., the whitish truffle *T. borchii* Vittad. and the white truffle *T. magnatum* Pico are of particular relevance.

The evaluation of the intraspecific genetic diversity and population genetic structure of a species is crucial to understand its biology and ascertain its origin, history and evolution. By using molecular markers and performing a wide geographical sampling, a fine assessment of the population genetic structure of *T. melanosporum*, *T. brumale*, *T. indicum*, *T. aestivum* and *T. magnatum* has been performed [3–9].

These studies highlighted the presence of geographically structured populations and phylogeographic patterns in these species.

Regarding *T. magnatum*, its genetic variability was initially investigated over a low number of specimens collected in Italy [10,11] or Italy and Croatia [12], using RAPD (Random Amplification of Polymorphic DNA), ITS-RFLP (Restriction Fragment Length Polymorphism of the Internal Transcribed Spacer of rDNA) and SNPs (Single Nucleotide Polymorphism) either in the ITS region, in the ß-tubulin gene or in a SCAR (Sequence Characterized Amplified Region). All these studies showed a very limited intraspecific polymorphism [13]. A broader investigation was carried out by Frizzi et al. [14] who analyzed the polymorphism of eleven isoenzymes on 139 specimens from 13 Italian populations. The low genetic variability across populations and the lack of an interpretable evolutionary trajectory were thought to be in agreement with a self-reproductive system, a restricted species endemism and a relatively recent differentiation of this taxon [14]. A few years later, by employing seven polymorphic simple sequence repeats (SSR) loci over 316 specimens from Italy and the Istrian peninsula (Croatia and Slovenia), Rubini and colleagues [4] disclosed for the first time an isolation by distance pattern and a phylogeographic structure in *T. magnatum*, with central Italy that likely represented a refugium for this species during the last ice age. In addition, this study has been instrumental for a deep reevaluation of the life cycle and the reproductive strategies of all *Tuber* spp. and to prove that truffle ascocarps are mainly made of the haploid, maternal tissue [15].

Around the late 90s of the last century, it became clear that *T. magnatum* can also be found in the Balkan peninsula and countries nearby [16] and, although more sporadically, in the south-east of France [17] and Switzerland [18]. The recent discovery of natural *T. magnatum* populations in areas ranging from Greece until Hungary and Romania ([13], and references therein), calls now for studies based on a larger sampling area than before. On these premises, here we employed SSR markers and an extensive sampling on most of the *T. magnatum* distributional range to shed more light on the genetic structure and phylogeography of this species. In particular, we aimed at evaluating whether Italian and south-east European populations are genetically differentiated and tracking the post-glacial expansion pattern of this species. The Balkan peninsula, like the Italian one, in fact could have represented a glacial refugia for *T. magnatum* during the last glaciation.

The assessment of the genetic diversity distribution could reveal important findings for aspects spanning from ecology, conservation and marketing of this prestigious fungus.

2. Materials and Methods

2.1. Sample Source and DNA Analysis

Tuber magnatum samples were collected in 2015–2016 with the help of local pickers. Sampling locations were mainly in Pannonian and Balkan countries and more marginally in central and southern Italy (Table 1). Genomic DNA was isolated from freeze-dried and fresh ascocarps (1–5 mg) according to Paolocci et al. [19]. DNA quantity and quality were evaluated using the spectrophotometer Nanodrop (MySpec, Wilmington, DE, USA). The isolated DNA was diluted to 20 ng/μL and stored at −20 °C. All samples were genotyped using eight microsatellite loci previously characterized in this species: the loci MA4, MA7, MA14, MA12, MA19 and MA13 derived from Rubini et al. [20] and the loci MA2-1 and MA5-1 derived from Rubini et al. [4]. The SSR loci were PCR-amplified using multiplex PCR [19]. To this purpose, two panels, each consisting of four loci, were defined (Table S1). PCR conditions were those reported in Rubini et al. [20]. Moreover, the locus MA13 was analyzed in all samples considered by Rubini et al. [4].

The SSR amplicons were analyzed by capillary electrophoresis using an ABI 3130 Genetic Analyzer in presence of the Genescan 500 LIZ size standard (Applied Biosystems, Foster City, CA, USA). Sizing of amplicons and allele scoring were performed using GeneMapper software version 3.7 (Applied Biosystems, Foster City, CA, USA).

Table 1. *Tuber magnatum* samples considered in this study.

Population No.	Population Name (Sampling Location)	Locality	Sample Size*	Mean No. of Alleles**	Expected Heterozygosity**	Allelic Richness**
1	Valle del Crati	Italy (Calabria)	3	1.5 ± 0.19	0.22 ± 0.08	ND^{++}
2	Potenza	Italy (Basilicata)	14 (12)	2.9 ± 0.35	0.37 ± 0.06	1.81 ± 0.15
3	Benevento, Avellino	Italy (Campania)	(5)	2.3 ± 0.37	0.39 ± 0.09	2.05 ± 0.29
4	Isernia, Campobasso	Italy (Molise)	(16)	3.4 ± 0.53	0.56 ± 0.06	2.27 ± 0.18
5	Agnone	Italy (Molise)	(10)	3.4 ± 0.73	0.46 ± 0.11	2.21 ± 0.33
6	Castel del Giudice	Italy (Molise)	(16)	2.9 ± 0.48	0.42 ± 0.09	1.95 ± 0.23
7	Ateleta, SP. Avellana	Italy (Abruzzo-Molise)	36 (18)	4.6 ± 0.89	0.50 ± 0.10	2.17 ± 0.26
8	Roio del Sangro	Italy (Abruzzo-Molise)	(10)	2.9 ± 0.48	0.41 ± 0.09	1.98 ± 0.23
9	Quadri, Rosello	Italy (Abruzzo)	(14)	2.9 ± 0.58	0.44 ± 0.11	2.04 ± 0.27
10	Valle Roveto	Italy (Abruzzo)	(7)	2.3 ± 0.41	0.34 ± 0.09	1.83 ± 0.25
11	Chieti	Italy (Abruzzo)	(5)	2.3 ± 0.41	0.37 ± 0.12	2.08 ± 0.35
12	Ascoli Piceno	Italy (Marche)	(10)	2.8 ± 0.56	0.37 ± 0.11	1.92 ± 0.30
13	Montecastrilli	Italy (Umbria)	(9)	2.8 ± 0.59	0.42 ± 0.10	2.06 ± 0.29
14	Fabro	Italy (Umbria)	(11)	2.8 ± 0.41	0.42 ± 0.07	1.96 ± 0.21
15	Appennino umbro	Italy (Umbria)	(9)	2.4 ± 0.50	0.36 ± 0.10	1.85 ± 0.27
16	Gubbio	Italy (Umbria)	(25)	3.8 ± 0.75	0.45 ± 0.09	1.97 ± 0.22
17	Città di Castello	Italy (Umbria)	(12)	2.9 ± 0.69	0.39 ± 0.10	1.93 ± 0.28
18	Montemaggiore	Italy (Umbria)	(12)	2.3 ± 0.53	0.29 ± 0.11	1.68 ± 0.27
19	Pietralunga	Italy (Umbria)	(7)	2.4 ± 0.42	0.36 ± 0.10	1.92 ± 0.28
20	Firenze	Italy (Toscana)	(32)	4.1 ± 0.90	0.47 ± 0.08	2.04 ± 0.21
21	Val di Zena	Italy (Emilia-Romagna)	(10)	2.9 ± 0.58	0.43 ± 0.10	2.02 ± 0.27
22	Pianoro	Italy (Emilia-Romagna)	(9)	2.5 ± 0.53	0.38 ± 0.11	1.94 ± 0.20
23	Alba	Italy (Piemonte)	(7)	2.8 ± 0.49	0.42 ± 0.09	2.12 ± 0.29
24	Asti	Italy (Piemonte)	(9)	2.5 ± 0.42	0.40 ± 0.10	1.98 ± 0.27
25	Langhe	Italy (Piemonte)	(15)	3.0 ± 0.76	0.42 ± 0.11	2.01 ± 0.30
26	Pavia	Italy (Lombardia)	(11)	2.4 ± 0.42	0.36 ± 0.10	1.85 ± 0.26
27	Istria	Croatia-Slovenia	(17)	3.0 ± 0.60	0.41 ± 0.10	1.93 ± 0.25
28	Southern Hungary	Hungary	14	3.1 ± 0.74	0.37 ± 0.11	1.91 ± 0.30
29	Šabac	Serbia	6	2.1 ± 0.44	0.28 ± 0.11	1.78 ± 0.31
30	Valjevo	Serbia	7	2.0 ± 0.33	0.33 ± 0.10	1.66 ± 0.27
31	Anina	Romania	3	1.4 ± 0.18	0.17 ± 0.08	ND^{++}
32	Craiova	Romania	22	3.6 ± 0.84	0.44 ± 0.12	2.04 ± 0.31
33	Oryahovo	Bulgaria	6	2.0 ± 0.27	0.39 ± 0.09	1.89 ± 0.22
34	Razgrad	Bulgaria	6	1.9 ± 0.40	0.29 ± 0.11	1.75 ± 0.32
35	Kastoria	Greece	18	3.6 ± 0.71	0.47 ± 0.11	2.13 ± 0.28
36	Elatochori	Greece	6	2.5 ± 0.46	0.40 ± 0.10	2.15 ± 0.33
Mean	Italy	-	12.46 ± 1.52	2.84 ± 0.12	0.40 ± 0.01	1.99 ± 0.03
Mean	Pannonia/Balkans	-	10.50 ± 2.09	2.52 ± 0.24	0.36 ± 0.03	1.92 ± 0.05
Mean	Total	-	11.92 ± 1.23	2.75 ± 0.11	0.39 ± 0.01	1.97 ± 0.02

* Samples derived from the dataset of Rubini et al. [4] are indicated in parentheses. ** Standard errors of the means are reported. ND^{++}: Not determined due to the small sample size and presence of missing data.

2.2. Genetic Diversity Data and Population Structure Analyses

The mean number of alleles (Na) and expected heterozygosity (He) for each locus and for each population were evaluated using GenAlEx v. 6.501 [21]. Allelic richness (Ar) was calculated with the software ADZE [22] using the rarefaction method [23] to correct for differences in sample size. The Ar was weighted to four individuals by excluding calculation for Population 1 and 31, which have a very small sample size. To avoid losing most of the information due to the small sample size of some populations, Ar was also calculated by sorting individuals into eight geographical groups according to their proximity (Figure S1). In this case, it was possible to consider a larger sample size, up to $n = 15$.

To estimate the degree of differentiation among populations, the analysis of molecular variance (AMOVA) and calculation of Fst and Rst values were performed using Arlequin software, version 3.5.1.2 [24]. These analyses were carried out both by comparing all 36 populations (sampling locations) and two regional groups of populations: Italy and Balkans/Pannonia.

Multilocus version 1.3b [25] was used to calculate the number of multilocus genotypes (MLGs) and genotypic diversity (i.e., the probability that two individuals taken at random have different genotypes). To evaluate if samples sharing the same MLG were true clones or resulted from random mating, the Psex (the probability of obtaining the same MLG from different sexual events) values and their significance levels were calculated with MLGsim software [26]. To test for the presence of an isolation by distance pattern, correlation between genetic and geographic distances was evaluated by performing a Mantel test according to Rousset [27] and using the software GenAlEx. The geographic distance matrix, consisting of the natural logarithm of the pairwise distance among populations, was calculated with GenAlEx considering the average geographic coordinates of each population. A genetic distance matrix consisting of pairwise Rst/(1−Rst) values was calculated using the software SPAGeDi version 1.5 [28].

The genetic structure of *T. magnatum* populations was evaluated by Bayesian analysis using the software STRUCTURE version 2.3.4 [29] and TESS version 2.3 [30,31]. Five independent runs of STRUCTURE for K (max number of estimated clusters) ranging from 2 to 10 were conducted. For each K, 2000,000 MCMC (Markov Chain Monte Carlo) and a burn-in of 200,000 iterations were performed, respectively. The admixture and no admixture models were tested considering correlated and uncorrelated allele frequencies. The optimal K was determined by comparing both the log-likelihood values and Evanno's ΔK [32] using the software Structure Harvester [33]. TESS analysis was performed both under the no admixture and the CAR admixture models by conducing 50 runs for each *K* ranging from 2 to 10 with 50,000 total MCMC steps and a burn-in of 10,000 sweeps. The spatial interaction parameter was set to the default value of 0.6 and the degree of trend to linear. To estimate the best K, the Deviance Information Criterion (DIC) was averaged across runs for each *K*. The smallest value before reaching a plateau was selected as the best K. Burn-in length and number of MCMC interactions for STRUCTURE and TESS were established by checking the convergence of summary statistics, and by evaluating consistence among runs of different lengths, following the recommendations in the software manuals.

STRUCTURE and TESS results were processed with the software CLUMPP [34] using the Greedy algorithm, random input order of runs and 1000 repeats. CLUMPP results were used to generate bar graphs using DISTRUCT [35]. The ancestry coefficients calculated with STRUCTURE were also plotted into a geographic map using the "POPSutilities" R script [36] according to the interpolation procedure described by Francois [37]. Interpolate values of ancestry coefficients among each pair of samples were calculated with R using the CLUMPP Q-matrix and a spatial grid obtained from the raster map of the area. For each sample, transition between one group (K) to another was displayed with a progressive change in the color intensity. The starting colors are those assigned to the different K.

3. Results

3.1. Genetic Diversity of *T. magnatum* Populations

SSR data were generated for 111 *T. magnatum* samples. Most of the samples were from the Balkan/Pannonian region (9 populations, 88 samples) and a few from central (Abruzzo-Molise, 18 samples) and southern Italy (Basilicata and Calabria, 5 samples) (Table S2). SSR analysis always showed the presence of a single allele per locus as expected for haploid organisms. In total, 49 alleles were detected and among them and 12 were new with respect to the alleles previously identified [4] (Table S2). The data obtained in this study were then merged with the data from Rubini et al. [4], which were relative to samples mainly collected in Italy, to end up with a dataset of 429 samples grouped into 36 populations covering almost all of the known *T. magnatum* distributional area (Table 1 and Figure 1).

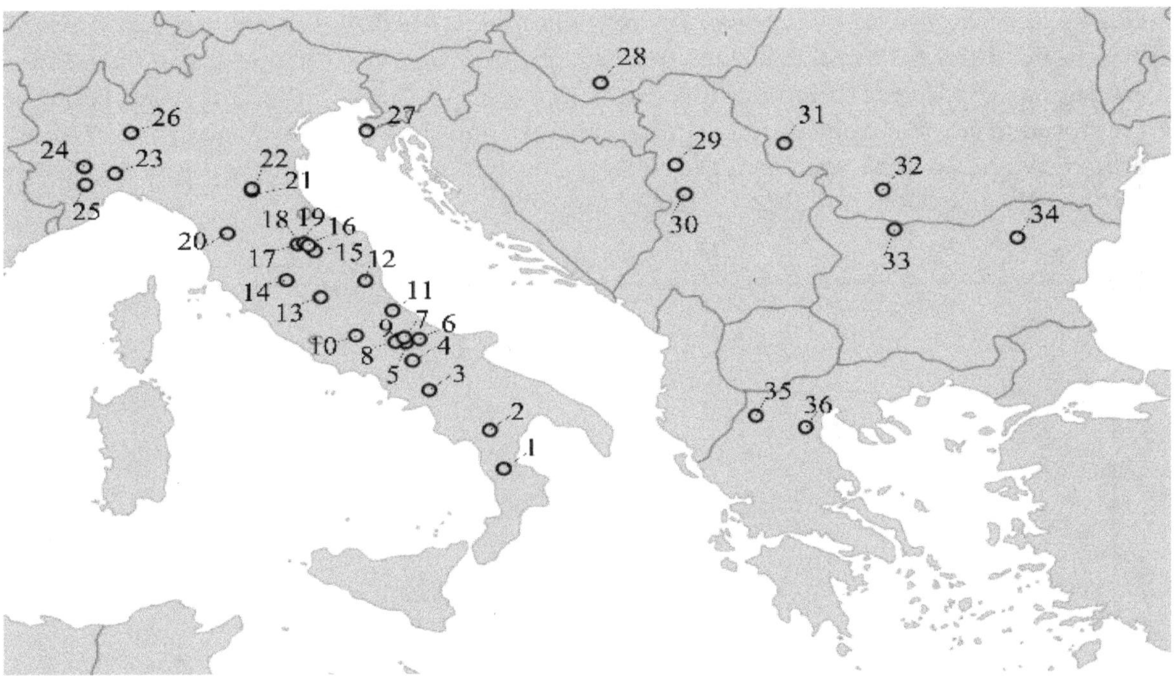

Figure 1. Map showing the geographical location of the *Tuber magnatum* samples analyzed. Populations are indicated with circles and numbered as in Table 1.

Considering this merged dataset, the number of alleles was 77 with a minimum of 3 alleles for the locus MA13 and a maximum of 19 for the locus MA4. The expected heterozygosity (He) ranged from 0.12 (MA13) to 0.84 (MA51) (Table S3). In each population the mean number of alleles ranged from 1.4 to 4.6 and the expected heterozygosity from 0.17 to 0.56 (Table 1). Comparison of allele distribution between samples from the Italian and Balkan/Pannonian regions revealed 10 private alleles specific to Italy and 28 to the Balkans/Pannonia. Some of these 38 alleles were also private for single populations, but their frequency was no higher than 0.041 (Table S4). The remaining 39 alleles were shared between the two regions, the most of them without relevant differences. Only a few showed a biased frequency, this was the case of the alleles 162 and 172 at MA4 and MA14 loci, respectively, being markedly more frequent among the Balkan/Pannonian samples, and the allele 121 at locus MA21 more frequent among Italian samples (Table S4). By combining the allelic profiles at the eight SSR loci, 362 MLGs were identified. Among these MLGs, 48 were shared at least between two individuals, but only a few (9) turned out to be clones as they had Psex values that were significantly lower, thus rejecting the hypothesis of their origin by sexual reproduction (Table S5). These putative clones were detected only within a population. Some MLGs were also shared among both close and distant populations but none of them showed significant Psex values, suggesting that they were generated by chance because of random mating (Table S5). The overall genotypic diversity was 0.998. When plotted against the number of loci, the genotypic diversity reached a value higher than 0.99 at six loci and also slightly increased up to eight loci (Figure S2).

The Ar for each single population ranged from 1.66 to 2.27. The highest values were observed for Abruzzo-Molise (Populations 4, 5 and 7) among the Italian populations and for Greece (Populations 35 and 36) among the Balkan/Pannonian populations (Table 1). Grouping individuals at a large geographical scale (i.e., 8 groups) allowed us to calculate the Ar by means of a rarefaction analysis based on a larger sample size (i.e., $n = 15$) and to show that the individuals from central-south Italy (Group 2) and those from the southern Balkans (Group 8) had the highest Ar values: 3.50 and 3.66, respectively (Figure S1).

3.2. The Balkan/Pannonian and Italian Populations Were Genetically Differentiated

The presence of a genetic structure was revealed by AMOVA, which showed a marked and significant genetic differentiation between the 36 populations (Fst 0.169, $p < 0.001$; Rst = 0.422, $p < 0.001$). A higher significant differentiation (Fst 0.212, $p < 0.001$; Rst = 0.427, $p < 0.001$) was detected when two regional groups of populations (Italian and Balkan/Pannonian) were considered (Table S6). Furthermore, the Mantel test showed that the among-populations differentiation significantly increased with geographic distance ($R^2 = 0.0232$, $p < 0.005$; Figure 2).

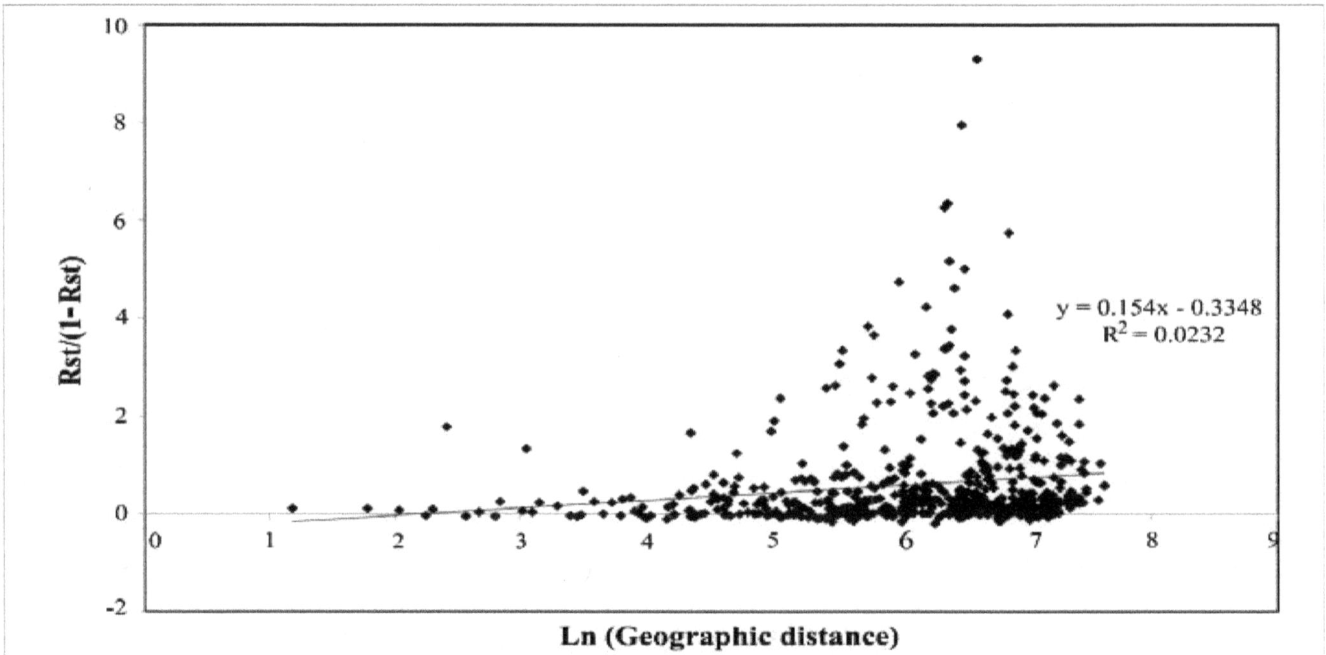

Figure 2. Mantel correlation between the genetic and geographic distances.

To better evaluate the genetic structure of *T. magnatum* populations, an admixture analysis was performed. Bayesian clustering using the software STRUCTURE, showed that the most probable number of clusters (K) was four (Figure S3A,B).

Three of these clusters were primarily associated with different geographical areas of sample acquisition, with one grouping most of the individuals from southern Italy, one those from central-northern Italy and Istria, and the third those from Balkans/Pannonia. Conversely, neither a specific nor a prevalent geographical provenance emerged within individuals of the fourth cluster (Figure 3A). We also noted that some individuals of Balkans/Pannonia shared common ancestry with those of central-northern Italy, Istria and with Population 1 from southern Italy. Plotting the ancestry coefficients obtained with STRUCTURE into the geographical map produced a better picture of the geographical distribution of individuals belonging to the four genetic clusters since they clearly matched the four distinct geographical areas: south Italy, central Italy, central-northern Italy and Istria as well as the Balkans/Pannonia (Figure 3B). No further relevant differentiation among individuals emerged when a higher value of K was considered (Figure 3A). Similar results were obtained using the no-admixture model considering either correlated or uncorrelated allele frequencies (data not shown). When STRUCTURE was run to analyze the Balkan/Pannonian regional group only, a further sub-structuration, with the differentiation of Greek Populations 35 and 36 starting from K = 3, emerged (Figure S4). Conversely, the same analysis on samples from Italy and Istria did not show any further sub-structuration (data not shown).

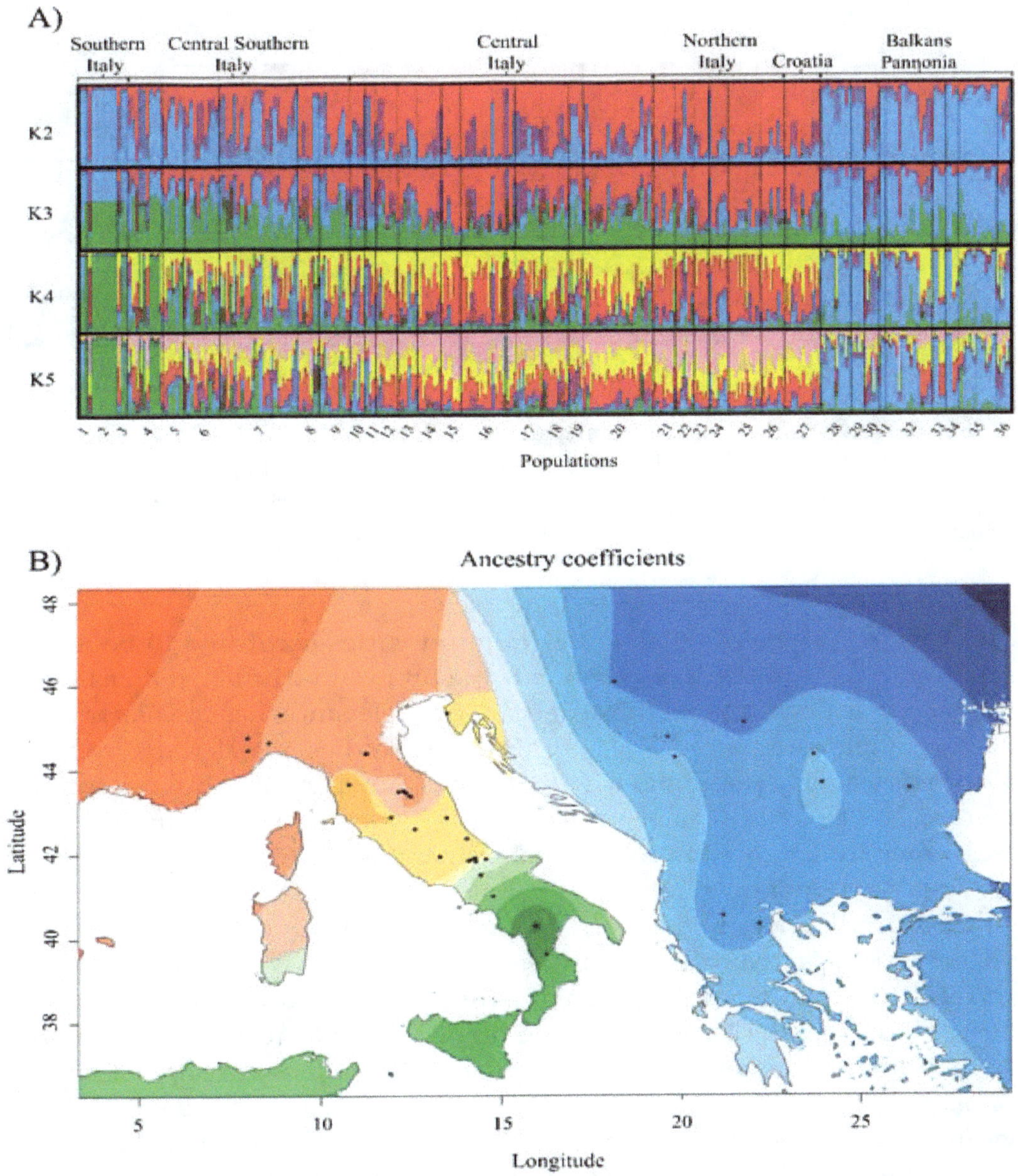

Figure 3. Bayesian analysis of genetic structure. (**A**) Plot of the ancestry coefficients of each single individual obtained with STRUCTURE based on admixture model and correlated allele frequencies. Each K is represented by a different color. Populations are indicated below the figure and their geographic origin above. (**B**) Spatial interpolation of population structure inferred for $K = 4$. Black dots represent the samples locations. The four K groups are represented by green, yellow, red and blue color gradients.

The genetic structure of populations was further explored using TESS. Unlike STRUCTURE, this software estimates the number of genetic clusters (K) by taking into account geographical coordinates of individuals to detect discontinuities in allele frequencies. Running TESS, under the admixture model there were three clusters matching; basically, the three main clusters found by STUCTURE (data not shown). The no-admixture model was more informative since, according to the DIC values, the most probable number of clusters was five (Figure 4A). In agreement with STRUCTURE, TESS evidenced a clear differentiation of southern Italian and Balkan/Pannonian populations. Moreover, the same analysis showed a further differentiation between populations of southern-central Italy and those of central and northern Italy. A fifth group was represented by the Balkan Population 35 from Greece and Population 1 from Calabria, which showed partial common ancestry (Figure 4B).

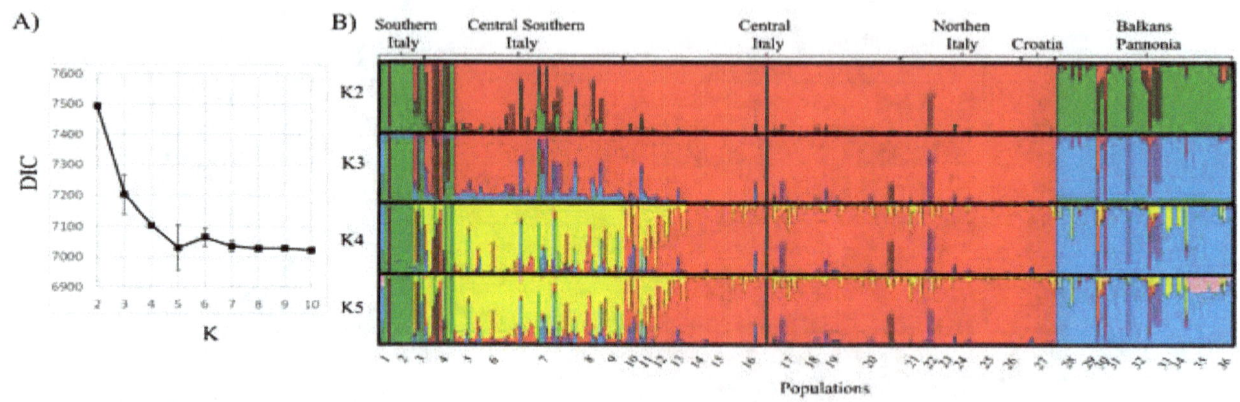

Figure 4. TESS analysis considering a no-admixture model. (**A**) Values of DIC in relation to the values of K. (**B**) Bar plots. Each individual is represented by vertical columns and the proportions of each K are represented by different colors. Populations and their geographic origin are given below and above the figure, respectively.

4. Discussion

When *T. magnatum* is evoked, what comes to mind varies according to people: According to the most gourmets and consumers, *T. magnatum* is the premium truffle, whereas for mycologists it represents the *Tuber* species that, among those of economic relevance, has the narrowest distributional range and the least understood autecology [13]. Previous studies revealed the presence of a genetic structure of *T. magnatum* Italian populations with southernmost and north-westernmost populations genetically differentiated from those of central Italy and Istria [4]. In these studies, only few samples other than those coming from Italy were considered; but, differently from what has been believed in the past, the *T. magnatum* distributional range is not indeed confined to the Italian peninsula only. Rather, it adds to the trans-Adriatic species as an increasing number of works have recently reported on the recovery of this fungus in different Balkan countries [16]. Thus, here we took advantage of *T. magnatum* specimens harvested from these new hotspots of white truffle production, spanning from Hungary, the northernmost, to Bulgaria, the easternmost, and down to Greece, to assess the genetic variability of this species and to ascertain whether genetic differentiation exists between populations of the two main regions of the *T. magnatum* distributional range: Italy and the Balkans/Pannonia. The analysis of more than 400 individuals, distributed into 36 populations located in Italy and the Balkan/Pannonian region and genotyped with eight SSR loci, unveils the presence of a high genetic variability and a clear genetic structure of *T. magnatum* populations.

4.1. New Insights into Genetic Diversity of White Truffle

The study by Rubini and colleagues [4] conducted over more than 300 *T. magnatum* samples harvested in Italy and Istria and genotyped by means of a few SSR markers was the first to show the presence of genetically structured populations and the occurrence of an extensive gene flow within and among *T. magnatum* populations, likely thanks to spore dispersal by mycophagist mammals. With respect to previous studies, not only the higher sample size but also the use of the more informative markers, that is, polymorphic SSRs vs. isoenzymes, RAPD or SNPs on highly conserved loci (i.e., β tubulin), were therefore crucial to these authors to gain first evidence on the abundance and distribution of the genetic variability in this truffle species. Yet, the use of a handful of polymorphic SSR loci coupled to a sampling of *T. melanosporum* all over the distributional range was sufficient to Riccioni et al. [5] to confute the thesis of a trifling genetic polymorphism in this species [38] and map its possible post glacial recolonization pattern. According to this observation, here we tested the hypothesis that even a relatively small number of polymorphic SSR loci would be sufficient to shed light into *T. magnatum* genetic diversity, if samples representative of the entire species distributional range are analyzed. Following this rationale, the first goal of the present study was to broaden the *T. magnatum* sampling areas. Thus, despite the widely known and hard to counteract secrecy of truffle hunters, an extensive sampling of *T. magnatum* fruitbodies from Bulgaria, Hungary, Romania, Serbia and Greece has been performed. Eighty-eight samples from these countries and a few samples from south Italy have been genotyped for eight SSR loci and these data merged with the SSR profiles retrieved from Rubini et al. [4]. By doing so, the number of alleles detected among the 429 specimens increased up to 77, with 12 new alleles being found, and with the number of alleles per locus (from 3 up to 19) in the same range found for other truffle species, such as *T. melanosporum* (2–18) and *T aestivum* (4–15) [39,40].

The value of the He over loci of 0.54 (Table S4) or 0.38, if the average value among populations is considered (Table 1), is much higher than that reported for this species (0.16) when sampling was confined to southern Italy only [41]. This high value of He suggests that at large geographical scale the level of genetic diversity in *T. magnatum* is comparable with that of other European species with a wider geographical distribution, such as *T. melanosporum* and *T. aestivum*, which showed He values of 0.41 and 0.50, respectively [7,39]. The high level of genetic diversity is also confirmed by the high number of MLGs: 332 over 429 fruit bodies analyzed and the overall genotypic diversity of 0.99, a value similar to that found in *T. melanosporum* and *T. aestivum* [7,39]. Moreover, the evidence that genotypic diversity reaches a plateau value of 0.99 with just six loci indicates that the number of loci used in this study is adequate to evaluate the genetic variability of *T. magnatum*.

If the entire sample set is split into two main regions, the Italian and the Balkan/Pannonian ones, then the presence of many private alleles specific to a single region emerges, although with a very low frequency. It is worth mentioning that a few of the shared alleles show a quite different frequency between the two regions. In sum, enlarging the sample size with individuals from the easternmost and southernmost *T. magnatum* distributional areas has allowed us to disclose new and private alleles and gain a closer look into the population genetics of this species.

4.2. Large Scale Sampling Reveals a Phylogeographic Structure in T. magnatum

The AMOVA analysis of the SSR data obtained from the 429 samples proves the presence of a genetic structure among populations. The values of Fst and Rst, in fact, are higher in the present research vs. the previous work by Rubini et al. [4], suggesting that the new populations considered, mainly from the Balkans/Pannonia, are genetically differentiated from those of Italy and Istria. This is confirmed by the Fst and Rst values that not only are significant but also increase when two regional groups of populations (Balkan/Pannonia and Italy) are compared. In keeping with this, the Mantel test depicts an isolation by distance pattern (Figure 2).

The presence of a genetic structure has been evaluated more in detail by performing Bayesian analysis. The STRUCTURE algorithm clearly reveals that Balkan/Pannonian and Italian populations are genetically differentiated. Moreover, Italian populations are split into tree genetic clusters: southern, central and central-northern Italy, in agreement with results of Rubini et al. [4].

Historical population expansions and restrictions resulting from climate changes have been reported for plant species, including truffle host species, which experienced population bottlenecks as a consequence of glaciations [42]. In concert with this, the geographical distribution of European truffle species has followed the population expansion and restriction processes of their hosts [38,43]. For example, within the black truffle clade, *T. melanosporum* survived in refugia located in the Iberian and Italian peninsulas, as inferred by ITS, ISSR and SSR markers [3,5,44]. Conversely the surviving pattern of *T. brumale* aggr., resulting from the phylogeny of ITS, LSU and PKC loci, was more complex: Within the *T. brumale* clade A, populations of haplotype I survived the last glaciation in Western Europe, those of haploytpe II in Eastern Europe whereas those of clade B in the Carpathian basin and Balkan region [6]. This latter clade was later proposed as representing a cryptic species, *T. cryptobrumale* [45]. Concerning the *T. aestivum* clade, according to the distribution and relatedness of the ITS haplotypes of samples all over Europe and Turkey, it has been suggested that this species survived in Turkey whereas European populations likely experienced a population bottleneck during the last glaciation [9]. The geographic structure of *T. magnatum* populations from the Italian peninsula, as per SSR analyses, was consistent with the occurrence a glacial refugium in central Italy from which the northernmost and southernmost populations originated [4]. Thanks to the large sampling performed, this hypothesis is here reinforced as among Italian populations, those from the central-southern area exhibit the highest levels of allelic richness. Our study also suggests that the Balkan peninsula may have represented a *T. magnatum* glacial refugium as well. Our inference stems from the following considerations: (i) Balkan/Pannonian populations, with a few exceptions, belong to a different genetic cluster with respect to Italian specimens, suggesting an independent evolutionary history; (ii) STRUCTURE analysis performed in Balkan/Pannonian populations shows that the individuals from the southernmost populations (Greece) tend to differentiate from the others and the He allelic richness is higher in these populations, a situation frequently expected in correspondence to putative glacial refugia [46]; and (iii) many truffle host plant species, including those that host *T. magnatum* (i.e., beech and hornbeam spp.) survived the last glaciation in three main Mediterranean peninsulas: the Iberian, Italian and Balkan ones [47–49].

It is noteworthy that, although a general phylogeographic pattern emerged from the STRUCTURE analysis, some individual from the Balkan/Pannonian region (e.g., individuals from Populations 30, 32 and 34) share common ancestry with individual from Italian populations. Moreover, analyses that take in account the geographical information (i.e., TESS algorithm), confirm clustering into four groups but also show that the individuals from Population 1 (southern Italy) share partial ancestry with those from Population 35, one of the southernmost populations from the Balkan area.

The finding that individuals across the two shores of the Adriatic sea partially share a common ancestry poses the question whether strain migration occurred from one side of the Adriatic shore to the other. In fact, the hypothesis that, in the past, strains from southern-central Italy moved to the Balkan region or vice versa, cannot be ruled out. Many Mediterranean taxa present disjunct distributions between the west and east Mediterranean, and these disjunct biogeographical patterns are the results of the complex paleogeographic history of the present Mediterranean region [50]. Land bridges between the Italian and Balkan shores of the Adriatic sea occurred in the Neogene through the formation of the Apulo-Dalmatic Realm [51] and likely during the Pleistocene glaciations [52]. Thus, it is conceivable that by enabling the migration of mycophagist animals these geological events may have favored truffle spore dispersal between the two peninsulas. Along the same reasoning, as the two shores of the Adriatic sea shared basically the same climate and soil (calcareous soil of cretaceous origin) conditions [53], it is more than conceivable that whatever the truffle migration direction was, the newly introduced truffle strains encountered host species and pedoclimatic conditions that have favored their settlement. Mycophagist-mediated spore dispersal across the Alps also appears to be the most conceivable explanation of the higher relatedness of Istrian specimens to those from Italy rather than to those from the Balkans. A more extensive sampling of specimens from both sides of the Adriatic sea covering, in particular, the latitudes spanning from 40° to 42° N, coupled with the use of phylogenetically informative functional markers, would help us to further test the "bridge" hypothesis.

4.3. Genetic Structure and Conservation Implications

Unearthing *T. magnatum* population genetic structure may have important implications for conservation of its biodiversity. Differently from black truffle species, *T. magnatum* cultivation is not yet established [13]; thus, propagation of this species is almost exclusively natural. In vitro isolation of mycelium strains is also very challenging as this species shows a very slow growth rate and optimal nutritional requirements have not been identified yet. Thus, the preservation of *T. magnatum* strains ex situ in genetic banks is currently unfeasible. Rather, the most affordable strategy for the conservation of *T. magnatum* biodiversity relies on the preservation of its natural habitats. On these premises, results of the present study may be of relevance to identify and preserve populations and strains specific and adapted to different environments.

5. Conclusions

Here we have shown that *T. magnatum* genetic diversity is higher than hitherto thought and geographically structured across the Italian peninsula and Balkan/Pannonia region. Our findings are of relevance to make inferences about the phylogeographic history of this species but also for marketing and conservation purposes. The price of white truffles is traditionally dictated by their geographic provenance; thus, by increasing the number of genetic markers it would be possible, in the near future, to trace the origin of these truffles. This is a prerequisite to promote and sustain local white truffle-linked ecosystem services and economies but also to evaluate if and to what extent genetic determinants concur to shape the aroma variability across white truffles of different provenance. From a conservation point-of-view, the presence of a phylogeographic structure led us to hypothesize that *T. magnatum* strains of different geographic areas might exhibit different adaptation traits. In the light of the difficulties in its cultivation/propagation and of a global warming scenario, preservation of *T. magnatum* natural habitats from both Italy and the Balkan/Pannonian countries is therefore crucial to prevent the erosion of its biodiversity.

Supplementary Materials: Table S1: Panels of SSR loci used, Table S2: List of the samples considered in this study, Table S3: Polymorphism levels of the 8 SSRs over the entire sample set, Table S4: Allele distribution and frequency in samples from Italian (I) and Balkan/Pannonian regions (B), Table S5: List of MLG and results of MLGsim analysis. Populations are indicated when identical MLG are found, Table S6: AMOVA analysis among all populations and considering two regional groups of populations (Italy, and Balkans/Pannonia), Figure S1: Allelic richness in eight geographical groups. The geographical groups are indicated below the figure. Each geographical group includes all individuals from populations (numbered as in Table 1) reported in parentheses, Figure S2: Average genotypic diversity in function of the number of loci, Figure S3: Estimation of the most probable k. (a) Mean log likelihood over 5 runs (error bars = standard deviations) and (b) ΔK, the second order rate of change in the likelihood at each K, Figure S4: STRUCTURE analysis performed on Balcan/Pannonian populations only, based on admixture model and correlated allele frequencies. Each K is represented by a different color. Populations are indicated below the figure and their geographic origin above.

Author Contributions: Conceptualization, A.R., G.P. and F.P.; methodology, A.R., B.B. and C.R.; investigation, data curation, B.B., V.D., M.L. and C.R.; formal analysis, A.R. and B.B.; resources, G.P. and M.L.; writing—original draft preparation, A.R., G.P., F.P.; writing—review and editing, B.B., V.D., C.R., M.L., G.P., F.P. and A.R. All authors have read and agree to the published version of the manuscript.

Acknowledgments: We are grateful to Pavlina Kladopoulou, Giorgio Konstanidis, Kiro Prodan, Kenan Kyose, Iordan Taralanski, Dejan Polic, Oszkar Fekete, Istvan Bagi, Silvio Guardiani, Gianni Miglietta, Mario Marchione, Marilena Oddis, Simona Ascione, Domenico Puntillo for providing us with fresh specimens of *Tuber magnatum* from the Balkans and Hungary and some areas of the Italian peninsula.

References

1. Smith, S.E.; Read, D.J. *Mycorrhizal Symbioses*, 3rd ed.; Academic Press: Cambridge, MA, USA, 2008.
2. Trappe, J.M.; Molina, R.; Luoma, D.L.; Cázares, E.; Pilz, D.; Smith, J.E.; Castellano, M.A.; Miller, S.L.; Trappe, M.J. *Diversity, Ecology and Conservation of Truffle Fungi in Forests of the Pacific Northwest*; Gen. Tech. Rep. PNW-GTR-772; USDA Forest Service, Pacific Northwest Research Station: Portland, OR, USA, 2009.

3. Murat, C.; Díez, J.; Luis, P.; Delaruelle, C.; Dupré, C.; Chevalier, G.; Bonfante, P.; Martin, F. Polymorphism at the ribosomal DNA ITS and its relation to postglacial re-colonization routes of the Perigord truffle *Tuber melanosporum*. *New Phytol.* **2004**, *164*, 401–411. [CrossRef]
4. Rubini, A.; Paolocci, F.; Riccioni, C.; Vendramin, G.G.; Arcioni, S. Genetic and phylogeographic structures of the symbiotic fungus *Tuber magnatum*. *Appl. Environ. Microbiol.* **2005**, *71*, 6584–6589. [CrossRef] [PubMed]
5. Riccioni, C.; Belfiori, B.; Rubini, A.; Passeri, V.; Arcioni, S.; Paolocci, F. *Tuber melanosporum* outcrosses: Analysis of the genetic diversity within and among its natural populations under this new scenario. *New Phytol.* **2008**, *180*, 466–478. [CrossRef] [PubMed]
6. Merényi, Z.; Varga, T.; Geml, J.; Orczán, Á.K.; Chevalier, G.; Bratek, Z. Phylogeny and phylogeography of the *Tuber brumale* aggr. *Mycorrhiza* **2014**, *24*, 101–113. [CrossRef] [PubMed]
7. Molinier, V.; Murat, C.; Peter, M.; Gollotte, A.; De la Varga, H.; Meier, B.; Egli, S.; Belfiori, B.; Paolocci, F.; Wipf, D. SSR-based identification of genetic groups within European populations of *Tuber aestivum* Vittad. *Mycorrhiza* **2016**, *26*, 99–110. [CrossRef] [PubMed]
8. Qiao, P.; Tian, W.; Liu, P.; Yu, F.; Chen, J.; Deng, X.; Wan, S.; Wang, R.; Wang, Y.; Guo, H. Phylogeography and population genetic analyses reveal the speciation of the *Tuber indicum* complex. *Fungal Genet. Biol.* **2018**, *113*, 14–23. [CrossRef]
9. Riccioni, C.; Rubini, A.; Türkoğlu, A.; Belfiori, B.; Paolocci, F. Ribosomal DNA polymorphisms reveal genetic structure and a phylogeographic pattern in the Burgundy truffle *Tuber aestivum* Vittad. *Mycologia* **2019**, *111*, 26–39. [CrossRef]
10. Gandeboeuf, D.; Dupré, C.; Roeckel-Drévet, P.; Nicolas, P.; Chevalier, G. Grouping and identification of *Tuber* species using RAPD markers. *Can. J. Bot.* **1997**, *75*, 36–45. [CrossRef]
11. Mello, A.; Fontana, A.; Meotto, F.; Comandini, O.; Bonfante, P. Molecular and morphological characterization of *T. magnatum* mycorrhizas in a long-term survey. *Microbiol. Res.* **2001**, *155*, 279–284. [CrossRef]
12. Mello, A.; Murat, C.; Vizzini, A.; Gavazza, V.; Bonfante, P. *Tuber magnatum* Pico, a species of limited geographical distribution: Its genetic diversity inside and outside a truffle ground. *Environ. Microbiol.* **2005**, *7*, 55–65. [CrossRef]
13. Riccioni, C.; Rubini, A.; Belfiori, B.; Gregori, G.; Paolocci, F. Tuber magnatum: The Special One. What Makes It so Different from the Other *Tuber* spp. In *True Truffle (Tuber spp.) in the World: Soil Ecology, Systematics and Biochemistry*; Zambonelli, A., Iotti, M., Murat, C., Eds.; Springer: Cham, Switzerland, 2016; pp. 87–103. [CrossRef]
14. Frizzi, G.; Lalli, G.; Miranda, M.; Pacioni, G. Intraspecific isozyme variability in Italian populations of the white truffle *Tuber magnatum*. *Mycol. Res.* **2001**, *105*, 365–369. [CrossRef]
15. Paolocci, F.; Rubini, A.; Riccioni, C.; Arcioni, S. Reevaluation of the life cycle of *Tuber magnatum*. *Appl. Environ. Microbiol.* **2006**, *72*, 2390–2393. [CrossRef] [PubMed]
16. Marjanović, Ž.; Grebenc, T.; Marković, M.; Glišić, A.; Milenković, M. Ecological specificities and molecular diversity of truffles (genus Tuber) originating from mid-west of the Balkan Peninsula. *Sydowia* **2010**, *62*, 67–87.
17. Tabouret, P. Exclusif! Description d'un site français producteur de truffe blanche d'Italie. *Le Trufficulteur* **2011**, *78*, 18.
18. Tabouret, P. Constitution d'une fédération Suisse de la truffe et de la trufficulture. *Le Trufficulteur* **2012**, *81*, 25.
19. Paolocci, F.; Rubini, A.; Granetti, B.; Arcioni, S. Rapid molecular approach for a reliable identification of *Tuber* spp. ectomycorrhizae. *FEMS Microbiol. Ecol.* **1999**, *28*, 23–30. [CrossRef]
20. Rubini, A.; Topini, F.; Riccioni, C.; Paolocci, F.; Arcioni, S. Isolation and characterization of polymorphic microsatellite loci in white truffle (*Tuber magnatum*). *Mol. Ecol. Notes* **2004**, *4*, 116–118. [CrossRef]
21. Peakall, R.; Smouse, P.E. GenAlEx 6.5: Genetic analysis in Excel. Population genetic software for teaching and research—An update. *Bioinformatics* **2012**, *28*, 2537–2539. [CrossRef]
22. Szpiech, Z.A.; Jakobsson, M.; Rosenberg, N.A. ADZE: A rarefaction approach for counting alleles private to combinations of populations. *Bioinformatics* **2008**, *24*, 2498–2504. [CrossRef]
23. El Mousadik, A.; Petit, R.J. High level of genetic differentiation for allelic richness among populations of the argan tree [*Argania spinosa* (L.) Skeels] endemic to Morocco. *Theor. Appl. Genet.* **1996**, *92*, 832–839. [CrossRef]
24. Excoffier, L.; Lischer, H.E. Arlequin suite ver 3.5: A new series of programs to perform population genetics analyses under Linux and Windows. *Mol. Ecol. Resour.* **2010** *10*, 564–567. [CrossRef] [PubMed]

25. Agapow, P.M.; Burt, A. Indices of multilocus linkage disequilibrium. *Mol. Ecol. Notes* **2011**, *1*, 101–102. [CrossRef]
26. Stenberg, P.; Lundmark, M.; Saura, A. MLGsim: A program for detecting clones using a simulation approach. *Mol. Ecol. Resour.* **2003**, *3*, 329–331. [CrossRef]
27. Rousset, F. Genetic differentiation and estimation of gene flow from F-statistics under isolation by distance. *Genetics* **1997**, *145*, 1219–1228. [PubMed]
28. Hardy, O.J.; Vekemans, X. SPAGEDI: A versatile computer program to analyze spatial genetic structure at the individual or population level. *Mol. Ecol. Notes* **2002**, *2*, 618–620. [CrossRef]
29. Pritchard, J.K.; Stephens, M.; Donnelly, P. Inference of population structure using multilocus genotype data. *Genetics* **2000**, *155*, 945–959.
30. Chen, C.; Durand, E.; Forbes, F.; Francois, O. Bayesian clustering algorithms ascertaining spatial population structure: A new computer program and a comparison study. *Mol. Ecol. Notes* **2007**, *7*, 747–756. [CrossRef]
31. Durand, E.; Jay, F.; Gaggiotti, O.E.; François, O. Spatial inference of admixture proportions and secondary contact zones. *Mol. Biol. Evol.* **2009**, *26*, 1963–1973. [CrossRef]
32. Evanno, G.; Regnaut, S.; Goudet, J. Detecting the number of clusters of individuals using the software STRUCTURE: A simulation study. *Mol. Ecol.* **2005**, *14*, 2611–2620. [CrossRef]
33. Dent, A.E.; von Holdt, B.M. STRUCTURE HARVESTER: A website and program for visualizing STRUCTURE output and implementing the Evanno method. *Conserv. Genet. Resour.* **2012**, *4*, 359–361.
34. Jakobsson, M.; Rosenberg, N.A. CLUMPP: A cluster matching and permutation program for dealing with label switching and multimodality in analysis of population structure. *Bioinformatics* **2007**, *23*, 1801–1806. [CrossRef] [PubMed]
35. Rosenberg, N.A. DISTRUCT: A program for the graphical display of population structure. *Mol. Ecol. Notes* **2004**, *4*, 137–138. [CrossRef]
36. Jay, F.; Manel, S.; Alvarez, N.; Durand, E.Y.; Thuiller, W.; Holderegger, R.; Taberlet, P.; François, O. Forecasting changes in population genetic structure of alpine plants in response to global warming. *Mol. Ecol.* **2012**, *21*, 2354–2368. [CrossRef] [PubMed]
37. How to Display Ancestry Coefficients (Q-matrix) on a Geographic Map. Available online: http://membres-timc.imag.fr/Olivier.Francois/TESS_Plot.html (accessed on 4 November 2019).
38. Bertault, G.; Raymond, M.; Berthomieu, A.; Callot, G.; Fernandez, D. Trifling variation in truffles. *Nature* **1998**, *394*, 734. [CrossRef]
39. Murat, C.; Riccioni, C.; Belfiori, B.; Cichocki, N.; Labbé, J.; Morin, E.; Tisserant, E.; Paolocci, F.; Rubini, A.; Martin, F. Distribution and localization of microsatellites in the Perigord black truffle genome and identification of new molecular markers. *Fungal Genet. Biol.* **2011**, *48*, 592–601. [CrossRef]
40. Molinier, V.; Murat, C.; Morin, E.; Gollotte, A.; Wipf, D.; Martin, F. First identification of polymorphic microsatellite markers in the Burgundy truffle, *Tuber aestivum* (Tuberaceae). *Appl. Plant. Sci.* **2013**, *1*, 1200220. [CrossRef]
41. Pomarico, M.; Figliuolo, G.; Rana, G.L. *Tuber* spp. Biodiversity in one of the southernmost European distribution areas. *Biodivers. Conserv.* **2007**, *16*, 3447–3461. [CrossRef]
42. Petit, R.J.; Brewer, S.; Bordács, S.; Burg, K.; Cheddadi, R.; Coart, E.; Cottrel, J.; Csaikl, U.M.; Van Dam, B.; Deans, J.D.; et al. Identification of refugia and post-glacial colonisation routes of European white oaks based on chloroplast DNA and fossil pollen evidence. *For Ecol. Manag.* **2002**, *156*, 49–74. [CrossRef]
43. Bertault, G.; Rousset, F.; Fernandez, D.; Berthomieu, A.; Hochberg, M.E.; Callot, G.; Raymond, M. Population genetics and dynamics of the black truffle in a man-made truffle field. *Heredity* **2001**, *86*, 451–458. [CrossRef]
44. García-Cunchillos, I.; Sánchez, S.; Barriuso, J.J.; Pérez-Collazos, E. Population genetics of the westernmost distribution of the glaciations-surviving black truffle *Tuber melanosporum*. *Mycorrhiza* **2014**, *24*, 89–100. [CrossRef]
45. Merényi, Z.; Varga, T.; Hubai, A.G.; Pitlik, P.; Erős, Á.; Trappe, J.M.; Bratek, Z. Challenges in the delimitation of morphologically similar species: A case study of *Tuber brumale* agg. (Ascomycota, Pezizales). *Mycol. Prog.* **2017**, *16*, 613. [CrossRef]
46. Widmer, A.; Lexer, C. Glacial refugia: Sanctuaries for allelic richness, but not for gene diversity. *Trends Ecol. Evol.* **2001**, *16*, 267–269. [CrossRef]

47. Taberlet, P.; Fumagalli, L.; Wust-Saucy, A.G.; Cosson, J.F. Comparative phylogeography and postglacial colonisation routes in Europe. *Mol. Ecol.* **1998**, *7*, 453–464. [CrossRef] [PubMed]
48. Grivet, D.; Petit, R.J. Chloroplast DNA phylogeography of the hornbeam in Europe: Evidence for a bottleneck at the outset of postglacial colonization. *Conserv. Genet.* **2003**, *4*, 47–56. [CrossRef]
49. Magri, D. Patterns of post-glacial spread and the extent of glacial refugia of European beech (*Fagus sylvatica*). *J. Biogeogr.* **2008**, *35*, 450–463. [CrossRef]
50. Sanmartín, I. Dispersal vs vicariance in the mediterranean: Historical biogeography of the Paleartic Pachydeminae (Coleoptera, Scarabeodidea). *J. Biogeog.* **2003**, *30*, 1883–1897. [CrossRef]
51. De Giuli, C.; Masini, F.; Torre, D. Island endemism in the Eastern Mediterranean mammalian paleofaunas patterns in the Gargano paleo-arcipelago. *Atti Accad. Naz. Lin.* **1990**, *85*, 247–262.
52. Kaiser, K. The climate of Europe during the Quaternary Ice Age. In *Quaternary Geology and Climate*; Wright, H.E., Jr., Ed.; National Academy of Sciences: Washington, DC, USA, 1969.
53. Gridelli, E. IL problema delle specie a diffusione attuale transadriatica con particolare riguardo ai Coleotteri. *Ital. J. Zool.* **1950**, *17*, 421–441.

15

Assessment of the Dyeing Properties of the Pigments Produced by *Talaromyces* spp.

Lourdes Morales-Oyervides [1,2], Jorge Oliveira [1], Maria Sousa-Gallagher [1], Alejandro Méndez-Zavala [2] and Julio Cesar Montañez [2,*]

1 School of Engineering, University College Cork, Cork, Ireland; lourdesmorales@uadec.edu.mx (L.M.-O.); j.oliveira@ucc.ie (J.O.); M.deSousaGallagher@ucc.ie (M.S.-G.)
2 Department of Chemical Engineering, Universidad Autónoma de Coahuila. Saltillo, Coahuila 25280, Mexico; alejandro.mendez@uadec.edu.mx
* Correspondence: julio.montanez@uadec.edu.mx.

Abstract: The high production yields of pigments by *Talaromyces* spp. and their high thermal stability have implied that industrial application interests may emerge in the food and textile industries, as they both involve subjecting the colourants to high temperatures. The present study aimed to assess the potential application of the pigments produced by *Talaromyces* spp. in the textile area by studying their dyeing properties. Dyeing studies were performed on wool. The dyeing process consisted of three stages: scouring, mordanting, and dyeing. Two different mordants (alum, A; ferric chloride, F) were tested at different concentrations on fabric weight (A: 5, 10, 15%; F: 10, 20, 30%). The mordanting process had a significant effect on the final colour of the dyed fabrics obtained. The values of dyeing rate constant (k), half-time of dyeing ($t_{1/2}$), and sorption kinetics behaviour were evaluated and discussed. The obtained results showed that pigments produced by *Talaromyces* spp. could serve as a source for the natural dyeing of wool textiles.

Keywords: fungal pigments; dyeing properties; textiles; *Talaromyces*

---

1. Introduction

Market trends to use natural colourants as food additives [1], natural dyes [2], functional foods [3], and cosmetic products [4] represent an opportunity for the application of natural pigments in several sectors of industry.

The possibility to exploit biological sources such as microorganisms for the production of natural pigments has been recommended [5]. However, the successful application of microbial pigments relies on high production yields, reasonable production costs and capital investment, regulatory approval, pigment characterisation, and stability to environmental factors such as temperature and light. Among the microorganisms with potential to produce a vast variety of pigments (*Monascus* homologues) is *Talaromyces* spp. (formerly *Penicillium* spp.) [6,7].

There has been a considerable effort in performing studies regarding the optimisation of the production of *Talaromyces* pigments [8–11]; however, there are only a few studies on the application of pigments produced by this strain [12,13].

Talaromyces pigments are thermally stable [14], implying that industrial application interests can emerge in the food and textile industries, as both processes involve subjecting the colourants to high temperatures.

Colourants used by the textile industry are predominantly synthetic; however, synthetic dyes are not environment-friendly, and recently the textile industry has been challenged to ensure compliance with environmental issues [15]. Natural colourants are environmentally friendly and biodegrade more quickly than synthetic dyes [16].

Additionally, natural pigments present properties of great interest in the textile industry, such as antibacterial properties. Fabrics can act as carriers of bacteria responsible for undesirable odours. It has been shown that *Talaromyces* pigment extracts possess antimicrobial properties [17]. These properties, along with the absence of toxicity [18,19], make them a valuable alternative as natural colourants in the textile industry.

However, natural colourants still face a huge disadvantage against synthetic colourants. Dyeing textile with natural colourants usually involves issues of limited shade range and lower fastness properties of the dyed fabrics. These problems have been overcome applying a pretreatment to the textile with mordants in order to create affinity between the fibre and the dye. Selection of mordant and concentration is important in natural dyeing processes, as mordants can increase the depth of shade or drastically alter the final colour of the dyed fabric [2].

The present study aimed to assess the potential application of the pigments produced by *Talaromyces* spp. in the textile area by studying their dyeing capacity.

2. Materials and Methods

2.1. Microorganism and Inoculum Preparation

Talaromyces spp. was used for the production of red pigments (Department of Food Science and Technology, Autonomous University of Coahuila, Saltillo, Coahuila). The purified strain had been previously isolated and characterised as *Penicillium purpurogenum* GH2 [20,21]. *Penicillium purpurogenum* has, however, been transferred to *Talaromyces* spp. [22]. The strain was maintained on PDA (Potato dextrose agar) slants at 4 °C and sub-cultured periodically. Inoculum was prepared in Erlenmeyer flasks (125 mL in capacity) containing 25 mL of Potato Dextrose Broth medium (PDB medium, ATCC medium:336), which were sterilised and inoculated with a spore suspension (1×10^5 spores/mL) of *Talaromyces* spp. previously incubated for 5 days. The flasks were then incubated at 30 °C for 84 h in an orbital shaker (Innova 94, New Brunswick Scientific, Edison, NJ, USA) at 200 rpm [9,23].

2.2. Culture Media

The PDA medium was prepared with a concentration of 39.0 g/L (Bioxon, Mexico). The medium PDB medium was prepared by finely boiling 0.3 kg of diced potatoes in 500 mL of water until thoroughly cooked; then the potatoes were filtered through cheesecloth and water was added to the filtrate to complete a volume of 1.0 L. Finally, 20.0 g of glucose was added before sterilisation. The Czapek-dox modified medium [24] consisted of (g/L): D-xylose 15.0, $NaNO_3$ 3.0, $MgSO_4{\cdot}7H_2O$ 0.5, $FeSO_4{\cdot}7H_2O$ 0.1, K_2HPO_4 1.0, KCl 1.0 and ethanol 20.0.

2.3. Cultivation Conditions for Pigment Production

The initial pH of the Czapek-dox modified medium was adjusted to 5.0 before sterilising by using 0.22 μm sterile membranes (Millipore, Billerica, MA, USA). A mycelial suspension of *Talaromyces* spp. was inoculated at 10% (*v/v*) in 125 mL Erlenmeyer flasks containing 25 mL of medium. The inoculated flasks were incubated at 30 ± 2 °C in an orbital shaker (Innova 94, New Brunswick Scientific, Edison, NJ, USA) at 200 rpm for 6 days.

2.4. Pigment Extraction

The pigment extraction was performed according to the methodology reported by Méndez-Zavala et al. (2011) [25]. The pigment extract was centrifuged at 8000 rpm and at 4 °C for 20 min (Sorvall, Primo R Biofuge Centrifugation Thermo, Waltham, MA, USA) and then filtered through a 0.45 μm cellulose membranes (Millipore, Billerica, MA, USA) for the subsequent analysis of pigments. In this study, only extracellular pigments were considered. The analysis of red pigment production was conducted by measuring the absorbance of the filtered extract at 500 nm using a spectrophotometer (Cary 50, UV-Visible Varian, Palo Alto, CA, USA). This wavelength was selected by scanning the maximum sensitivity for the presence of the pigment (that is, the pigment absorbs maximum light at this wavelength). Red pigment extracts were stored in the dark at 4 °C before being used for the subsequent test.

2.5. Wool Dyeing Process

Wool fabric was bought at a local textile company. The dyeing process consisted of three stages: scouring [26], mordanting, and dyeing [12].

2.5.1. Fabric Scouring

The fabric was first scoured to remove any impurities so that they would not interfere with the dyeing process [26]. Wool was rinsed with 100 mL of hot distilled water (60 °C) per gram of wool; then, 18 mL of neutral soap per litre of water was added, along with 25% w/v of sodium carbonate, taking into consideration the wool weight. The mixture was heated to 60 °C and the wool was constantly moved from top to bottom. Then, the fabric was washed with water at ambient temperature (six times).

2.5.2. Fabric Mordanting

The fabric mordanting process was carried out using the pre-mordanting technique [12]. To assess the most appropriate mordant for *Talaromyces* spp. pigments, two different mordants were tested at different concentrations on the weight of fabric (Table 1). Concentrations were selected from literature according to recommended concentrations for these two mordants [27]. Wool was heated in mordant solutions at 70 °C for 1 h. Subsequently, the fabric was squeezed to remove excess liquid and then air dried at room temperature overnight.

Table 1. Mordants studied for the pre-mordanting process and selected concentrations. Codes were assigned.

Mordant	Concentration, % w/w	Code
Ferric chloride	10	F1
	20	F2
	30	F3
Alum	5	A1
	10	A2
	15	A3

2.6. Dyeing

The pre-mordanted wool was dyed with 40 mL of pigment extract (40:1, pigment extract per gramme of fabric) in a conical flask at 80 °C for 90 min. pH was not controlled. The colour of fabrics after dyeing was determinate by CIELAB colour coordinates using ColorEye XTS colorimeter (GretagMacbeth, Grand Rapids, MI, USA).

Pigment uptake was determined by measuring the optical density of the dye solution samples at a wavelength of 500 nm [27]. Percentage of pigment uptake (q, %) was calculated using the following equation:

$$q = \frac{OD_o - OD_i}{OD_o} * 100 \tag{1}$$

where OD_o is the initial optical density (500 nm) of the dye bath, and OD_i is the optical density after dyeing at different sampling times (i min).

The first-order rate equation of Lagergren, which is one of the most widely used equations for the sorption of solute from a liquid solution, was employed to describe the pigment sorption kinetics.

Rearrangement of the Lagergren model [28] was used for the variation of the adsorbed pigment as a function of time:

$$\frac{q_t}{q_r} = 1 - \frac{exp[(-kt_r)(t/t_r)]}{1 - exp(-kt_r)} \tag{2}$$

where q_t and q_r are the amount of adsorbed pigment at time t (min) and at equilibrium (%), respectively, k is the first-order rate constant (min^{-1}), and t_r is the longest time of the sorption process (min).

2.7. Data Analysis

The model parameters were estimated by non-linear regression analysis [29]. Results were analysed statistically by factorial ANOVA to test statistical differences ($p < 0.05$), followed by Tukey's test at 5% probability for comparisons. Statistical and regression analyses were made with STATISTICA 7.1. (StatSoft, Inc., Tulsa, OK, USA, 2005,).

3. Results and Discussion

Dying Properties of Pigments

The effect of the mordant used on L^*, a^*, and b^* parameters are given in Table 2; L^* represents a lightness value (a higher lightness value represents a lower colour yield). a^* and b^* represent the tone of the colour; positive values of a^* and b^* represent redder and yellowish tones, respectively.

Total colour difference ΔE^* between the dyed fabrics and the undyed wool was calculated as:

$$\Delta E^* = \sqrt{(L^*_U - L^*_D)^2 + (a^*_U - a^*_D)^2 + (b^*_U - b^*_D)^2} \tag{3}$$

where sub-index U and D represent undyed and dyed wool, respectively, for L^*, a^*, and b^* values.

Fabrics were qualitatively perceived as different; the dyeing process using mordant F presented a red colour while the dyeing process using mordant A showed a red tending to brown colour. Colorimetric studies indicated that the tested mordants significantly affected the colour exhibited by the fabric. The dyed wool using mordant F presented a higher total colour difference from the undyed wool than the dyed wool using mordant A (Table 2). Higher values of a^* were obtained with mordant F in comparison with mordant A, indicating that more reddish tones can be obtained using mordant F. Meanwhile, b^* values indicated that more yellowish tones can be obtained with mordant A. These results were more evident with the hue values obtained; hue values near to 0 indicated the degree of redness while values near to 90 represented the level of yellowness. Moreover, the wool dyed using mordant F presented a higher saturation of colour (Chroma) and a higher yield (L^*) than that dyed with mordant A.

Dyed fabrics obtained here presented a stronger shade (red colour) than the optimum conditions reported for the dyeing of cotton using pigments produced by different fungal strains (*Monascus purpureus*, *Isaria farinose*, *Emericella nidulans*, *Fusarium verticilliodes* and *Penicillium purpurogenum*) [12].

Similarly, a^* and b^* values were higher than those reported for the dyeing of leather and silk with *Talaromyces* pigments [11,30]; however, those results were attained without the addition of mordants.

Mordant concentration also affected the colour exhibited by the dyed fabric. When mordant F was used, the colour yield obtained increased with concentration. However, there was not a statistical difference between F2 and F3. It is noted that a more reddish colour (a^*) was obtained by increasing the mordant concentration while there was not a statistical difference achieving a yellow tone (b^*) between F2 and F3 or between F1 and F2. In terms of colour saturation, hue, and total colour difference, there was not statistical difference between F2 and F3 indicating that if an F concentration between 20 and 30% *w/w* is used during the mordanting process, the same colour can be obtained.

Different results were obtained using mordant A; colour yield increased by increasing mordant concentration. A more reddish colour (a^*) was obtained using the highest concentration while the yellowish tone (b^*) was not affected by concentration. Furthermore, the concentration did not affect colour saturation (Chroma). A higher degree of redness (Hue) and higher total colour difference (ΔE^*) was attained using the highest concentration of mordant A.

These results are consistent with those reported by Arroyo-Figueroa et al. (2011) [26], who stated that a colour variation in terms of CIELAB scale is obtained as function of mordant concentration during the dyeing process using natural colourants.

In technical dyeing using synthetic dyes, a total colour difference of 1 is accepted as a tolerable colour difference between dyeings. However, with the introduction of natural dyes into the textile dyeing process, a wider total colour difference can be accepted ($\Delta E^* = 2$) [31]. Results demonstrated that, independent of the mordant and concentration tested, an acceptable colour difference between dyeings was obtained ($\Delta E^* = 0.71$–1.70).

The effect of the dyeing process time on pigment uptake using different mordants at different concentrations is shown in Figure 1. It can be seen that dyeings with mordant F presented higher values of pigment uptake (42.54–82.44%) than mordant A (33.98–41.98%). In Table 3 are listed the kinetic parameters obtained with the regression analysis of Equation (2). When mordant F was used, pigment uptake at equilibrium (q_r) increased as mordant concentration increased. Mordant A did not show a significant difference between concentrations A1 and A2, reaching the maximum pigment uptake with concentration A3. The higher pigment uptake here obtained (81.33 ± 0.43%) is similar to the maximum (80%) reported by Velmurugan et al. (2010) [12], which was achieved dyeing cotton with red pigments produced by *Monascus purpureus*.

Table 2. L^*, a^*, b^* values of dyed wool.

Experiment	Colour Coordinates					
	L^*	a^*	b^*	Chroma	Hue	ΔE^*
Wool	93.89 ± 1.16	0.44 ± 0.22	5.15 ± 1.47	5.17 ± 1.47	84.94 ± 2.63	0.00
F1	31.39 ± 1.11 ^{b}	24.35 ± 0.44 ^{c}	21.01 ± 0.20 ^{c}	32.16 ± 0.23 ^{b}	40.79 ± 0.76 ^{b}	68.78 ± 0.96 ^{b}
F2	28.20 ± 0.64 ^{a}	27.62 ± 0.68 ^{b}	20.61 ± 0.41 ^{bc}	34.47 ± 0.64 ^{a}	36.73 ± 0.78 ^{a}	72.76 ± 1.70 ^{a}
F3	27.53 ± 0.65 ^{a}	28.91 ± 0.35 ^{a}	21.87 ± 0.81 ^{b}	36.25 ± 0.25 ^{a}	37.10 ± 1.34 ^{a}	74.14 ± 0.90 ^{a}
A1	44.47 ± 0.68 ^{e}	14.84 ± 0.15 ^{e}	27.34 ± 0.32 ^{a}	31.11 ± 0.34 ^{d}	61.50 ± 0.21 ^{d}	56.07 ± 0.71 ^{d}
A2	42.48 ± 0.43 ^{d}	15.07 ± 0.24 ^{e}	27.83 ± 0.25 ^{a}	31.65 ± 0.13 ^{d}	61.56 ± 0.57 ^{d}	58.09 ± 1.65 ^{d}
A3	37.08 ± 0.56 ^{c}	17.14 ± 0.35 ^{d}	27.76 ± 0.14 ^{a}	32.63 ± 0.18 ^{d}	58.32 ± 0.57 ^{c}	63.40 ± 1.52 ^{c}

Different letters in each column indicate significant differences (Tukey's post-hoc comparison, $p < 0.05$).

Results showed that there was no statistical difference in terms of exponential rate (k) using mordant F at all concentrations and mordant A at concentrations A2 and A3. Only mordant A showed the lowest exponential rate at the lowest concentration studied.

Table 3. Kinetic parameters and goodness of fit of the data.

Experiment	q_r, %	k, min^{-1}	R^2
F1	50.56 ± 0.57 ^{c}	0.0504 ± 0.0028 ^{a}	0.98
F2	76.99 ± 0.55 ^{b}	0.0542 ± 0.0021 ^{a}	0.99
F3	81.33 ± 0.43 ^{a}	0.0537 ± 0.0015 ^{a}	0.99
A1	36.91 ± 0.82 ^{e}	0.0218 ± 0.0011 ^{b}	0.98
A2	35.40 ± 0.33 ^{e}	0.0497 ± 0.0023 ^{a}	0.99
A3	41.02 ± 0.39 ^{d}	0.0569 ± 0.0030 ^{a}	0.99

Different letters in each column indicate significant differences (Tukey's post-hoc comparison, $p < 0.05$).

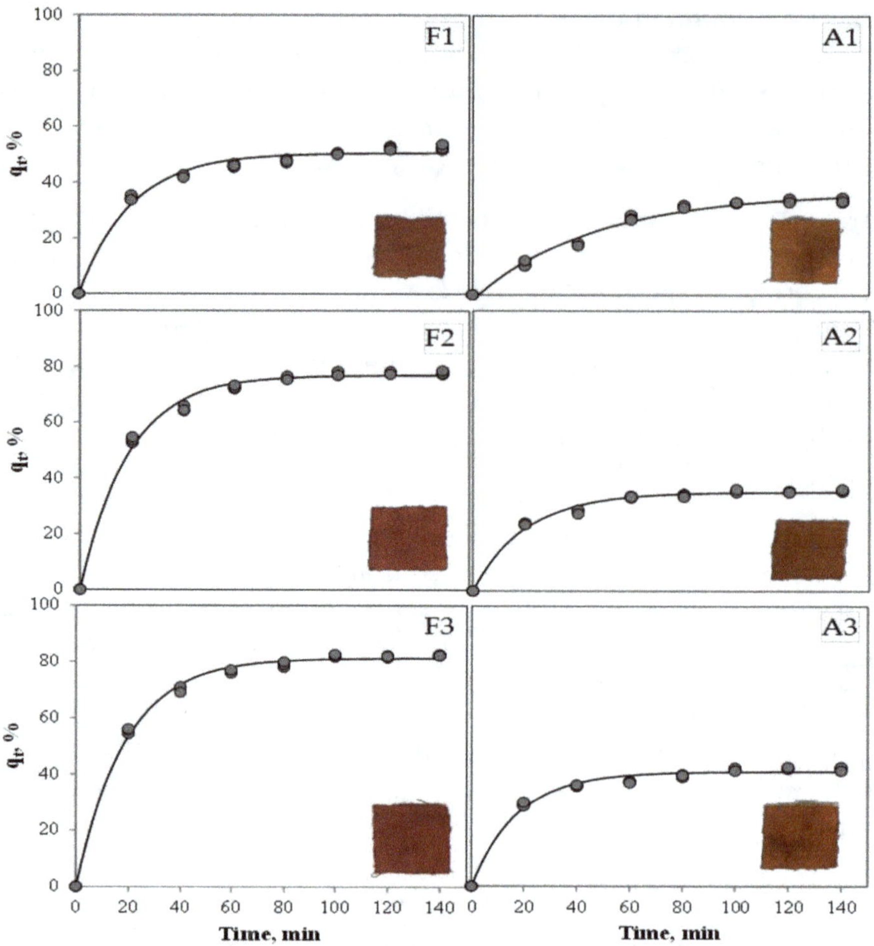

Figure 1. Effect of mordant type and concentration on pigment sorption on wool. Circles represent experimental data points and continuous lines represent the Lagergren model.

In literature, sorption curves are classified into four zones (0 to 4) according to the kt_r value; the kinetic behavior goes from no adsorption to drastic (Please see sorption kinetic behavior reported by Tseng et al. [28].

The time of half dyeing $t_{1/2}$ is also used to express the time required for a fabric to adsorb half of the amount of the dye adsorbed at equilibrium ($q_t/q_r = 0.5$), calculated as:

$$t_{1/2} = \frac{ln(2)}{k} \tag{4}$$

Table 4 lists the dimensionless rate constants (kt_r), the half dyeing time values ($t_{1/2}$), and the zone of kinetic sorption behavior presented for the mordants studied at three concentration levels.

Table 4. Sorption kinetic parameter (kt_r), $t_{1/2}$ and zone of the sorption behavior.

Experiment	kt_r	$t_{1/2}$, min	Zone
F1	7.06	13.74	III
F2	7.58	12.79	III
F3	7.51	12.92	III
A1	3.04	31.84	II
A2	6.95	13.95	III
A3	7.97	12.17	III

Results showed that most of sorption kinetic curves fell into zone III (Kinetic curve type, rapidly rise; kinetic behavior, quick), while only the dyeing process using A1 fell into zone II (Kinetic curve type, continuously rising; kinetic behavior, good).

The $t_{1/2}$ values obtained were in range with those reported for the optimised conditions for dying wool ($t_{1/2}$ = 13.1 min) with natural anthraquinones dyes produced by the fungus *Fusarium oxysporum* [32]. The $t_{1/2}$ values are relatively short in comparison with the longest time of the sorption process (t_r), which could result in undesired colour yields. Thus, the above kinetics are useful to establish a target value ($q_t/q_r > 0.5$) to assist the engineering design and optimisation of the dyeing process of wool by *Talaromyces* pigments.

4. Conclusions

The above results showed the feasible potential applications of the pigments produced by *Talaromyces* spp.

These pigments could serve as a useful alternative source for the natural dyeing of wool textiles. High values of pigment uptake on the fabric were obtained. The values of dyeing rate constant (k), half-time of dyeing ($t_{1/2}$) and sorption kinetics behaviour compare well with other natural dyes used for dyeing textile. However, the process still needs to be optimised in terms of dying process conditions (pigment concentration, pH, temperature).

It is concluded that the mordant used plays an important role in the dying process. Colorimetric studies showed that it is possible to attain a strong red shade; therefore, these pigments can compete with other synthetic or natural red dyes.

From this point onward, full characterisation of the molecules, toxicity studies, and more in-depth analyses such as molecular interactions between the dye and the fabric are worth tfurther investigation.

Acknowledgments: Author Lourdes Morales-Oyervides acknowledges CONACyT-México for the financial support provided for conducting her PhD studies (215490-2011).

Author Contributions: Julio Cesar Montañez and Jorge Oliveira conceived and designed the experiments; Lourdes Morales-Oyervides performed the experiments; Jorge Oliveira and Maria Sousa-Gallagher analysed the data; Alejandro Méndez-Zavala and Julio Cesar Montañez contributed reagents/materials/analysis tools; Lourdes Morales-Oyervides wrote the paper.

References

1. Carocho, M.; Morales, P.; Ferreira, I.C.F.R. Natural food additives: Quo vadis? *Trends Food Sci. Technol.* **2015**, *45*, 284–295. [CrossRef]
2. Shahid, M.; Mohammad, F. Recent advancements in natural dye applications: A review. *J. Clean. Prod.* **2013**, *53*, 310–331. [CrossRef]
3. Fernández-García, E.; Carvajal-Lérida, I.; Jarén-Galán, M.; Garrido-Fernández, J.; Pérez-Gálvez, A.; Hornero-Méndez, D. Carotenoids bioavailability from foods: From plant pigments to efficient biological activities. *Food Res. Int.* **2012**, *46*, 438–450. [CrossRef]
4. Kostick, R.; Wang, S.; Wang, J. Color Cosmetics—Coloring/Staining the Skin from Fruit and Plant Color Pigments. Patent US20060280762, 2006.
5. Venil, C.K.; Zakaria, Z.A.; Ahmad, W.A. Bacterial pigments and their applications. *Process Biochem.* **2013**, *48*, 1065–1079. [CrossRef]
6. Arai, T.; Kojima, R.; Motegi, Y.; Kato, J.; Kasumi, T.; Ogihara, J. PP-O and PP-V, *Monascus* pigment homologs, production and phylogenetic analysis in *Penicillium purpurogenum*. *Fungal Biol.* **2015**, *119*, 1226–1236. [CrossRef] [PubMed]
7. Mapari, S.A.S.; Hansen, M.E.; Meyer, A.S.; Thrane, U. Computerized screening for novel producers of *Monascus*-like food pigments in *Penicillium* species. *J. Agric. Food Chem.* **2008**, *56*, 9981–9989. [CrossRef] [PubMed]

8. Santos-Ebinuma, V.C.; Roberto, I.C.; Teixeira, M.F.S.; Pessoa, A., Jr. Improvement of submerged culture conditions to produce colorants by *Penicillium purpurogenum*. *Braz. J. Microbiol.* **2014**, *45*, 731–742. [CrossRef] [PubMed]
9. Morales-Oyervides, L.; Oliveira, J.C.; Sousa-Gallagher, M.J.; Méndez-Zavala, A.; Montañez, J.C. Selection of best conditions of inoculum preparation for optimum performance of the pigment production process by *Talaromyces* spp. using the Taguchi method. *Biotechnol. Prog.* **2017**. [CrossRef] [PubMed]
10. Padmapriya, C.; Murugesan, R. Optimization of physical parameters on polyketide red pigment production from *Penicillium purpurogenum*. *Trends Biosci.* **2014**, *7*, 1664–1667.
11. Sudha; Gupta, C.; Aggarwal, S. Optimization and extraction of extra and intracellular color from *Penicillium minioluteum* for application on protein fibers. *Fibers Polym.* **2017**, *18*, 741–748. [CrossRef]
12. Velmurugan, P.; Kim, M.J.; Park, J.S.; Karthikeyan, K.; Lakshmanaperumalsamy, P.; Lee, K.J.; Park, Y.J.; Oh, B.T. Dyeing of cotton yarn with five water soluble fungal pigments obtained from five fungi. *Fibers Polym.* **2010**, *11*, 598–605. [CrossRef]
13. Dhale, M.A.; Vijay-Raj, A.S. Pigment and amylase production in *Penicillium* sp. NIOM-02 and its radical scavenging activity. *Int. J. Food Sci. Technol.* **2009**, *44*, 2424–2430. [CrossRef]
14. Morales-Oyervides, L.; Oliveira, J.C.; Sousa-Gallagher, M.J.; Méndez-Zavala, A.; Montañez, J.C. Effect of heat exposure on the colour intensity of red pigments produced by *Penicillium purpurogenum* GH2. *J. Food Eng.* **2015**, *164*, 21–29. [CrossRef]
15. Arora, S. Textile Dyes: It's impact on environment and its treatment. *J. Bioremediation Biodegrad.* **2014**, *5*, 1. [CrossRef]
16. Borges, M.E.; Tejera, R.L.; Díaz, L.; Esparza, P.; Ibáñez, E. Natural dyes extraction from cochineal (*Dactylopius coccus*). New extraction methods. *Food Chem.* **2012**, *132*, 1855–1860. [CrossRef]
17. Patil, S.; Sivanandhan, G.; Thakare, D. Effect of physical and chemical parameters on the production of Red exopigment from *Penicillium purpurogenum* isolated from spoilt onion and study of its antimicrobial activity. *Int. J. Curr. Microbiol. Appl. Sci.* **2015**, *4*, 599–609.
18. Chadni, Z.; Rahaman, M.H.; Jerin, I.; Hoque, K.M.; Reza, M.A. Extraction and optimisation of red pigment production as secondary metabolites from *Talaromyces verruculosus* and its potential use in textile industries. *Mycology* **2017**, *8*, 48–57. [CrossRef]
19. Sopandi, T.; Wardah, W. Sub-Acute toxicity of pigment derived from *Penicillium resticulosum* in mice. *Microbiol. Indones.* **2012**, *6*, 35–41. [CrossRef]
20. Cruz-Hernández, M.; Rodráguez, R. Aislamiento y caracterización morfológica de cepas microbianas degradadotas de taninos. In Proceedings of the XXII Annual Meeting of the Mexican Academy of Chemical Engineering (AMIDIQ), Mazatlan, Mexico, 9 September 2001.
21. Espinoza-Hernández, T. Caracterización Morfológica, Fisiológica y Molecular de tres Cepas Fúngicas Productoras de Pigmentos. Bachelor's Thesis, Universidad Autónoma de Coahuila, Saltillo, México, 2004.
22. Frisvad, J.C.; Yilmaz, N.; Thrane, U.; Rasmussen, K.B.; Houbraken, J.; Samson, R.A. Talaromyces atroroseus, a new species efficiently producing industrially relevant red pigments. *PLoS ONE* **2013**, *8*, e84102. [CrossRef] [PubMed]
23. Morales-Oyervides, L.; Oliveira, J.C.; Sousa-Gallagher, M.J.; Méndez-Zavala, A.; Montañez, J.C. Quantitative assessment of the impact of the type of inoculum on the kinetics of cell growth, substrate consumption and pigment productivity by *Penicillium purpurogenum* GH2 in liquid culture with an integrated stochastic approach. *Food Bioprod. Process.* **2015**, *96*, 221–231. [CrossRef]
24. Méndez-Zavala, A. Identificación de Factores que Afectan la Producción de Pigmentos por Penicillium purpurogenum GH2 y Obtención de Perfiles Cromatográficos. Master's Thesis, Universidad Autónoma de Coahuila, Saltillo, México, 2011.
25. Méndez-Zavala, A.; Pérez, C.; Montañez, J.C.; Martínez, G.; Aguilar, C.N. Red pigment production by *Penicillium purpurogenum* GH2 is influenced by pH and temperature. *J. Zhejiang Univ. Sci. B* **2011**, *12*, 961–968. [CrossRef] [PubMed]
26. Arroyo-Figueroa, G.; Ruiz-Aguilar, G.M.L.; Cuevas-Rodriguez, G.; Sanchez, G.G. Cotton fabric dyeing with cochineal extract: Influence of mordant concentration. *Color. Technol.* **2011**, *127*, 39–46. [CrossRef]
27. Velmurugan, P. Studies on the Production and Dyeing Properties of Water Soluble Pigments from Filamentous Fungi. Ph.D. Thesis, Bharathiar University, Coimbatore, India, 2008.

28. Tseng, R.-L.; Wu, F.-C.; Juang, R.-S. Characteristics and applications of the Lagergren's first-order equation for adsorption kinetics. *J. Taiwan Inst. Chem. Eng.* **2010**, *41*, 661–669. [CrossRef]
29. Hill, C.G.J.; Grieger-Block, R.A. Kinetic data: Generation, interpretation, and use. *Food Technol.* **1980**, *34*, 55–66.
30. Gupta, C.S.; Aggarwal, S. Dyeing wet blue goat nappa skin with a microbial colorant obtained from *Penicillium minioluteum*. *J. Clean. Prod.* **2016**, *127*, 585–590.
31. Bechtold, T.; Mahmudali, A.; Mussak, R. Natural dyes for textile dyeing: A comparison of methods to assess the quality of Canadian golden rod plant material. *Dyes Pigments* **2007**, *75*, 287–293. [CrossRef]
32. Nagia, F.A.; El-Mohamedy, R.S.R. Dyeing of wool with natural anthraquinone dyes from *Fusarium oxysporum*. *Dyes Pigment.* **2007**, *75*, 550–555. [CrossRef]

16

Phylogenetic Characterization of *Botryosphaeria* Strains Associated with *Asphondylia* Galls on Species of Lamiaceae

Beata Zimowska [1], Sylwia Okoń [2], Andrea Becchimanzi [3], Ewa Dorota Krol [1] and Rosario Nicoletti [3,4,*]

1 Department of Plant Protection, University of Life Sciences, 20-068 Lublin, Poland; beata.zimowska@up.lublin.pl (B.Z.); ewa.krol@up.lublin.pl (E.D.K.)
2 Institute of Plant Genetics, Breeding and Biotechnology, University of Life Sciences, 20-068 Lublin, Poland; sylwia.okon@up.lublin.pl
3 Department of Agricultural Sciences, University of Naples Federico II, 80055 Portici, Italy; andrea.becchimanzi@unina.it
4 Council for Agricultural Research and Economics, Research Centre for Olive, Citrus and Tree Fruit, 81100 Caserta, Italy
* Correspondence: rosario.nicoletti@crea.gov.it

Abstract: In the last decade, *Botryosphaeria dothidea* has been steadily reported as an associate of gall midges (Diptera, Cecidomyiidae) in a variety of host plants and ecological settings. This cosmopolitan fungus is well-known for its ability to colonize many plant species, as both a pathogen and an endophyte. Thus, the shift from this general habit to a lifestyle involving a strict symbiotic relationship with an insect introduces expectancy for possible strain specialization which could reflect separated phylogenetic lineages. Considering the recent taxonomic revision concerning species of *Botryosphaeria*, we evaluated the phylogenetic relationships among strains recovered from *Asphondylia* galls collected on several species of Lamiaceae in Poland and in Italy, and all the currently accepted species in this genus. A number of strains previously characterized from gall samples from Australia and South Africa, whose genetic marker sequences are deposited in GenBank, were also included in the analysis. As a result, full identity as *B. dothidea* is confirmed for our isolates, while strains from the southern hemisphere grouped separately, indicating the existence of genetic variation related to the geographic origin in the association with gall midges.

Keywords: *Asphondylia*; *Botryosphaeria*; *B. dothidea*; DNA sequencing; gall-associated fungi; Lamiaceae; phylogenetic relationships; symbiosis

---

1. Introduction

Although the nature of their symbiotic relationship has not been clearly ascertained, the occurrence of *Botryosphaeria dothidea* as an associate of many gall midge species (Diptera, Cecidomyiidae) is steadily reported, regardless of host plants and ecological contexts. For a long time, the identity of the fungal symbiont has been controversial, by reason of inherent difficulties in the isolation procedure, and of several taxonomic reassessments. In fact, several other fungi, such as *Cladosporium* spp. and *Alternaria* spp., have been frequently reported as gall associates, basically in connection with their saprophytic aptitude, which occasionally makes them conceal the real symbiont during the isolation procedure [1–4]. On the other hand, nomenclatural inconsistency, which only recently has been resolved after the epitypification of *B. dothidea* [5], may account for some previous incorrect reports referring to *Macrophoma*, *Diplodia*, and *Dothiorella* [2,3,6,7]. Indeed, taxonomists recommend a careful interpretation of past literature concerning this fungus [8,9].

Like several species in the Botryosphaeriaceae, *B. dothidea* is well-known for its cosmopolitan distribution and ability to colonize a high number of plants, either as a pathogen or as an endophyte [8–10]. The involvement in cecidomyid-gall formation on a variety of plants confirms it as a very adaptive species. Although the shift from association with plants to a lifestyle characterized by a strict symbiotic relationship with an insect suggests possible strain specialization, which could reflect separated phylogenetic lineages, data resulting from previous studies did not provide evidence for this hypothesis [1,11]. However, in the last decade, the evolution in fungal taxonomy boosted by the application of DNA sequencing has provided a remarkable contribution in view of a better resolution of the *Botryosphaeria* species aggregate. Six species presenting a *Fusicoccum* anamorph, namely *B. corticis*, *B. dothidea*, *B. fabicerciana*, *B. fusispora*, *B. ramosa*, and *B. scharifii*, were recognized based on a detailed phylogenetic analysis; another species, *B. agaves*, is included in this group, although its anamorphic stage has never been described so far [8]. In addition to this basic set, several new species have been more recently identified, mostly based on isolations from tree plants in China (Table 1).

Species of Lamiaceae are widespread in the Mediterranean region, where they seem to represent a diversity hotspot for gall midges. In fact, two new species of the genus *Asphondylia* have been recently described from galls collected on host plants such as *Coridothymus capitatus* [12] and *Clinopodium nepeta* [4], and two more are in course of characterization from *Micromeria graeca* and *Clinopodium vulgare* (Viggiani, personal communication). However, their distribution appears to reach Central Europe, following the geographical spread of some hosts, as documented in the case of *A. serpylli* and *A. hornigi*, respectively associated with *Thymus* spp. and *Origanum vulgare* [3,13]. In the course of our investigations on a number of species of Lamiaceae, we had the opportunity to collect isolates from galls of *Asphondylia* spp. in two European countries with different climatic conditions. Despite a certain variation in biometric characteristics, sequence homology of internal transcribed spacers of ribosomal DNA (rDNA-ITS) confirmed *B. dothidea* as the fungal symbiont of gall midges in both contexts. However, the relative unreliability of data available in GenBank for this species, which is to be taken into account after the recent rearrangements within *Botryosphaeria*, prompted us to undertake a more accurate study of the phylogenetic relationships with the new taxa and other isolates from *Asphondylia* galls from other plant species/countries, in order to assess whether or not strains adapted to this particular symbiotic relationship are homogeneous in taxonomic terms.

Table 1. *Botryosphaeria* species with *Fusicoccum* anamorph described after 2013.

Species	Main Hosts	Country	Reference
B. auasmontanum	*Acacia*	Namibia	[14]
B. kawatsukai	*Malus, Pyrus*	China	[15]
B. minutispermatia	dead wood	China	[16]
B. pseudoramosa	*Eucalyptus, Melastoma*	China	[17]
B. qinlingensis	*Quercus*	China	[18]
B. qingyuanensis	*Eucalyptus*	China	[17]
B. rosaceae	*Amygdalus, Malus, Pyrus*	China	[19]
B. sinensia	*Juglans, Malus, Morus, Populus*	China	[20]
B. wangensis	*Cedrus*	China	[17]

2. Materials and Methods

2.1. Isolation and Morphological Observations

Isolations of fungal associates of gall midges were carried out on potato dextrose agar (PDA) amended with 85% lactic acid (1 mL·L^{-1}) in 90 mm diameter Petri dishes. After removing the outer residues of the flower calyx, fragments from gall walls were cut and transferred onto the agar medium. Plates were incubated in darkness at 25 °C. Hyphal tips from the emerging fungal colonies of the botryosphaeriaceous morphotype were transferred to fresh PDA plates, for morphological observations and storage of pure cultures at 4 °C. Production of pycnidia was induced in cultures prepared in plates containing 2% water agar (WA), topped with sterilized pine needles, which were kept at room temperature under near-UV illumination [21]. Observations were carried out under a Motic BA 210 microscope (Xiamen, China), and images were taken through a 1 MP Motic camera and Scopelmage 9.0 software. Minimum, maximum, mean values, and the length/width (L/W) ratios of 50 conidia from each isolate were measured for a comparison with previously annotated reference sizes for *Botryosphaeria* species [8,17].

2.2. DNA Sequencing and Phylogenetic Analysis

Selected strains recovered from galls on several species of Lamiaceae collected in Italy and Poland (Table 2) were sampled from the surface of PDA cultures with a scalpel. The mycelial matter was transferred to 1.5 mL Eppendorf tubes, for DNA extraction. DNA isolation was performed by using a DNA easy plant and fungi isolation kit (EurX, Gdańsk, Poland), according to manufacturer's protocol. DNA concentration was estimated on 1.5% agarose gel, compared with GeneRulerTM DNA Ladder Plus (Thermo Scientific, Waltham, MA, USA), and measured by using a NanoDrop 2000 spectrophotometer (Thermo Scientific, Waltham, MA, USA). DNA samples were diluted to a concentration of 20 ng·μL^{-1} and stored at 20 °C. Amplification of loci currently considered in taxonomy of *Botryosphaeria* [8] was carried out, using primers ITS1 and ITS4 for the rDNA-ITS region [22], and primers EF1-728F and EF1-986R for the translation elongation factor 1-alpha (TEF1) region [23]. PCR reaction mixtures contained 20 ng of genomic DNA, 0.2 mM of dNTP, 0.2 mM of each primer, 1 × Taq buffer mM buffer (10 mM of Tris-HCl, 1.5 mM of $MgCl_2$, and 50 mM of KCl), and 1 U of Taq polymerase, and were adjusted to a final volume of 25 µL with sterile distilled water. PCR was conducted in a Biometra T1 thermocycler (Analytik Jena, Jena, Germany). The following reaction profile was applied: 95 °C—5 min, 35 cycles (95 °C—45 s, 52 °C—45 s, and 72 °C—45 s), with final elongation at 72 °C—5 min. PCR products were separated in 1.5% agarose gels containing EtBr in TBE buffer, at 140 V, for 1 h.

Table 2. Isolates of *B. dothidea* from *Asphondylia* galls collected on Lamiaceae used in this study.

Number	Host	Location	GenBank Accession	
			ITS	TEF1
SG3	*Clinopodium nepeta*	San Giorgio a Cremano, Italy	MN731265	MN737437
AcE3	*C. nepeta*	Astroni Nature Reserve, Italy	MN731266	MN737438
AcAs2	*C. nepeta*	Astroni Nature Reserve, Italy	MN731267	MN737439
AcSe1	*C. nepeta*	Serino, Italy	MN731268	MN737440
CLRi2	*Clinopodium vulgare*	Rivello, Italy	MN731272	MN737444
Mp26j	*Mentha piperita*	Konopnica, Poland	MN731273	MN737445
Mp30p	*M. piperita*	Konopnica, Poland	MN731274	MN737446
MgBt1	*Micromeria graeca*	Boscotrecase, Italy	MN731269	MN737441
MgPC6	*M. graeca*	Palma Campania, Italy	MN731270	MN737442
MgPC7	*M. graeca*	Palma Campania, Italy	MN731271	MN737443
OvdF3e	*Origanum vulgare*	Fajsławice, Poland	MN731275	MN737447
OvFs/g	*O. vulgare*	Fajsławice, Poland	MN731276	MN737448
Th/g2017	*Thymus vulgaris*	Fajsławice, Poland	MN731277	MN737449
ThgI/10	*T. vulgaris*	Fajsławice, Poland	MN731278	MN737450

ITS: internal transcribed spacer; TEF1: translation elongation factor 1-alpha.

After checking and determining the size of the resulting PCR products, we submitted samples to Genomed (Warsaw, Poland), for sequencing. The obtained nucleotide sequences were compared with reference strains of *Botryosphaeria* spp. from GenBank. All sequences were checked and manually edited by using CLC Main Workbench 8.1.2 software (QIAGEN, Aarhus, Denmark) where necessary. Besides our original sequences, additional sequences of *Botryosphaeria* isolates from *Asphondylia* galls were searched in GenBank for inclusion in the phylogenetic analysis, where a strain of the species *Botryobambusa fusicoccum* was used as outgroup (Table 3). The combined and single ITS and TEF1 sequences were aligned by using Muscle [24] and manually adjusted with AliView software [25], where necessary. The phylogenetic analysis was conformed to a recent protocol [26]. Congruence between the different datasets was tested by using the partition homogeneity test in PAUP software version 4.0b10 [27]. Gaps were treated as missing characters.

Phylogenetic analyses of the concatenated and single-sequence data for maximum likelihood (ML) were performed by using RAxML software version 8.2.12 [28] with GTR+G model of nucleotide substitution and 1000 bootstrap replications. Concatenated sequences were also analyzed for maximum parsimony (MP) by using PAUP, under the heuristic search parameters with tree bisection reconnection branch swapping, 100 random sequence additions, maxtrees set up to 1000, and 1000 bootstrap. Posterior probabilities of the concatenated dataset were determined by Markov Chain Monte Carlo (MCMC) sampling in MrBayes version 3.0b4 [29]. MCMC chains were run for 4000,000 generations, sampling every 100, with a 25% burn-in discarded. Phylogenetic trees were drawn by using FigTree software [30]. Both the alignments and the trees of concatenated dataset were deposited in TreeBase (http://purl.org/phylo/treebase/phylows/study/TB2:S25558).

Table 3. Reference strains used in the phylogenetic analysis.

Species	Number	GenBank Accession	
		ITS	TEF1
Botryobambusa fusicoccum	MFLUCC 11-0143	JX646792	JX646857
Botryosphaeria agaves	MFLUCC 10-0051	JX646790	JX646855
	MFLUCC 11-0125	JX646791	JX646856
Botryosphaeria auasmontanum	CBS 121769	KF766167	EU101348
	MFLUCC 15-0923	MF398858	MF398910
	MFLUCC 17-1071	MF398863	MF398915
Botryosphaeria corticis	CBS 119047	DQ299245	EU017539
	ATCC 22927	DQ299247	EU673291
Botryosphaeria dothidea	CBS 110302	AY259092	AY573218
	CBS 115476	AY236949	AY236898
	3161	EF614924	EF614940
	3179	EF614917	EF614933
	3241	EF614920	EF614937
	3242	EF614916	EF614936
	3247	EF614923	EF614941
	3253	EF614921	EF614938
	3261	EF614926	EF614943
	3275	EF614925	EF614942
	3278	EF614919	EF614934
	3279	EF614918	EF614935
Botryosphaeria fabicerciana	CMW 27094	HQ332197	HQ332213
	CMW 27108	HQ332200	HQ332216
Botryosphaeria fusispora	MFLUCC 10-0098	JX646789	JX646854
	MFLUCC 11-0507	JX646788	JX646853
Botryosphaeria kawatsukai	PGZH18	MG637267	MG637243
	PGZH19	MG637266	MG637242
Botryosphaeria minutispermatia	GZCC 16-0013	KX447675	KX447678
	GZCC 16-0014	KX447676	KX447679
Botryosphaeria pseudoramosa	CERC 2001	KX277989	KX278094
	CERC 3455	KX277997	KX278102
Botryosphaeria qinlingensis	CFCC 52984	MK434301	MK425020
	CFCC 52985	MK434302	MK425021
Botryosphaeria qingyuanensis	CERC 2946	KX278000	KX278105
	CERC 2947	KX278001	KX278106
Botryosphaeria ramosa	CBS 122069	EU144055	EU144070
Botryosphaeria rosaceae	CFCC 82350	KX197079	KX197097
	DZP B	KX197076	KX197096
Botryosphaeria scharifii	IRAN1529C	JQ772020	JQ772057
	IRAN1543C	JQ772019	JQ772056
Botryosphaeria sinensia	BJFU DZP141005-06	KT343254	KU221233
	BJFU DZP141111-10	KT343256	KU221234
Botryosphaeria wangensis	CERC 2298	KX278002	KX278107
	CERC 2300	KX278004	KX278109

3. Results

Cultures on PDA of *Botryosphaeria* isolates recovered from *Asphondylia* galls on Lamiaceae displayed a sparse to moderately dense aerial mycelium, with diverse colors, from white-cream to gray to olivaceous-black, darkening with age, occasionally with narrow or wider columns of mycelium (Figure 1A). Pycnidial conidiomata developed after 10–14 days on PDA, or 8–12 days on pine needles in WA (Figure 1B). These fruiting bodies released buff, and, respectively, black (Figure 1C) or cream masses of spores containing typical *Fusicoccum* conidia, one-celled or with one septum (Figure 1D). They were smooth, hyaline, mostly with granular content, fusiform or irregularly fusiform, wider in the middle to upper third, base-truncate or subtruncate with rounded apex, and quite variable in size (Table 4). Muriform conidia referable to the synanamorphic stage *Dichomera* were never observed, unlike what previously resulted in subcultures of strains from galls collected on *T. vulgaris* directly prepared from the isolation plates [3]. This failure was assumed to possibly derive from prolonged storage at 4 °C of the stock cultures of our strains. Likewise, no isolate produced ascomata throughout the observation period.

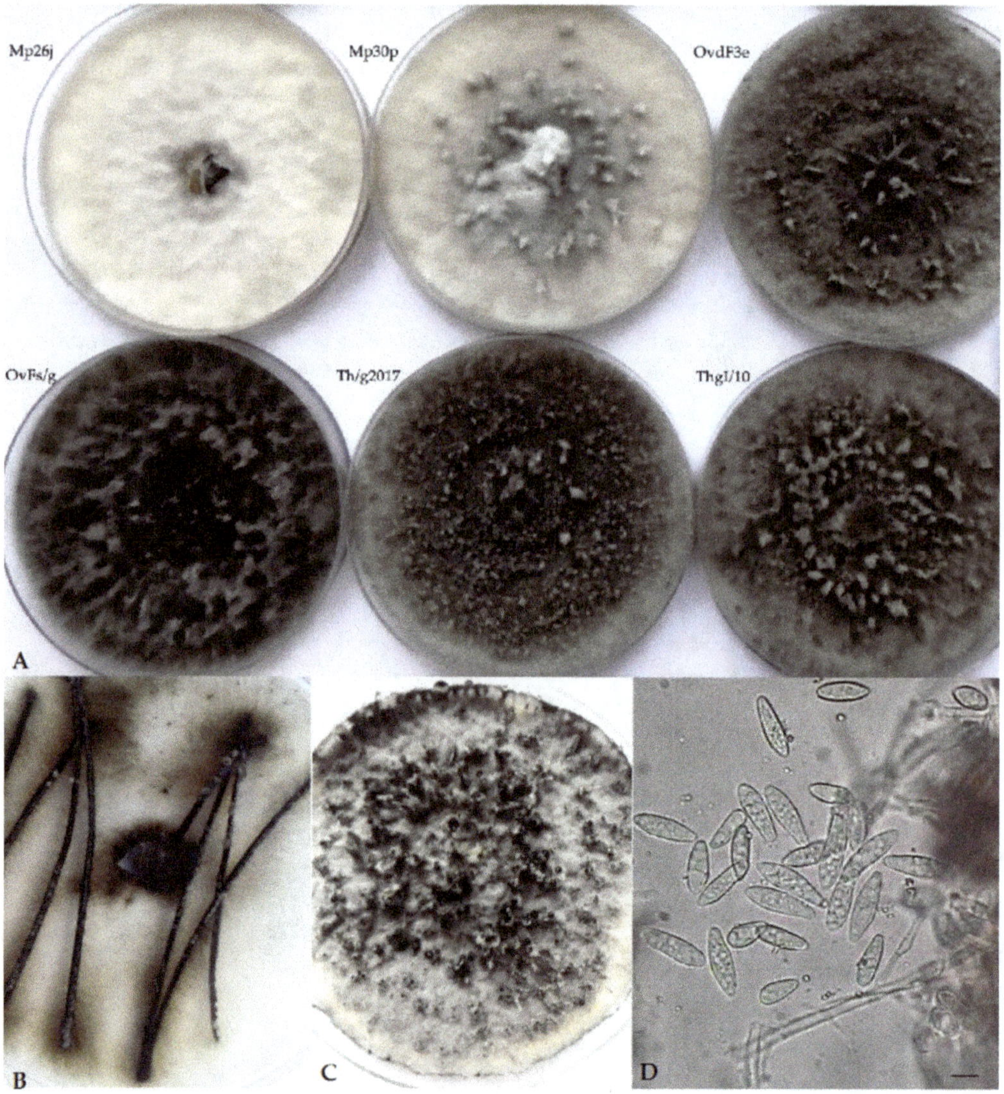

Figure 1. (**A**) Variable morphology of isolates from *Asphondylia* galls collected in our study. (**B**,**C**) Production of pycnidia, respectively, on pine needles on WA (water agar) and on PDA (potato dextrose agar). (**D**) *Fusicoccum* conidia.

Table 4. Morphological features of *B. dothidea* isolates from Lamiaceae examined in this study.

Strain Number	Conidial Dimensions (Length × Width, μm)		
	Range (50 Conidia)	Mean	Length/Width Ratio
SG3	(14–)16.8–21.8 × (3.5–)4.2–5.6	18.9 × 4.9	3.2–5.2
AcAs2	15.9–21 × (3.9–)4.3-5.8(–6.6)	18.7 × 5.2	2.6–4.7
AcE3	(14.2–)16.4–21.6 × (3.4–)4.6–5.4	19.4 × 5.1	3.1–5.4
AcSe1	(13.4–)15.3–21(–22.9) × 3.8–5.7	18.2 × 5.1	2.3–5.1
CLRi2	14.9–21.8 × (3.9–)4.3–5.8(–6.6)	18.7 × 5.2	2.6–5.4
Mp26j	(17.5–)18.5–29.7 × 3.7–7(–7.5)	22.3 × 5.1	3.5–6.1
Mp30p	(16.5–)19.0–28.7 × (3.2–)4.0–7.3	22.3 × 5.1	3.7–6.1
MgBt1	(13–)16.8-21.2 × 4.6–5.5(–6.2)	17.3 × 5.2	2.4–4.6
MgPC6	(12.8–)15.3–21 × 3.8–5.7	15.6 × 4.9	2.1–5.5
MgPC7	(13.4–)15.5–21(–22.9) × 3.3–5.7	18.9 × 4.4	2.3–6.0
OvdF3e	(17–)18.5–25.9 × (3.0–)3.5–7.4	24.3 × 5.3	3.0–6.1
OvFs/g	(17.5–)18.5–28.7 × (2.8–)3.7–7.4	23.5 × 5.1	3.5–6.0
Th/g2017	(15.2–)16.5–25.7 × (2.7–)3.9–6.8	20.3 × 4.8	2.9–5.5
ThgI/10	(16.0–)18.5–28.4 × (3.1–)3.5–7.0	22.0 × 5.1	3.4–6.0

Because of the above unreliability of morphological characters, DNA sequence homology was fundamental for an accurate taxonomic identification. PCR products of approximately 560 bp for ITS region and 300 bp for TEF1 region were amplified and successfully sequenced in 14 isolates from *Asphondylia* galls considered in this study. All the nucleotide sequences obtained were deposited in GenBank (Table 2), and blasted against the ex-epitype strain of *B. dothidea* (CMW8000/CBS115476) [5]. Identity was 99.79% for all strains but one (MgBt1, 98.97%) for ITS sequences, while for TEF1 sequence identity was 100% except four isolates (Th/g2017, OvFs/g, OvdF3e, and again MgBt1) matching at 99.16%.

This identity was highlighted in the subsequent phylogenetic analysis considering reference strains of all the recognized *Botryosphaeria* spp. producing *Fusicoccum* conidia, along with previously identified strains of *B. dothidea* collected from *Asphondylia* galls, in other contexts, worldwide. Although several contributions have been published in recent years on the subject of gall midges and associated fungi, a search in GenBank showed that sequences of both ITS and TEF1 are only available for some isolates from *Acacia* spp., collected in Australia and South Africa, which were the subject of a previously mentioned study [1].

The trimmed and manually adjusted alignment of concatenated locus contained 58 strains (including the outgroup) and consisted of 490 and 255 bp for ITS and TEF1, respectively. The best scoring RAxML tree (Figure 2) had a final likelihood value of −1961.235515. The matrix had 172 distinct alignment patterns, with 5.74% of undetermined characters or gaps. The ML tree of ITS alone showed poor resolution compared to ML trees based on TEF1 and concatenated sequences (Figure S1). ML trees of TEF1 and ITS + TEF1 showed almost the same topology, except for *B. qingyuanensis*, which is included in the same clade of *B. wangensis*-*B. sinensis*-*B. qinlingensis* in the phylogram based on TEF1 alone (Figure S2). Parsimony analysis yielded 1000 equally parsimonious trees (tree length = 174 steps; consistency index = 0.902; retention index = 0.915; relative consistency index = 0.826; homoplasy index = 0.098). Of the 745 characters used, 76 were parsimony-informative, 65 were variable and parsimony-uninformative, and 604 were constant. The same clades were supported in MP and ML analyses, except for *B. minutispermatia*, which resulted in being less closely related to the Australian isolates in the MP tree (Figure S3).

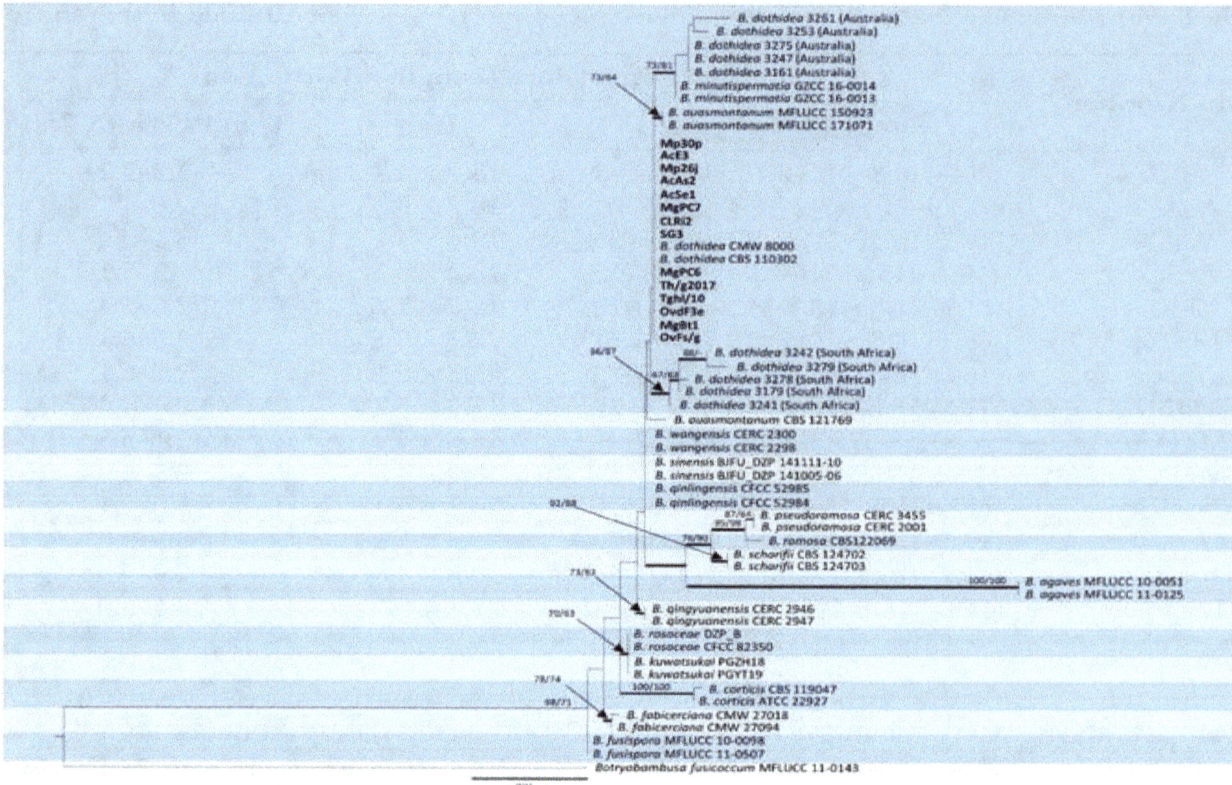

Figure 2. Phylogenetic tree based on maximum likelihood (ML) analyses of concatenated ITS and TEF1 sequences from strains considered in this study. Original isolates from Lamiaceae are in bold. Bootstrap support values ≥ 60% for ML and maximum parsimony (MP) are presented above branches as follows: ML/MP, bootstrap support values < 60% are marked with '-'. Branches in bold are supported by Bayesian analysis (posterior probability ≥ 90%). *Botryobambusa fusicoccum* MFLUCC 11-0143 was used as the outgroup reference.

4. Discussion

The wide morphological variation observed in our study is in line with recent findings that biometric data of *Botryosphaeria* species may overlap and are no longer relevant for the identification of *B. dothidea* [8,9,17]. Even *Dichomera* conidia are infrequently observed in this species [8,14,21], indicating that this character cannot be reliably taken into account for taxonomic purposes. Hence, reports from *Asphondylia* galls primarily referring to this anamorphic stage [1] should be considered with caution, because of the common co-occurrence of saprophytic *Alternaria* producing similar phaeodictyospores.

Therefore, nucleotide sequencing has become the primary identification method for *B. dothidea* through the assessment of homologies with the ex-epitype strain (CMW8000/CBS115476) [5], particularly considering the loci of ITS and TEF1. In fact, although ITS was shown to clearly distinguish *B. dothidea* from its closest relatives, the recent discovery of several cryptic species within the Botryosphaeriaceae makes it necessary to combine with TEF1 for a more reliable identification [8,31,32].

Phylogenetic analysis disclosed a clear identity with *B. dothidea* of our heterogeneous strain sample from Lamiaceae, regardless of their origin from two different climatic contexts. In fact, both Polish and Italian isolates grouped together with the type strains of this species. Conversely, Australian and South African strains formed two distinct groupings, which indicated a divergence from the most recent common ancestor. Particularly, the former was associated to type strains of the species *B. minutispermatia* in a more comprehensive clade, while the latter exhibited a slightly higher phylogenetic distance. Considering that both Australian and South African isolates were collected in association with gall midges on *Acacia* spp., this finding could be interpreted as being in agreement with a previous analysis of the bulk *B. dothidea* sequences deposited in GenBank, showing a population structure which is shaped by geographical distance rather than host-plant preference [9]. In fact, lower identity with *B. dothidea* strains in the GenBank database resulted for these isolates, particularly those from South Africa whose TEF1 sequence homology was not higher than 96.83%. The observed variation particularly concerning this genetic marker requires further assessments to establish if these clusters should be interpreted as separated lineages within *B. dothidea*, or if they may represent distinct species.

The case of *B. auasmontanum*, a species characterized from a single strain (CBS121769) collected in Namibia [14], deserves further consideration. In fact, besides the holotype, our analysis included two more isolates from Italy ascribed to this species [26], whose ITS and TEF1 sequences are available in GenBank. Unexpectedly, they did not cluster with the holotype in the phylogenetic tree (Figure 2). A BLAST search in GenBank with sequences by these two strains clearly shows their 100% identity with *B. dothidea*, while ITS and TEF1 sequence homology with CBS121769 is lower (95.21% and 90.28%, respectively). The phylogenetic separation of the holotype of *B. auasmontanum* is supported by two large gaps resulting from the sequence alignment (Figure 3), which means that the claimed evidence of possible synonymy between the two species [26] applies only to the two Italian isolates. Thus, their inclusion in phylogenetic studies as representatives of *B. auasmontanum* should be avoided, and their incorrect taxonomic identification should be taken into account. Figure 3 also describes the substantial similarity among all strains of *B. dothidea*, including Australian and South African isolates, and the closely related *B. minutispermatia*, which essentially differ for nucleotide substitutions at single definite positions.

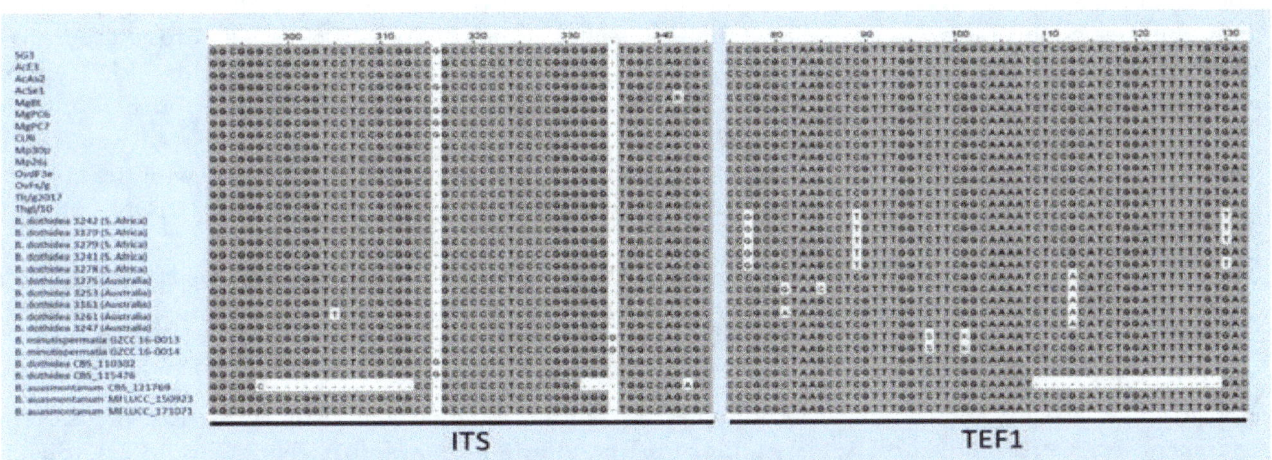

Figure 3. Alignment of ITS and TEF1 sequences of strains of *B. dothidea*, *B. minutispermatia*, and *B. auasmontanum*. Gaps in the sequences of the holotype of the latter species (CBS121769) are evident, along with single nucleotide substitutions in TEF1 sequences of Australian and South African isolates, and the type strains of *B. minutispermatia*.

5. Conclusions

Characterization of *Botryosphaeria* strains from different plant species and environmental contexts goes through the finding of novel species, with the expectation that additional undescribed taxa may be discovered in the future [17]. This concept involves strains associated with cecidomyids, inducing gall formation on a wide variety of plant species, which so far have been overlooked in assessments concerning taxonomy of such a highly adaptive and widespread fungus. Results of our investigation showed full identity with *B. dothidea* of isolates from galls collected from Lamiaceae, while a possible separation from this species should be verified for isolates previously recovered from *Acacia* in Australia and, particularly, South Africa. Indeed, a more adequate definition could be obtained by integrating these findings with data concerning cecidomyid-associated *Botryosphaeria* strains from other countries.

Therefore, a more active cooperation among researchers currently working on this topic worldwide is to be encouraged, in order to shed further light on this unique biological association.

Author Contributions: Conceptualization, B.Z. and R.N.; methodology, A.B., B.Z., E.D.K., and S.O.; formal analysis, A.B. and S.O.; writing—original draft preparation, B.Z., R.N., and S.O.; writing—review and editing, B.Z., E.D.K., and R.N.; funding acquisition, E.D.K. All authors have read and agreed to the published version of the manuscript.

Acknowledgments: Authors thank Sarah Lucchesi (University of Southern Maine, Portland, USA) for revising the English style of this paper.

References

1. Adair, R.J.; Burgess, T.; Serdani, M.; Barber, P. Fungal associations in *Asphondylia* (Diptera: Cecidomyiidae) galls from Australia and South Africa: Implications for biological control of invasive acacias. *Fungal Ecol.* **2009**, *2*, 121–134. [CrossRef]
2. Lebel, T.; Peele, C.; Veenstra, A. Fungi associated with *Asphondylia* (Diptera: Cecidomyiidae) galls on *Sarcocornia quinqueflora* and *Tecticornia arbuscula* (Chenopodiaceae). *Fungal Divers.* **2012**, *55*, 143–154. [CrossRef]
3. Zimowska, B.; Viggiani, G.; Nicoletti, R.; Furmańczyk, A.; Becchimanzi, A.; Kot, I. First report of the gall midge *Asphondylia serpylli* on thyme (*Thymus vulgaris*), and identification of the associated fungal symbiont. *Ann. Appl. Biol.* **2017**, *171*, 89–94. [CrossRef]
4. Bernardo, U.; Nugnes, F.; Gualtieri, L.; Nicoletti, R.; Varricchio, P.; Sasso, R.; Viggiani, G. A new gall midge species of *Asphondylia* (Diptera: Cecidomyiidae) inducing flower galls on *Clinopodium nepeta* (Lamiaceae) from Europe, its phenology, and associated fungi. *Environ. Entomol.* **2018**, *47*, 609–622. [CrossRef] [PubMed]
5. Slippers, B.; Crous, P.W.; Denman, S.; Coutinho, T.A.; Wingfield, B.D.; Wingfield, M.J. Combined multiple gene genealogies and phenotypic characters differentiate several species previously identified as *Botryosphaeria dothidea*. *Mycologia* **2004**, *96*, 83–101. [CrossRef]
6. Denman, S.; Crous, P.W.; Taylor, J.E.; Kang, J.-C.; Pascoe, I.; Wingfield, M.J. An overview of the taxonomic history of *Botryosphaeria*, and a re-evaluation of its anamorphs based on morphology and ITS rDNA phylogeny. *Stud. Mycol.* **2000**, *45*, 129–140.
7. Park, I.; Sanogo, S.; Hanson, S.F.; Thompson, D.C. Molecular identification of *Botryosphaeria dothidea* as a fungal associate of the gall midge *Asphondylia prosopidis* on mesquite in the United States. *BioControl* **2019**, *64*, 209–219. [CrossRef]
8. Phillips, A.J.L.; Alves, A.; Abdollahzadeh, J.; Slippers, B.; Wingfield, M.J.; Groenewald, J.Z.; Crous, P.W. The Botryosphaeriaceae: genera and species known from culture. *Stud. Mycol.* **2013**, *76*, 51–167. [CrossRef]
9. Marsberg, A.; Kemler, M.; Jami, F.; Nagel, J.H.; Postma-Smidt, A.; Naidoo, S.; Wingfield, M.J.; Crous, P.W.; Spatafora, J.W.; Hesse, C.N.; et al. *Botryosphaeria dothidea*: a latent pathogen of global importance to woody plant health. *Mol. Plant Pathol.* **2017**, *18*, 477–488. [CrossRef]
10. Slippers, B.; Wingfield, M.J. Botryosphaeriaceae as endophytes and latent pathogens of woody plants: diversity, ecology and impact. *Fungal Biol. Rev.* **2007**, *21*, 90–106. [CrossRef]
11. Janson, E.M.; Peeden, E.R.; Stireman, J.O.; Abbot, P. Symbiont-mediated phenotypic variation without co-evolution in an insect–fungus association. *J. Evol. Biol.* **2010**, *23*, 2212–2228. [CrossRef] [PubMed]
12. Malagaris, P. Biology and ecology of *Asphondylia coridothymi* (Diptera: Cecidomyiidae) inducing galls on *Coridothymus capitatus* on the island of Samos, Greece. *Acta Soc. Zool. Bohemicae* **2011**, *75*, 239–251.
13. Viggiani, G.; Stinca, A. Infestations on oregano by the gall midge *Asphondylia hornigi* Wachtl. *Prot. Colture* **2018**, *4*, 29–31.
14. Slippers, B.; Roux, J.; Wingfield, M.J.; van der Walt, F.J.J.; Jami, F.; Mehl, J.W.M.; Marais, G.J. Confronting the constraints of morphological taxonomy in the Botryosphaeriales. *Persoonia* **2014**, *33*, 155–168. [CrossRef]
15. Xu, C.; Wang, C.S.; Ju, L.L.; Zhang, R.; Biggs, A.R.; Tanaka, E.; Li, B.; Sun, G.Y. Multiple locus genealogies and phenotypic characters reappraise the causal agent of apple ring rot in China. *Fungal Divers.* **2015**, *71*, 215–231. [CrossRef]
16. Ariyawansa, H.A.; Hyde, K.D.; Liu, J.K.; Wu, S.P.; Liu, Z.Y. Additions to Karst Fungi 1: *Botryosphaeria minutispermatia* sp. nov., from Guizhou Province, China. *Phytotaxa* **2016**, *275*, 35–44. [CrossRef]

17. Li, G.Q.; Liu, F.F.; Li, J.Q.; Liu, Q.L.; Chen, S.F. Botryosphaeriaceae from *Eucalyptus* plantations and adjacent plants in China. *Persoonia* **2018**, *40*, 63–95. [CrossRef]
18. Liang, L.Y.; Jiang, N.; Chen, W.Y.; Liang, Y.M.; Tian, C.M. *Botryosphaeria qinlingensis* sp. nov. causing oak frogeye leaf spot in China. *Mycotaxon* **2019**, *134*, 463–473. [CrossRef]
19. Zhou, Y.P.; Zhang, M.; Dou, Z.P.; Zhang, Y. *Botryosphaeria rosaceae* sp. nov. and *B. ramosa*, new botryosphaeriaceous taxa from China. *Mycosphere* **2017**, *8*, 162–171. [CrossRef]
20. Zhou, Y.; Dou, Z.; He, W.; Zhang, X.; Zhang, Y. *Botryosphaeria sinensia* sp nov., a new species from China. *Phytotaxa* **2016**, *245*, 43–50. [CrossRef]
21. Crous, P.W.; Slippers, B.; Wingfield, M.J.; Rheeder, J.; Marasas, W.F.; Philips, A.J.; Alves, A.; Burgess, T.; Barber, P.; Groenewald, J.Z. Phylogenetic lineages in the Botryosphaeriaceae. *Stud. Mycol.* **2006**, *55*, 235–253. [CrossRef] [PubMed]
22. White, T.J.; Bruns, T.; Lee, S.J.W.T.; Taylor, J. Amplification and direct sequencing of fungal ribosomal RNA genes for phylogenetic. In *PCR Protocols, A Guide to Methods and Applications*; Innis, M.A., Gelfand, D.H., Sninsky, J.J., White, T.J., Eds.; Academic Press: New York, NY, USA, 1990; Volume 18, pp. 315–322.
23. Carbone, I.; Kohn, L.M. A method for designing primer sets for speciation studies in filamentous ascomycetes. *Mycologia* **1999**, *91*, 553–556. [CrossRef]
24. Edgar, R.C. MUSCLE: Multiple sequence alignment with high accuracy and high throughput. *Nucleic Acids Res.* **2004**, *32*, 1792–1797. [CrossRef] [PubMed]
25. Larsson, A. AliView: A fast and lightweight alignment viewer and editor for large datasets. *Bioinformatics* **2014**, *30*, 3276–3278. [CrossRef]
26. Dissanayake, A.J.; Camporesi, E.; Hyde, K.D.; Yan, J.Y.; Li, X.H. Saprobic Botryosphaeriaceae, including *Dothiorella italica* sp. nov., associated with urban and forest trees in Italy. *Mycosphere* **2017**, *8*, 1157–1176. [CrossRef]
27. Wilgenbusch, J.C.; Swofford, D. Inferring evolutionary trees with PAUP*. *Curr. Prot. Bioinformatics* **2003**, *1*, 6.4.1–6.4.28. [CrossRef]
28. Stamatakis, A. RAxML-VI-HPC: maximum likelihood-based phylogenetic analyses with thousands of taxa and mixed models. *Bioinformatics* **2006**, *22*, 2688–2690. [CrossRef]
29. Huelsenbeck, J.P.; Ronquist, F. MRBAYES: Bayesian inference of phylogenetic trees. *Bioinformatics* **2001**, *17*, 754–755. [CrossRef]
30. Molecular Evolution, Phylogenetics and Epidemiology. Available online: http://tree.bio.ed.ac.uk/software/figtree/ (accessed on 25 November 2018).
31. Chen, S.; Pavlic, D.; Roux, J.; Slippers, B.; Xie, Y.; Wingfield, M.J.; Zhou, X.D. Characterization of Botryosphaeriaceae from plantation-grown *Eucalyptus* species in South China. *Plant Pathol.* **2011**, *60*, 739–751. [CrossRef]
32. Abdollahzadeh, J.; Zare, R.; Phillips, A.J. Phylogeny and taxonomy of *Botryosphaeria* and *Neofusicoccum* species in Iran, with description of *Botryosphaeria scharifii* sp. nov. *Mycologia* **2013**, *105*, 210–220. [CrossRef]

PERMISSIONS

All chapters in this book were first published in MDPI; hereby published with permission under the Creative Commons Attribution License or equivalent. Every chapter published in this book has been scrutinized by our experts. Their significance has been extensively debated. The topics covered herein carry significant findings which will fuel the growth of the discipline. They may even be implemented as practical applications or may be referred to as a beginning point for another development.

The contributors of this book come from diverse backgrounds, making this book a truly international effort. This book will bring forth new frontiers with its revolutionizing research information and detailed analysis of the nascent developments around the world.

We would like to thank all the contributing authors for lending their expertise to make the book truly unique. They have played a crucial role in the development of this book. Without their invaluable contributions this book wouldn't have been possible. They have made vital efforts to compile up to date information on the varied aspects of this subject to make this book a valuable addition to the collection of many professionals and students.

This book was conceptualized with the vision of imparting up-to-date information and advanced data in this field. To ensure the same, a matchless editorial board was set up. Every individual on the board went through rigorous rounds of assessment to prove their worth. After which they invested a large part of their time researching and compiling the most relevant data for our readers.

The editorial board has been involved in producing this book since its inception. They have spent rigorous hours researching and exploring the diverse topics which have resulted in the successful publishing of this book. They have passed on their knowledge of decades through this book. To expedite this challenging task, the publisher supported the team at every step. A small team of assistant editors was also appointed to further simplify the editing procedure and attain best results for the readers.

Apart from the editorial board, the designing team has also invested a significant amount of their time in understanding the subject and creating the most relevant covers. They scrutinized every image to scout for the most suitable representation of the subject and create an appropriate cover for the book.

The publishing team has been an ardent support to the editorial, designing and production team. Their endless efforts to recruit the best for this project, has resulted in the accomplishment of this book. They are a veteran in the field of academics and their pool of knowledge is as vast as their experience in printing. Their expertise and guidance has proved useful at every step. Their uncompromising quality standards have made this book an exceptional effort. Their encouragement from time to time has been an inspiration for everyone.

The publisher and the editorial board hope that this book will prove to be a valuable piece of knowledge for researchers, students, practitioners and scholars across the globe.

LIST OF CONTRIBUTORS

Jorge Oliveira and Maria Sousa-Gallagher
School of Engineering, University College Cork, Cork, Ireland

Alejandro Méndez-Zavala and Julio Cesar Montañez
Department of Chemical Engineering, Universidad Autónoma de Coahuila, Saltillo 25280, Mexico

Lourdes Morales-Oyervides
School of Engineering, University College Cork, Cork, Ireland
Department of Chemical Engineering, Universidad Autónoma de Coahuila, Saltillo 25280, Mexico

Carolina Elena Girometta, Rebecca Michela Baiguera, Francesco Bracco, Simone Buratti, Anna Maria Picco and Elena Savino
Department of Earth and Environmental Sciences (DSTA), University of Pavia, 27100 Pavia, Italy

Annarosa Bernicchia
School of Agriculture and Veterinary Medicine, University of Bolognavia, Guidotti 39, 40134 Bologna, Italy

Marco Cartabia
Department of Earth and Environmental Sciences (DSTA), University of Pavia, 27100 Pavia, Italy
MOGU S.r.l. Via San Francesco d'Assisi 62, 21020 Inarzo (VA), Italy

Sarath M. Vega Gutierrez and Sara C. Robinson
Department of Wood Science & Engineering, 119 Richardson Hall, Oregon State University, Corvallis, OR 97331, USA

Elias Polemis, Vassiliki Fryssouli and Vassileios Daskalopoulos
Laboratory of General and Agricultural Microbiology, Agricultural University of Athens, 11855 Athens, Greece

Mark Busman
Mycotoxin Prevention and Applied Microbiology Research Unit, National Center for Agricultural Utilization Research, Agricultural Research Service, United States Department of Agriculture, Peoria, IL 61604, USA

Maria Letizia Gargano
Department of Agricultural and Environmental Science, University of Bari Aldo Moro, Via Amendola 165/A, I-70126 Bari, Italy

Georgios I. Zervakis
Laboratory of General and Agricultural Microbiology, Agricultural University of Athens, 11855 Athens, Greece

Omoanghe S. Isikhuemhen
Mushroom Biology and Fungal Biotechnology Laboratory, Department of Natural Resource & Environmental Design, North Carolina A&T State University, Greensboro, NC 27411, USA

Giuseppe Venturella, Roberta Calvo and Valeria Ferraro
Department of Agricultural, Food and Forest Sciences, University of Palermo, Viale delle Scienze, Bldg. 5, I-90128 Palermo, Italy

Anna Giammanco and Teresa Fasciana
Department of Health Promotion, Mother and Child Care, Internal Medicine and Medical Specialities, University of Palermo, 90127 Palermo, Italy

Mekala Venkatachalam and Melissa Llorente
Laboratoire de Chimie des Substances Naturelles et des Sciences des Aliments—LCSNSA EA 2212, Université de La Réunion, 15 Avenue René Cassin, CS 92003, F-97744 Saint-Denis CEDEX 9, Ile de La Réunion, France

Mireille Fouillaud and Laurent Dufossé
Laboratoire de Chimie des Substances Naturelles et des Sciences des Aliments—LCSNSA EA 2212, Université de La Réunion, 15 Avenue René Cassin, CS 92003, F-97744 Saint-Denis CEDEX 9, Ile de La Réunion, France
Ecole Supérieure d'Ingénieurs Réunion Océan Indien—ESIROI, 2 Rue Joseph Wetzell, F-97490 Sainte-Clotilde, Ile de La Réunion, France

Helene Magalon and Pascale Cuet
UMR ENTROPIE and LabEx CORAIL, Université de La Réunion, 15 Avenue René Cassin, CS 92003, F-97744 Saint-Denis CEDEX 9, Ile de La Réunion, France

Neji Mahmoudi and Mohamed Mars
Unité de Recherche Biodiversité et Valorisation des Bio-ressources en Zones Arides (BVBZA), Faculté des Sciences de Gabès, Cité Erriadh Zrig 6072, Tunisie

Teresa Dias and Cristina Cruz
Centre for Ecology, Evolution and Environmental Changes (cE3c), Faculdade de Ciências da Universidade de Lisboa, Edifício C2, Piso 5, Sala 2.5.03 Campo Grande, 1749-016 Lisboa, Portugal

Mosbah Mahdhi
Center for Environmental Research and Studies, Jazan University, Jazan 82817, Saudi Arabia

Maria F. Caeiro
Centro de Estudos do Ambiente e do Mar (CESAM), Faculdade de Ciências da Universidade de Lisboa, Edifício C2, Piso 2, Campo Grande, 1749-016 Lisboa, Portugal

Javier Avalos, Javier Pardo-Medina, Obdulia Parra-Rivero, Macarena Ruger-Herreros and María Carmen Limón
Departamento de Genética, Facultad de Biología, Universidad de Sevilla, 41012 Sevilla, Spain

Roberto Rodríguez-Ortiz
Departamento de Genética, Facultad de Biología, Universidad de Sevilla, 41012 Sevilla, Spain
Present Address: CONACYT-Instituto de Neurobiología-UNAM, Juriquilla, Querétaro 076230, Mexico

Dámaso Hornero-Méndez
Departamento de Fitoquímica de los Alimentos, Instituto de la Grasa, CSIC, Campus Universidad Pablo de Olavide, 41013 Sevilla, Spain

Jelena Lazarević
Biotechnical faculty, University of Montenegro, Mihaila Lali´ca 15, 81 000 Podgorica, Montenegro

Audrius Menkis
Department of Forest Mycology and Plant Pathology, Uppsala BioCenter, Swedish University of Agricultural Sciences, SE-75007 Uppsala, Sweden

Juliana Lebeau and Yanis Caro
Laboratoire de Chimie des Substances Naturelles et des Sciences des Aliments (LCSNSA), Université de la Réunion, F-97490 Sainte-Clotilde, Ile de la Réunion, France

Thomas Petit
UMR QualiSud, Université de la Réunion, IUT, F-97410 Saint-Pierre, Ile de la Réunion, France

Francesco Vinale
Istituto per la Protezione Sostenibile delle Piante (IPSP-CNR) and Dipartimento di Agraria, Università degli Studi di Napoli Federico II, 80055 Portici, NA, Italy

Anna Poli, Elena Bovio, Lucrezia Ranieri, Giovanna Cristina Varese and Valeria Prigione
Mycotheca Universitatis Taurinensis, Department of Life Sciences and Systems Biology, University of Torino, Viale Mattioli 25, 10125 Torino, Italy

Hendrik Pollmann, Jürgen Breitenbach and Gerhard Sandmann
Biosynthesis Group, Molecular Biosciences, Fachbereich Biowissenschaften, Goethe Universität Frankfurt, Frankfurt am Main 60438, Germany

Hendrik Wolff
Merck Stiftungsprofessur für Molekulare Biotechnologie, Fachbereich Biowissenschaften, Goethe Universität Frankfurt, Frankfurt am Main 60438, Germany

Helge B. Bode
Merck Stiftungsprofessur für Molekulare Biotechnologie, Fachbereich Biowissenschaften, Goethe Universität Frankfurt, Frankfurt am Main 60438, Germany
Buchmann Institute for Molecular Life Sciences (BMLS), Goethe Universität Frankfurt, Max-von-Laue-Strasse 15, Frankfurt am Main 60438, Germany

Beatrice Belfiori, Claudia Riccioni, Francesco Paolocci and Andrea Rubini
National Research Council, Institute of Biosciences and Bioresources–Perugia Division, Via Madonna Alta n. 130, 06128 Perugia, Italy

Valentina D'Angelo, Marco Leonardi and Giovanni Pacioni
Department of Life, Health & Environmental Sciences, University of L'Aquila, 67010 Coppito-L'Aquila, Italy

Beata Zimowska and Ewa Dorota Krol
Department of Plant Protection, University of Life Sciences, 20-068 Lublin, Poland

Sylwia Okoń
Institute of Plant Genetics, Breeding and Biotechnology, University of Life Sciences, 20-068 Lublin, Poland

Andrea Becchimanzi
Department of Agricultural Sciences, University of Naples Federico II, 80055 Portici, Italy

Rosario Nicoletti
Department of Agricultural Sciences, University of Naples Federico II, 80055 Portici, Italy
Council for Agricultural Research and Economics, Research Centre for Olive, Citrus and Tree Fruit, 81100 Caserta, Italy

Index

A
Adsorption, 226, 229
Aging, 24, 30, 133
Amino Acid, 174
Anthocyanins, 159

B
Bifurcations, 81
Bioavailability, 112, 122, 227
Biodiversity, 10, 14, 23, 26, 28-29, 31, 89-90, 92, 103, 106-107, 143, 153, 157, 194, 217, 219
Biofilm, 80, 84-87
Biomass, 2-8, 11, 14, 28, 30, 93-94, 96, 98, 102, 105, 114, 118, 120, 122-125, 160-165, 168, 173-175
Bootstrapping, 184

C
Calibration Curve, 71, 166
Cancer, 70, 86, 89, 105
Carbohydrates, 28, 84, 86
Cell Lines, 70, 76
Cell Membrane, 5, 7
Cellulose, 3, 23, 28, 44, 46, 79, 92, 162, 222
Centrifugation, 3, 114, 175, 222
Chemical Composition, 83, 87
Confocal Microscopy, 34, 37-38, 43-44, 46
Consumers, 214
Correlation Coefficient, 120
Cytotoxicity, 70

D
Dehydration, 104
Dichotomy, 25
Divergence, 186, 236
Documentation, 124

E
Ecosystem, 14, 68, 88, 110, 112-113, 120-124, 217
Ecosystem Services, 217
Electron Microscopy, 33-34, 44
Enzyme, 11, 24, 27, 106, 120, 125, 128-130, 139, 141, 205-206
Escherichia Coli, 79, 128, 198, 204-206
Etiology, 157

F
Fragmentation, 193, 197, 202-203

G
Gene Expression, 132, 205
Genotypes, 210
Glacial Refugium, 65, 216
Global Warming, 47, 217, 219
Gram-negative Bacteria, 79

H
Habitat, 47-49, 51, 57-58, 60-61, 63-64, 67, 78, 81, 153, 176
Health-promoting Properties, 30
Histograms, 164
Hydrological Cycle, 47
Hydroxyl Groups, 200, 204
Hyperactivity, 159

I
Interpolation, 210, 213
Investment, 221
Iodine, 78, 80

L
Liberation, 168, 175
Lycopene, 128

M
Magnetic Resonance, 44, 75, 106
Marine Ecosystems, 89, 92, 108, 195
Marine Environments, 87, 89-90, 103-106
Mass Spectrometry, 33, 69-71, 106, 158, 162, 197, 202, 204, 206
Meta-analysis, 124
Metabolites, 5, 22, 24, 30, 76, 89-90, 103-107, 109-110, 126, 128, 131, 135-137, 141, 149, 152, 156, 161-162, 165, 168, 170, 172, 175-178, 228
Methodologies, 79
Methodology, 3, 13, 27, 65, 86, 123, 128, 153, 195, 217, 222, 238
Multidimensional Scaling, 148, 150

N
Next Generation Sequencing, 65
Niacin, 84

O

Online Service, 66
Optimization, 228
Oxidation, 88, 90, 129-130, 160

P

Pathogenicity, 151, 156
Polysaccharide, 27
Private Sector, 78
Probability, 51, 210, 224, 236
Production Level, 5
Production Process, 10, 228
Protein Synthesis, 130
Proteins, 2, 84, 134-137, 141, 165, 174, 176
Pseudomonas Aeruginosa, 77, 79

Q

Quality Control, 146

R

Raw Material, 2
Regression, 111, 115, 123, 162, 224-225
Rehabilitation, 107, 123
Resistance, 32, 43, 45, 76, 86-87, 90, 140-141

S

Seasonality, 124
Simulation, 50, 219
Staphylococcus Aureus, 77, 79, 87
Sustainability, 121, 159, 175

T

Taxonomy, 26, 57, 65, 109, 150, 186, 231-232, 237-239
Topology, 79, 186, 235

U

Ubiquitin, 135

V

Vegetation, 15, 64, 78, 112-113, 121, 142, 153
Visualization, 66, 86, 123, 195

www.ingramcontent.com/pod-product-compliance
Lightning Source LLC
LaVergne TN
LVHW081012210926
848485LV00004B/844